£27

Thermodynamics, Combustion and Engines

Brian E. Milton

School of Mechanical and
Manufacturing Engineering
University of New South Wales
Kensington
Australia

CHAPMAN & HALL

London · Glasgow · Weinheim · New York · Tokyo · Melbourne · Madras

Published by Chapman & Hall, 2–6 Boundary Row, London SE1 8HN, UK

Chapman & Hall, 2–6 Boundary Row, London SE1 8HN, UK

Blackie Academic & Professional, Wester Cleddens Road, Bishopbriggs,
Glasgow G64 2NZ, UK

Chapman & Hall GmbH, Pappelallee 3, 69469 Weinheim, Germany

Chapman & Hall USA, 115 Fifth Avenue, New York, NY 10003, USA

Chapman & Hall Japan, ITP-Japan, Kyowa Building, 3F, 2-2-1 Hirakawacho,
Chiyoda-ku, Tokyo 102, Japan

Chapman & Hall Australia, 102 Dodds Street, South Melbourne,
Victoria 3205, Australia

Chapman & Hall India, R. Seshadri, 32 Second Main Road, CIT East,
Madras 600 035, India

First edition 1995

© 1995, Brian E. Milton

Typeset in 10/12 pt Times by Thomson Press (India) Ltd, Madras
Printed in Great Britain by St. Edmundsbury Press, Bury St Edmunds, Suffolk

ISBN 0 412 53840 7

A catalogue record for this book is available from the British Library.

Library of Congress Catalog Card Number: 94-69924

∞ Printed on permanent acid-free text paper, manufactured in accordance with
ANSI/NISO Z39.48-1992 and ANSI/NISO Z39.48-1984 (Permanence of Paper).

Contents

Preface

The sciences of thermodynamics, heat transfer and fluid mechanics play an important role in many applied areas where they cannot in practice be considered in isolation from each other. However, this is often not immediately apparent because they are usually taught, discussed and researched as separate subjects, an approach which sometimes inhibits both the fundamental understanding of and the ability to apply them. To overcome this, a thorough grounding in at least one practical application is desirable in order to obtain a good understanding of their interaction and integration. If this happens to be an area where the appropriate machinery is readily available for use and examination, so much the better, particularly if it is one which many find interesting and stimulating. One such application is in the field of internal combustion engines which is itself a subset of the broad and fascinating area covering the production of useful work and power. Work producing machinery (engines and power plants) cannot be properly understood, evaluated and optimized without a detailed knowledge of the above thermofluid subjects and the way in which they are linked. In addition, while these engines may obtain their energy from a variety of sources, e.g. from nuclear processes or from solar radiation, the vast majority, at the present stage of our technological development, derive their power from combustion, internal or external, principally that of fossil fuels. For these, the fundamentals of combustion, often linked only with chemical engineering, must also be considered in conjuction with the above three sciences (thermodynamics, heat transfer, and fluid mechanics) which are regarded as fundamental to both mechanical and chemical engineering. In this text, concepts from all of these sciences are developed together so that the reader can have, in the one book, the means of integrating them and understanding the way in which each impinges on and interacts with the others. While this should provide a background for engine design, analysis and modelling, it is hoped that the reader will be able to translate this approach to a great many other thermodynamic devices.

Combustion engines operate using either internal or external combustion, this terminology referring respectively to whether or not the combustion processes take place within or separate from the engine working fluid. Both types are common. The first covers most of the small to medium-size engines used in transport and stationary applications, these being spark ignition (SI), compression ignition (CI or '*diesel*'), reciprocating engines and most gas turbines. The second, external combustion is now largely confined to electrical power generation which utilizes large steam turbines for the initial conversion of the primary energy source to useful work. With these, the working substance is continuously recycled. Other external combustion types with which the reader may be familiar are some (rarely used) forms of large, stationary gas turbines, *Stirling cycle* engines and the reciprocating, locomotive type steam engines. In this text, only internal combustion engines will be considered, and these include both the 'steady' flow gas turbine and the 'unsteady', intermittent flow reciprocating types. Together, they provide adequate scope for the development of a comprehensive understanding of the above sciences. They are readily available and well understood mechanically in their role as propulsion units for aircraft and automobiles respectively.

Some of the most demanding fluid flow, heat transfer and combustion processes occur in the operation of internal combustion engines because of the constraints imposed by the working fluid and the wide range of operating conditions. Reciprocating internal combustion engines in particular provide a major challenge to combustion researchers and computer modellists because of the short time duration available for the gas exchange processes, fuel addition and its preparation as a mixture for combustion, the ignition of the mixture and the combustion process itself. While many of the physical concepts have been known for many years, the modern emphasis on clean, energy-efficient engines which still retain excellent response characteristics has placed great pressure on researchers, designers and manufacturers to improve their understanding of the fundamental engine processes. Thus, there is substantial continuing research and development in this area. Internal combustion engines are therefore excellent devices to use as examples for an advanced study of the thermal/fluid sciences.

A normal learning approach to any of the above fundamental areas is linear in the sense that the theory is developed from the basic ideas through to more complex ones with a few applications usually being dealt with at the end of a course. Thermodynamic texts are typical of this approach. A usual approach provides basic material on ideal gas properties and a tabular approach to the liquid–vapour transition states together with a fairly thorough study of the first and second law of thermodynamics using the property values so obtained. This is followed by a more detailed discussion of properties and some simple combustion calculations approached from a chemical viewpoint. Finally, a section on basic cycle analysis of engines and heat pumps completes the book. The thermodynamics texts referenced at the end of this book tend to follow this approach. Engine texts on the other hand often assume that a wide, fundamental knowledge of thermodynamics, gas dynamics and, sometimes, heat transfer and combustion is available and the average student may have to make up the deficiency by isolating the appropriate sections in more fundamental works. The present text is aimed at providing a link between these approaches. While it assumes that the very basic material has been covered in a first course, it refers frequently to these fundamentals so that they are revised in the reader's mind. In its structure, it reverses the normal logic by immediately examining cycle analysis in detail, pointing out the advantages of a simple procedure and where it must be improved in order to reach the stage where advanced development can occur. In doing so, the appropriate knowledge in the relevant fundamental engineering sciences is highlighted. In summary, this text assumes that the reader has a basic (i.e. a first course) knowledge of fluid mechanics, thermodynamics and heat transfer with perhaps a very rudimentary feel for the chemistry of combustion. The objective of the book is to develop a more sophisticated awareness of these subjects while at the same time introducing the reader to a fundamental understanding of internal combustion engines.

Once the link between fundamentals and application is perceived in one device, a similar association can be more easily established for many other applications. In this book, the '*engine*' serves as a vehicle for relating advanced theory to engineering practice, while the book serves in part as a fundamental treatment of engines in particular and the applied thermal sciences in general. It is not intended as a replacement for either the many very good thermodynamics, combustion and gas dynamic texts such as those referred to above which give a full treatment from the very beginning of their particular subject or for the excellent engine text books which assume a great deal of specialized knowledge at the outset. Indeed, it would be desirable if many of these are studied simultaneously with this text. The intention of this work is rather to give an eclectic approach to the individual subjects with applications focused on engines, these being a well-known group of devices.

B. E. Milton
May 1993

Useful data

(a) General

Universal gas constant	$R_0 = 8314.3 \, \text{J kmol}^{-1} \, \text{K}^{-1}$
Gravitational constant	$g = 9.81 \, \text{m s}^{-2}$
Avogadro number	$N_a = 6.023 \times 10^{26} \, \text{kmol}^{-1}$
Planck constant	$h = 6.626 \times 10^{-34} \, \text{J s}$
Boltzmann constant	$k = 1.380 \times 10^{-23} \, \text{J K}^{-1}$
Velocity of light	$c = 2.998 \times 10^8 \, \text{m s}^{-1}$
Stefan–Boltzmann constant	$\sigma = 5.67 \times 10^{-8} \, \text{W m}^{-2} \, \text{K}^{-4}$

(b) For air

Specific gas constant	$R = 287 \, \text{J kg}^{-1} \, \text{K}^{-1}$
Molecular mass	$\bar{M} = 28.97 \, \text{kg kmol}^{-1}$

Standard atmospheric conditions 101.3 kPa, 298 K

Specific heat at constant pressure	$c_p = 1004.5 \, \text{J kg}^{-1} \, \text{K}^{-1}$
Specific heat at constant volume	$c_v = 717.5 \, \text{J kg}^{-1} \, \text{K}^{-1}$
Isentropic index	$\gamma = 1.400$
Absolute viscosity (at 1 atmos.)	$\mu = 184.6 \times 10^{-7} \, \text{kg m}^{-1} \text{s}^{-1}$
Thermal conductivity	$k = 26.3 \times 10^{-3} \, \text{W m}^{-1} \text{K}^{-1}$

500 K

Specific heat at constant pressure	$c_p = 1028.6 \, \text{J kg}^{-1} \, \text{K}^{-1}$
Specific heat at constant volume	$c_v = 741.6 \, \text{J kg}^{-1} \, \text{K}^{-1}$
Isentropic index	$\gamma = 1.387$
Absolute viscosity (at 1 atmos.)	$\mu = 270.1 \times 10^{-7} \, \text{kg m}^{-1} \text{s}^{-1}$
Thermal conductivity	$k = 40.7 \times 10^{-3} \, \text{W m}^{-1} \text{K}^{-1}$

1000 K

Specific heat at constant pressure	$c_p = 1141.2 \, \text{J kg}^{-1} \, \text{K}^{-1}$
Specific heat at constant volume	$c_v = 854.2 \, \text{J kg}^{-1} \, \text{K}^{-1}$
Isentropic index	$\gamma = 1.336$
Absolute viscosity (at 1 atmos.)	$\mu = 424 \times 10^{-7} \, \text{kg m}^{-1} \text{s}^{-1}$
Thermal conductivity	$k = 66.7 \times 10^{-3} \, \text{W m}^{-1} \text{K}^{-1}$

(c) Atomic masses of common substances used in engines

Substance	Atomic Mass
Carbon, C	12.01
Hydrogen, H	1.008
Oxygen, O	16.00
Nitrogen, N	14.007

(d) Typical higher and lower heating values of some common fuels

	Higher (MJ kg^{-1})	Lower (MJ kg^{-1})
Petrol (gasoline)[1]	47.3	44
Distillate (light diesel)[1]	44.8	42.5
Fuel Oil (heavy diesel)[1]	43.8	41.4
Aviation Fuel (kerosene)[1]	44.0	42.0
Natural Gas[2]	51.7	46.7
LPG[3]	50.2	46.3
Methane	55.5	50.0
Propane	50.3	46.4
n-Butane	49.5	45.7
n-Octane (liq.)	47.9	44.4
Methyl Alcohol (liq.)	22.7	19.9
Ethyl Alcohol (liq.)	29.7	26.8

[1] Values may vary slightly due to composition of the fuel. They are representative values only. [2] Composition varies considerably, typical sample for above data: 90.9% methane, 5.0% ethane, 1.1% propane, 0.3% butane, 2.4% carbon dioxide, 0.3% nitrogen. [3] Composition varies considerably, typical sample for above data: 90.0% propane, 5.0% butane, 5.0% propylene.

Introduction to internal combustion engines

1

1.1 THERMODYNAMICS, COMBUSTION AND ENGINES

The topics, **thermodynamics, combustion** and **engines**, are important elements of the engineering background of either a generalist or a specialist engineer because they examine the principal way in which energy is used in society. While they are studied in many engineering courses they are often areas in which both students and practitioners do not feel comfortable. This particularly applies to the first two in that the concepts of thermodynamics are perhaps considered too abstract for a practical approach while the area of combustion is still regarded as ill-defined and intuitive. Engines, however, are concrete examples of practical devices well known to all but which are currently evolving rapidly because of the requirements of greater fuel efficiency, low emission levels and possible future changes in their fuels and they therefore now require a much better understanding of the fundamental sciences on which their design and development is based. So by studying both simultaneously, an understanding of both the more abstract science of thermodynamics and the broad technology of combustion can be enhanced by illustration from the well known mechanical devices, engines. This book is intended to be used for that purpose while simultaneously providing a scientific background sufficient to provide progression to further more specialized engine studies.

A few definitions are required.

Thermodynamics is the science which studies the conversion and interchange of energy in its various forms with particular regard to work and heat transfer. It requires the understanding of both the energy-related aspects of the properties of matter and the laws which govern the interchanges, these being designated as the (four) laws of thermodynamics. From an engineering perspective, the amount of useful work and power which can be obtained (or used) via a machine can be evaluated by first law studies whilst the possible efficiencies of the interchange are governed by the second law. It is assumed that the reader has undertaken a basic thermodynamics course, is familiar with these laws and has a cursory understanding of the thermodynamic properties of such substances as ideal gases and steam. This text starts from that point and undertakes a thorough thermodynamic analysis of engine cycles with the necessary additional background being given for the evaluation of high temperature and non-ideal gas properties relevant to them. Some revision of basic thermodynamics is required and the reader is referred to the books listed at the end of this text. These are only a selection of the many excellent texts on the subject and wide reading is recommended.

Combustion is the process of oxidation of molecules, usually carbon, hydrogen or hydrocarbons which occurs readily at high temperature with the release of energy. That is, the process is exothermic. If the energy release is sufficient, the process is selfsustaining and may liberate further energy in the form of heat and light for useful purposes. The importance of combustion can be understood when it is realized that the vast majority of the energy currently consumed worldwide is obtained from the combustion of fossil fuels and, even if renewable sources eventually provide much of the world's energy, it is probable that a high proportion of these will continue to be utilized through combustion because of the need for energy storage and the advantage of combustible substances in that respect. While combustion in the form of simple fires has been an important fact of human technology since the beginning of civilization, it is surprisingly poorly understood. This is because a thorough knowledge of a great many factors involved in the process is required and some, such as the interaction of combustion and fluid mechanics, are still being refined. Factors such as the fuel composition, the chemical reactions, the flow and mixing of the fuel and air, the heat transfer and the time and space available for the reaction, are important and these are quite variable between different devices and the different operating regimes of any one device. Combustion is therefore a complex topic and has in the past been viewed by many engineers as, at best, a cross between an art and a science. However, modern research and numerical modelling have removed much of its intuitive nature, making the design of combustion devices more amenable to a fully scientific approach. Using the steady flow gas turbine and the unsteady flow reciprocating engines as a background, this text builds up ideas on the intricacies of the combustion process. Several basic texts on combustion are listed at the end of this book.

Engines are the machines which are used for the production of useful work and power. Strictly, a single process machine, e.g. the devices for hurling rocks at castle walls in mediaeval times, can be called an engine but in modern parlance, the word is generally reserved for a machine with a continuous output for at least a short period of time. Thus, the prime focus is on its power. There is some overlap in the use of the words engine, motor and power plant and they are, to some extent, synonymous. **Engine** is the term most likely to be used for combustion machines in the small to intermediate power ranges (from less than 1 kW to, say, around 400 kW), **motor** is used for devices using a secondary energy carrier such as electricity as an energy source (although in transport it is also in common usage for mobile combustion units) and **power plant** for large stationary units. In general, all combustion engines work on a cyclical basis, taking in energy at high temperature and rejecting it at a lower value. A cycle of this type is termed a **thermodynamic** one with a series of processes being repeated continuously implying that the same thermodynamic states are reached repetitively. Cycle analysis is therefore a fundamental part of the thermodynamics of engines and will be examined here with the aim of giving an appreciation of the complexities that are required to closely simulate engine operation. It should be noted that this text concentrates on **internal combustion engines** and a further general discussion on them will be given in the later sections of this chapter. Again, a number of excellent texts are available either on internal combustion engines as a whole or on aspects of their operation and these are noted at the end of the book.

1.2 COMBUSTION ENGINES

The majority of engines use combustion for their energy supply although other types do exist, e.g. those powered by nuclear, solar or wind energy sources.

Combustion engines may be immediately classified in two groups, these being as follows.

- *External combustion engines* In these, the working substance and the hot combustion products are different fluids and are separated by a conducting wall. That is, the air, fuel and combustion products which are continuously exchanged do not at any stage contact the moving parts of the engine. This is the function of the separate working substance which is continuously recycled through the various components of the machine. External combustion engines include steam engines of various types (reciprocating and turbine) and Stirling cycle (air or other light gas) engines.
- *Internal combustion engines* Here, the hot combustion gases themselves become the working substance and energy is transferred directly from them as work. They can be reciprocating types, of either the spark ignition (SI) or compression ignition (called either 'diesel' or CI) which have intermittent flow and combustion processes or a variety of continuous types, such as gas turbines and rocket engines.

Although at first glance, it may seem simpler to combine the functions of the combustion and working fluids, the problems with internal combustion engines are to some extent greater than external combustion because, in the former, the moving parts are themselves subject to the high temperature gases at some stage during the cycle. However, their advantage lies in their having fewer components (although not necessarily moving parts) and therefore lower weight and size for a given power output. In the continuous flow types, the temperatures of the moving components can become severe and much of the progress in their development is limited by this aspect. In the reciprocating types, combustion occurs during a limited part of the cycle only. Thus many problems exist in continually initiating combustion and effectively transferring the heat energy to mechanical work in the short time period available. This text deals with air breathing internal combustion engines and examines the thermodynamics of the cycles, combustion processes, working substance (gas) exchange, heat transfer and other aspects which are important in maximizing work output and efficiency and minimizing exhaust emissions.

1.3 BRIEF HISTORICAL REVIEW OF INTERNAL COMBUSTION ENGINE DEVELOPMENT

1.3.1 Reciprocating engines

It is not the intention of this section to discuss fully the history of the reciprocating internal combustion engine. This is an immense task and the interested reader is referred to the book, by Cummins (1976) which deals with this in full. What follows is a simple overview which will perhaps give the reader an introductory assessment of the problems which engine designers faced, what sort of problems were overcome and what concepts still need further consideration. In the future, new ideas or development of old ones will emerge which will further challenge designers.

Reciprocating internal combustion engines were developed before the arguably simpler concept of rotodynamic (turbine) types and became successful units in the latter half of the nineteenth century. Some attempts at creating a gas turbine did occur about the beginning of the twentieth century but successful types did not appear for almost fifty years. While this was undoubtedly due to some difficult aspects of their design which will be explained later, the initial concentration on reciprocating types was probably due to their likeness to the cannon which was a well-developed early combustion-driven apparatus. Since its inception, the reciprocating internal combustion engine has undergone considerable development although its overall basis has remained substantially similar for a long period. It exists in the spark ignition (SI) form in the smaller range and the compression ignition (CI) in the larger although these two overlap to a large extent. While development has in fact proceeded continuously over the whole period, it is useful to view it in three stages because of the prevailing emphasis in each particular period. These can perhaps be grouped as a first stage of **initial concepts and invention** when the basic understanding of engine operation was underway to produce reliable machines similar in overall design, if not in detail, to those currently in operation, an intermediate period of **consolidation** when the requirements of high specific power and high reliability were at the forefront and a final period of **refinement** driven by the need for the improved emission levels and efficiency necessary for the constraints of the highly populated energy-hungry modern world.

Initial concepts and inventions – The early period

External combustion engines, in the form of the steam engine, predated the first successful internal combustion engine by several centuries. This was partially due to the comparative ease of the continuous combustion process combined with the high specific heat capacity of steam making heat transfer from the burner to the working fluid the obvious choice. However, these engines were large and the small power unit so conveniently obtained from internal combustion was delayed for a very long time because the processes for creating a successful engine of that type were not understood. It is not surprising that, when the first internal combustion engines appeared, they followed the then current steam engine practice of using reciprocating pistons in cylinders. This also tied in with the concept of an engine based on the operation of cannons which could, with some licence, themselves be decribed as the first successful internal combustion engines and which were certainly the devices on which the earliest IC engine designs were founded.

Based on this line of thought, Frenchman Denis Papin in 1690, produced a device in which exploding gunpowder was used to compress air but it was a further century until the next development, more akin to the workable approaches of later engines, by Englishman Robert Street who, in 1794, patented an engine which was to use a mixture of vapourized turpentine and air. This apparently was not built but, in the early nineteenth century, the combustible gas then being obtained from coal stimulated the invention of several similar engine concepts. Other patents for liquid fuel engines also followed but none were exploited commercially, possibly because they could not compete with the, by then, well-developed steam engine or the developing air engine.

The major developmental stage should be dated from 1857 with the design by two Italians, Eugenio Barsanti and Felice Matteucci, who built a free-piston engine in which the explosion drove a heavy piston freely up a vertical cylinder. On falling under the influence of both gravity and the vacuum created by its own overrun during its ascent, it engaged a ratchet which turned a shaft. This was, in hindsight, obviously not likely to be a very successful system mechanically. It was followed very soon afterwards (1860) by the first machine which, although less efficient, was successful commercially, this being the gas engine of Belgian Etienne Lenoir. Here, gas and air were inducted at atmospheric pressure for the first half of the stroke at which point ignition occurred, providing a pressure rise which was sufficient to produce the work output, the expulsion of the burned gases, the induction of the next charge and the return of the piston to the position for ignition. Efficiency was very low being perhaps about 5% (Cummins 1976) but it fulfilled a need for a small engine in workshops and some hundreds were constructed. A number of other similar 'atmospheric' engines followed. In 1867, Germans Nikolaus Otto and Eugen Langen produced their first engine which, although possibly an independent invention, worked on the same principle as that of Barsanti and Matteucci. It was limited in power to about 4 kW and was extremely noisy but proved to be noticeably more efficient than Lenoir's engine because its burnt gas was expanded over a greater ratio. This was an important factor not realized at the time and it allowed efficiencies of up to 11%. It further established that a small engine was a marketable and sought-after commodity and a number were produced.

During this period, the basic theoretical concepts of the operation of an internal combustion engine were being grasped due to the work of French physicist Sadi Carnot in general and Beau de Rochas, a French railway engineer, in particular. The latter, between the years 1851 and 1888 published a paper in which he defined the basic four stroke cycle in common use today but this did not receive much publicity until many years later. The most important point was that he noted that compression before ignition was needed in order to give an adequate expansion ratio for the working stroke so that a good output could be obtained.

It is not clear whether the necessary knowledge to progress to a 'modern' engine was commonly understood or whether progress stemmed from practical development. However, in 1876, the first engine following these principles was produced. This was the Otto 'Silent' engine which worked as a four-stroke and used a compression ratio of 2.5:1. Mechanically, it dispensed with the crosshead which, following steam engine practice had been used on all previous engines, and connected the piston directly to the crank. The engine had a noticeably higher efficiency (of about 14%) than any prior to it but nevertheless, it appears that Otto himself did not fully appreciate the value of compression and argued that this was due to the stratification of the charge. A controversy, stimulated by court cases over patent rights, raged on this point with Scottish engineer Dugald Clerk who was the first to use an air standard cycle analysis, supporting the case of compression before ignition. Clerk made other notable contributions to the internal combustion engine, being one of the developers of the two-stroke cycle, possibly to circumvent patents already in existence. These engines have provided an important alternative to the four-stroke, particularly in very small petrol and in very large diesel engines and may make further inroads into a wider range of sizes in the future.

The initial Otto engines rotated quite slowly and consequently produced little power, by modern standards, for their size. Eventually, the engine was refined by German engineer Gottlieb Daimler who in 1883, using more precise engineering, produced what he called the 'high speed' internal combustion engine which operated at almost 100 rpm using a hot bulb type of ignition device. This now had a sufficient power to weight ratio for vehicle use and was adapted for that purpose by Daimler in 1886, a few months after the first internal combustion engined vehicle–the two-stroke engined tricycle of another German Karl Benz.

The peripheral equipment essential for engine operation also underwent considerable development. It should be noted that the engines predominantly in use up to this time used coal gas as the principal fuel but, with the advent of the motor vehicle, the advantage of gasoline (petrol) was obvious because of its high energy density. Hence, during the 1880s, many methods of vapourizing fuel or introducing it into the cylinder during compression were tried. These basically relied on 'surface' methods where air was bubbled through the liquid. Eventually German Wehelm Maybach's design of the carburettor in 1892 solved the problems. This drew a fuel spray into the air stream as it passed through a venturi, a principle that became the major method for fuel introduction for almost a century and one which is still in use today. Ignition systems also needed considerable development. The early ignition systems were primitive and either a red-hot tube or bulb heated by a flame or a low tension wiper type spark was used for a considerable period. Eventually, the invention of the high tension system by Robert Bosch in the 1890s allowed the development of the modern electric spark plug.

About 1880, an additional type of combustion engine was being developed by William Priestman and Herbert Ackroyd–Stuart in England. While what could be termed the 'light oil' (i.e. petrol) engines used fuel vapourized before induction and were therefore similar to gas engines, these used the heavier fractions vapourized by heating within the cylinder itself. Ackroyd–Stuart developed a system to inject fuel into a vapourizer chamber with a restricted throat to the main chamber to give better combustion. However, the full implication of the turbulent mixing in a prechamber does not appear to have been understood. Nevertheless, his designs were the basis of the modern oil engine and, although they did not rely on compression ignition, the high temperature of the vapourizer chamber allowed them to continue running without other forms of ignition. Compression ignition, the final fundamental development in reciprocating engines came with the concept of the compression ignition engine by German Rudolf Diesel. Diesel, an engineer by training, was the first of the inventors to use a scientific approach to engine design, his initial idea being to create one with a cycle as close to the Carnot cycle as possible. Practicalities, such as the amount of work produced per cycle, forced him to depart from this but the overall thrust of his ideas remained scientific. Efficiencies of up to 25% were achieved early in the development. One of the major hurdles to be overcome to produce a workable device was that of fuel injection as a finely atomized spray into the highly compressed air. Diesel's original designs used an air-blast fuel injection system. These required a large complicated high-pressure compressor (to about 7 MPa) driven from the crankshaft and therefore the CI principle was confined to very large engines. Although Ackroyd–Stuart had invented a fuel only (solid) injection system several years before Diesel's original patent, it was not until 1910 that l'Orange at Benz with his prechamber design successfully converted an air-blast engine to solid injection. Soon afterwards, Hawkes, in conjunction with Vickers Ltd in the UK managed a similar conversion (Walshaw, 1950). These improvements paved the way for the development of the light high speed CI engine of today.

The powerful, reliable engine – A consolidation of ideas

The period immediately following the early development (i.e from about 1900 for SI engines and 1910 for CI ones) saw basically a consolidation of the principles and ideas of the previous era. By about 1920, most of the basic principles that are in use today in engine analysis were known although they were not able to be applied with the thoroughness that modern computational techniques allow. These known effects included such things as the burning rate, compression ratio, air/fuel ratio and pumping losses on engine efficiency and concepts as advanced in a modern sense as stratified charge combustion in SI engines had been put forward by Englishman Harry Ricardo. Engines of that time were already impressive. The work of Midgeley (1920) on combustion and knock (i.e. the so-called 'detonation') in SI engines and on additives to the fuel which would minimize it took place in the early 1920s (Boyd, 1950) and it is to him that the discovery, good or bad depending on the stance taken, of the addition of tetra-ethyl-lead as a knock suppressant was due. Another example of pioneering work on the details of engine processes of that time is the work of Mock (1920) which initially described the separated fuel air flows which can exist in inlet manifolds. Details of the ignition delay in CI engines and its relationship to 'diesel' knock were reported as early as 1930, fuel sprays in CI engine combustion chambers were photographed in the early 1930s and combustion bomb experiments quantifying the ignition delay of fuels at high temperatures and pressures were obtained in the mid 1930s. A thorough method of cycle analysis (subsequently termed **fuel air cycle analysis**), replacing the traditional air standard cycle, was described by Goodenough and Baker (1927) and thermodynamic charts developed by Hottel, Williams and Satterfield (1936) allowed easy cycle calculations with the correct properties at different temperatures, different fuel/air mixtures and combustion including dissociation effects to be performed. Heat transfer calculations relevant to engines were introduced by German Wilhelm Nusselt as early as 1914 and considerable work in that area was carried out by both he and Eichelberg (1939) during the next twenty years. Of considerable interest is the research and development work of Ricardo on many aspects of both SI and CI engines in the 1920s and 1930s as is the research carried out in the Sloan Laboratories at the Massachusetts Institute of Technology (MIT) after about 1930 originally under the direction of C. F. Taylor.

By about 1920, the SI engine had become a reliable machine with a reasonable power output. For example, the Continental engine was typical of the USA automobile engines of the period. It was a six cylinder in-line engine of 3.7 litre capacity, compression ratio about 5:1, weighing 260 kg and developing 41 kW at 2600 rev min^{-1}, a low power-to-weight ratio at a low speed by modern standards but nevertheless an impressive engine. It used the then almost universal (in automotive applications) side-valve arrangement although the advantages of overhead valves were already known. For example, in their aeroengines for World War I, Rolls–Royce had opted for a V12, with a crossflow cylinder head with overhead valves and camshafts (Robins, 1986). The 1914 Mercedes Grand Prix car used a similar design. However, because of its simplicity, the side-valve engine persisted in some automobiles even into the 1950s. Specific power (power per unit mass) gradually increased over the next thirty to forty years with improvements in fuel octane numbers and combustion chamber design allowing higher compression ratios to be used and improved carburettion and breathing giving higher engine speeds. The downdraft carburettor replaced the earlier updraft version and the airvalve, constant depression sidedraft type was introduced for high-performance vehicles.

During World War II, the most powerful SI engines ever built were developed for aircraft propulsion, one of the best known being the 27 litre, V12 Rolls–Royce Merlin although its later larger (by 10 litre) version, the Griffon, was even more powerful. The Merlin was originally a 760 kW (1030 HP) engine when it entered service in 1937 but, with the introduction of two-stage intercooled supercharging it eventually produced 1530 kW (2050 HP) and could maintain power equal to that of the original version to 11 000 m (36 000 ft). These changes occurred without increase in size or alterations to the compression ratio or valve timing. However, the advent of the gas turbine brought a halt to further development along these lines and the SI engine reverted to first of all a prime mover for automobiles. Fuel injection, desmodromic (springless controlled closure) inlet and exhaust valves and turbocharging had been tried in racing car engines and, by the 1950s, it was perhaps felt that the major period in SI engines research and development was over. This, however, was not the case.

The pollution and fuel crises – The modern period of refinement

By 1960, the air quality in the city of Los Angeles in California had deteriorated to such a level that drastic action was required. While there were (and are) a number of causes, one of the major sources was found to be the carbon monoxide and unburned hydrocarbons from the exhaust gases of automobiles. Emission controls became mandatory, firstly in California beginning with the introduction of the positive crankcase ventilation (PCV) valve in 1961, (which was not a new device as it had been used previously by British manufacturers) as it allowed the unoxidized or partially oxidized gases which leak past the piston and accumulate in the crankcase to be returned to the combustion chamber. Full emission control legislation followed almost immediately with consequent adverse effects on the fuel consumption of the current vehicles. Similar legislation followed throughout the USA and, more slowly, in many other parts of the world. In some European countries, tight emission controls were introduced soon after the USA with the EEC legislating for major universal controls in 1986. Japan, Canada and Australia introduced controls from about the mid-1970s onwards.

Photochemical smog was soon perceived as a major environmental problem. It was known that the nitrogen–oxygen compounds (NO_x) in conjunction with unburned hydrocarbons were instrumental in its formation and this added a new dimension to the emission control problem. Both carbon monoxide and hydrocarbon (HC) compounds could be removed by oxidation but the NO_x compounds required reduction. Hence the multiple control of the different exhaust emissions required a renewed approach to fundamental research in engine combustion. In the USA, new research programs sprang up based both on experimental work and engine modelling. There were many contributors in industry, universities and research establishments. Such things as the variation in unburned hydrocarbon emissions during an engine cycle were at least partially explained and the amount of NO_x during operation was quantified. Much work, however, remains to be done on all these topics.

The oil crisis beginning in 1973 further emphasized the necessity of a continuing high level of fundamental research into engine operation. This occurred at a time when many of the emission problems, if not fully understood, were at least readily controlled in practice. The reduction of emissions had generally increased fuel consumption and it was now necessary not only to retain and further improve exhaust cleanliness but to revert to pre-emission control or even better levels of engine efficiency. In addition, alternative fuels and engine types needed study. The importance of engine fluid mechanics, as it effected both the gas exchange processes, mixing within the cylinders and the combustion phenomenon itself, now needed to be emphasized to a greater extent. This applied particularly to the role of turbulence. Engine modelling therefore became a dominant factor in research and remains so to the present. New problems continue to present themselves and those that are currently the most pressing (in 1994) are the need to meet even more stringent clean air legislation (the ultralow and the zero emission vehicle concepts in California), the minimization of 'greenhouse' gases and the further improvement in fuel economy.

1.3.2 The gas turbine

Turbine type devices were familiar from very early in the history of machinery as the waterwheel, windmill and, eventually, pumps and hydraulic turbines. The topic of turbomachinery development is too extensive for details to be given here and the reader is referred to more specialist texts on the subject. A brief but more comprehensive treatise is given in the introduction to Wilson (1984). What is perhaps of importance is the development of the rotodynamic air compressor which is an essential component of the gas turbine but which predated it. The first commercially successful compressors were produced by Englishman Charles Parsons in the latter part of the nineteenth century. These consisted of both axial and radial flow devices although the latter were the more successful. At the turn of the century, Auguste Rateau produced compressors which Wilson quotes as having pressure ratios exceeding 1.5:1 and isentropic efficiencies above 56%. Also, the steam turbine is itself of significance because of the similarity in some of the flows associated with it to the gas turbine. The concepts of reaction where the blade is driven by the increase in velocity across it and impulse where the change in direction of the fluid provides the motive force was elucidated. Compounding of moving rotor blade rows with fixed stator rows between them to redirect the fluid was introduced. In addition, the concept of the supersonic nozzle with its limitation on mass flow rate due to choking became understood. The basic knowledge of turbomachinery from which a gas turbine could be developed was therefore in existence by the beginning of the twentieth century.

Due largely to the pioneering work of Parsons, the steam turbine started to replace reciprocating steam engines in the last decade of the nineteenth century. It was natural then to also look towards a gas turbine as a successor to reciprocating internal combustion engines. However, the development of this engine type was slow. Many of the problems of high rotational speeds had been overcome with the steam turbine development but two other factors kept the gas turbine from equivalent progress. These were first of all metallurgical in that, when the continuous internal combustion processes of the gas turbine are compared to the intermittent ones of the reciprocating engine, it becomes evident that for the same gas temperatures parts of the engine structure of the former will become extremely hot. This is because there is no cooling inflow over the hot surfaces and no time relief from the combustion process. The problem is particularly evident at the combustor and the first row of turbine blades. Thus low temperatures were necessary to maintain metal integrity giving potentially low work outputs and efficiencies. The second factor was the efficiency of the rotodynamic machinery. It will be shown later that very high compressor and turbine component efficiencies are required to produce a gas turbine with an acceptable overall efficiency. That is, a low turbine efficiency uses too much of the output to run the compressor, a low compressor efficiency uses up too much turbine work. Development of the gas turbine was therefore delayed until both highly efficient rotodynamic machinery was developed and strong high temperature alloys were available.

The earliest gas turbine concept also emanated from Parsons who, in his 1884 patent of the steam turbine, commented that the possibility of a gas turbine also existed. In 1905, a gas turbine burning oil was built by the Société Anonyme des Turbomoteurs in Paris to the design of René Armengaud and Charles Lemale. They used a centrifugal compressor with a pressure ratio of about 3:1, continuous combustion with an atomized oil spray with water cooling of the turbine. A high combustion temperature was necessary to overcome the low isentropic efficiency of the compressor. Steam raised in the turbine cooling was used with the combustion products as they entered convergent–divergent nozzles ahead of the turbine. This machine ran at around 4000 rpm and produced power, although some reports indicate that at times it was barely sufficient to drive its own compressor. Wilson quotes its thermal efficiency as 3.5%.

In 1908 Hans Holzworth in Hanover produced a more successful design. The combustion was at constant volume which was controlled by valves in the combustion chamber. Discharge then took place through convergent–divergent nozzles to a turbine rotor which was driven by a succession of pulses rather than continuously. The first unit did not compress the mixture before combustion but later versions added this process. Generally, mixed gaseous fuel was used although some units sprayed oil into the air into the combustion chamber before ignition. These engines were built up until about 1920 and produced significant power (hundreds of kilowatts) although at low efficiencies of less than 13%. Development of other engines, particularly the CI engine, caused them to become superseded at about this stage. Exhaust gas turbines for aircraft engine supercharging then remained the only useful rotodynamic devices with a major power producing role in internal combustion engines for some years.

Advances in aircraft design stimulated the revival of the gas turbine in the 1930s. Here, turbine rotors became less of a problem because only a small proportion of the work, that for the compressor, was extracted in the turbine, the remainder being obtained from expansion in the jet nozzle which was far simpler to design. In 1930, Frank Whittle in England patented a design using continuous, constant pressure combustion between the compressor and turbine. His design used a pressure ratio of 4 obtained from a single stage centrifugal compressor. Excess air from the compressor which bypassed the initial combustion zone was fed back in downstream into the combustion chamber as a coolant. The hot gases flowed in a fully annular convergent nozzle system to the small turbine before expansion in a jet nozzle. This engine first came into production in 1945. At about the same time, Pabst von Ohain in Germany produced a similar gas turbine which used a centrifugal compressor and a radial inflow turbine. This proved effective in initial performance and the world's first jet powered aircraft flight was carried out with it in a Heinkel. In modern engines, most high-performance types now use axial flow compressors although the robust nature of the centrifugal type still makes them useful. The basic principles, however, have remained the same and improvements have come through high pressure ratios, high turbine entry temperatures obtained by improving their cooling and better metal alloys. Also, in aircraft propulsion, the use of bypass air to improve the thrust efficiency of the jet has made a marked improvement.

1.4 CURRENT AND FUTURE ENGINE DEVELOPMENT

The modern engine, SI, CI or gas turbine, now in use is a technically sophisticated unit. Nevertheless, continuing pressure on fuel resources, the need to implement alternative fuel strategies in the future and increasing population levels, energy use and the consequent environmental problems mean that a high rate of development will continue for some considerable time. In order to assess current technology and its future direction, it is perhaps better to consider engine development for efficiency and modifications for alternative fuels separately. It is emphasized, however, that there is a strong interrelationship between the two in practice.

1.4.1 Modern IC engines and their alternatives

Various engine types and their potential developments are discussed below. In many cases, reference will be made to features that improve their performance and efficiency which may be obscure to the reader. These will be dealt with in detail in the later text. It is important here to give an overall impression of engines and their development so that some of the following material assumes its proper relevance. The engine types to be discussed are homogeneous charge spark ignition engines, stratified charge spark ignition engines, compression ignition engines, turbocharged engines, and gas turbines.

Homogeneous charge spark ignition engines

The modern SI engine is predominantly a homogeneous charge type operating on the same general principles as did those at the turn of the century. However, its fuel efficiency, power output and exhaust cleanliness are now much superior. Engine efficiency improves with a number of factors, the most important being:

- **High compression ratio** The beneficial effect of a high compression ratio has been well understood since the debates of Clerk and Otto last century. However, an increase in compression ratio above certain limits does not continue to raise the efficiency due to the greater heat transfer from the combustion chamber and greater friction which offset the thermodynamic gains. This was originally shown in the work of D. F. Caris and E. E. Nelson from General Motors who found for the type of combustion chamber tested, that an optimum value was about 17:1. Engine knock has generally limited compression ratios to about 10:1. Recent work on different combustion chamber designs (May, Ricardo) have shown that higher compression ratios are feasible given good chamber design. In particular, lean operation, compact chambers with high swirl and cool end-gas all improve the knock resistance allowing higher compression ratios to be used.

- **Fast burn** Cycle analysis highlights the beneficial effect of faster heat addition rates although these can be excessive, causing roughness and higher heat transfer losses. The correct amount of turbulence needs to be designed into the head and, as turbulent flame propagation is still not well understood, considerable research is currently underway in this area. Lean mixtures burn more slowly than stoichiometric ones and require greater turbulence and higher ignition energies. The higher temperatures of high compression ratio increases the burning rate.

- **Lean burn** A lean burn engine has effectively a higher ratio of specific heats during the cycle which can again be shown to be beneficial as long as the burn rate can be maintained. There are also advantageous exhaust emission effects (for CO and HC) which can replace alternative emission control strategies which are fuel inefficient. There are limits on the extent of lean burn due to misfire and cycle-by-cycle variability which can negate the advantages.

- **Improved control and optimization of the engine** This applies to air/fuel ratios particularly between cylinders and during transient operation. Any maldistribution results in an efficiency deterioration. Also important, although less complex, is the optimization of spark timing and recirculated exhaust gas. The simultaneous control of all these parameters to maintain maximum engine efficiency under the constraint of meeting the necessary exhaust emission standards is a challenging problem.

- **Reduced engine size** For the same maximum power, a smaller engine will be operating at a higher mean effective pressure. Here the pumping and friction losses become a smaller proportion of useful power. Gains are particularly noticeable at part throttle. The smaller displacement can be realized by improved engine design to allow better breathing at higher engine speed and by turbocharging.

The modern SI engine incorporates most of these features. Engines now becoming available may have high turbulence and swirl, multiple valve heads and multipoint fuel injection. Obviously, while these all can be advantageous, they raise the cost of an engine considerably.

Stratified charge spark ignition engines

The concept of a stratified charge SI engine that used a very lean mixture in the end-gas (last to combust) region but a stoichiometric or richer mixture near the spark plug so that reliable ignition occurs dates from Otto's ideas in 1872. This type of engine could operate on near optimum compression ratios because the end-gas would not be prone to autoignition which causes knock and its control would be possible by varying the fuel/air ratio in the main, lean part of the combustion chamber thereby reducing part-load throttling losses. In addition, in relation to the more modern problem of exhaust emission, it would be possible to have an overall lean mixture to minimize CO and a flame which burned from a slightly rich into a very lean mixture avoiding the near stoichiometric condition at which NO_x formation is high. Unfortunately, it has been hard to develop because it requires a direct in-cylinder injection system with rapid mixing but a retention of the stratification in the desired location. Long-term development has occurred with the Ford PROCO engine and the Texaco TCCS engines since the early 1970s but neither is yet in production. Both use a deep bowl in the piston together with high inlet swirl but the Ford engine has a low pressure early (in the compression stroke) injection while the Texaco engine uses high pressure late injection. Thus the Texaco engine operates more like a diesel engine but has neither cetane nor octane requirements for its fuel. The Honda Motor Corporation developed a prechamber type of stratified charge engine, the CVCC, in which the 'jet' issuing from the spark ignited rich prechamber ignited the fuel in the lean main chamber. This engine was used principally to meet USA emission standards and was not universally adopted. Finally, the recent Australian development of the three-cylinder, two-stroke direct in-cylinder injection engine by the Orbital Engine Company should be noted. This injection system is an airblast type but of lower pressure than the old diesel engine concept. Very good atomization is claimed and, if the stratification can be retained for a wide range of operating conditions particularly during transients, it could be the most promising development of this type. Whatever stratified charge engine will prove to be best is not yet clear but, during the 1990s, the progressive amalgamation of the best ideas from the homogeneous and stratified charge concepts seems likely.

Compression ignition engines

Compression ignition engines are even more varied in detailed design than SI ones in that the largest have cylinder bores approaching 1 m diameter and rotate at about 100 rev min^{-1}. while the smallest have a capacity of around 0.4 to 0.5 l per cylinder and engine speeds up to about $4000 \text{ rev min}^{-1}$. Cylinder head designs also vary commensurably. The major subdivision is between direct injection (DI) where fuel is sprayed into the single main chamber and indirect injection (IDI) where the spray is into a prechamber which generates high swirl or turbulence and is fuel rich. The burning mixture issues through a throat into the main chamber where final oxidation occurs. The IDI engine is less efficient than the DI engine but has wider speed/load characteristics and is mostly used on small automotive engines. The major changes are that smaller DI engines are being produced and that, in both small DI and IDI engines, optimization of chamber shape and fuel injection spray patterns is occurring. This is largely a result of better simulation of in-cylinder flow patterns. Another major change is that turbo-charged diesel engines are commanding an increasing proportion of the market.

Turbocharged engines

There is substantial energy remaining in the exhaust gas of an engine and harnessing it is an attractive proposition. Turbocharging was first proposed and patented by Alfred Buchi of Switzerland in 1905 but it was many years before it became commonplace even on CI engines to which it is most adaptable. For many years, mechanically driven superchargers were more commonly used, particularly on SI aircraft engines in order to obtain alti-tude compensation. This is because the matching of turbine and compressor to the engine with problems of lag during transients and excess pressure rise at high loads had not been solved. Also units were large slowly rotating devices suited much more to very large diesels. Substantial developments have occurred in recent years in both SI and CI engines. Turbine and compressor speeds have risen considerably and improved component isen-tropic efficiencies have made the total system more viable. The introduction of the waste gate whereby excess exhaust gas bypasses the turbine thereby preventing turbine overspeed has improved their viability for SI engines.

For CI engines, there are few disadvantages other than cost. The higher pressures and temperatures reduce the ignition delay of any given cetane number fuel and an adequate power-to-weight ratio can be obtained with-out pushing the fuelling to the limit. Thus, smoke levels at peak loads are reduced. Any disadvantages lie in the higher mechanical and thermal load-ings and, for vehicle engines, the peakier torque curves caused by turbo-charger speed and hence engine intake pressure falling as engine speed reduces. Turbocharged trucks require gearboxes with many more ratios than do those with equivalent maximum power normally aspirated engines. Attempts are underway to overcome these problems by use of variable area volutes for the turbine. More truck engines are now turbocharged than not and most manufacturers offer three versions of the one engine, normally aspirated, turbocharged and turbocharged/intercooled. An important devel-opment in CI engines is the use of ceramic materials to minimize heat flow from the engine, the ultimate possibility being the 'adiabatic diesel' which does not require coolant. Here, the additional energy retained in the gas because of the substantially reduced heat transfer can, in theory, be used not as additional power to the piston because of expansion ratio limitations but by an additional power turbine in the exhaust. Combining its output to that of the crankshaft poses considerable problems and it is likely to be a long time if at all before turbocompounded ceramic engines are in use.

For SI engines, the picture is more complex. There is a misconception that simply turbocharging a given engine will increase efficiency because of its use of the exhaust energy. However, turbocharging only increases efficiency if the engine size is reduced, otherwise for the same load as the original engine, it must run at part load with greater throttling causing increased pumping losses. In addition, the higher cylinder temperatures and pressures do not reduce combustion knock as in CI engines, but increase it. Thus, the compression ratio needs to be reduced causing a further drop in efficiency. The lag which occurs on any turbocharged engine due to the time for the turbine/compressor unit to build up speed can be exacerbated by additional fuel buildup on inlet manifold walls. The most successful SI turbocharged engines have therefore used port fuel injection.

Gas turbines

These are most widely used by far in aircraft and are, in fact, totally dominant in that market except for very small aeroplanes. In addition, they have specific applications in power generation, pumping and marine propulsion. They have been considered for automotive use but not to date with great success. This is because they cannot match the CI engine for efficiency and flexibility under road conditions although high specific power (power per unit mass) is available. As mentioned previously, considerable improvement in efficiency is still possible and these developments depend on improvements in the details of the gas flows in the various components, better design for cooling the turbine blades and better high-temperature materials. Currently, in aircraft engines, the pressure ratios used are at maximum about 30:1 and peak temperatures (turbine entry temperatures, TET) about 1650 K. It has been projected that future engines will have pressure ratios approaching 75:1 with TET values of 2300 K. For stationary gas turbines where weight is not an important consideration, there is potential for water cooling of blades and the use of regeneration to allow very high thermal efficiencies at low pressure ratios. The use of regenerators could make the gas turbine viable as an automotive power plant but this depends on the development of regenerator units with very high effectiveness. Ceramic, rotating units have been tried but it remains to be seen how well they operate in practice before any predictions on future use can be made.

For very large gas turbine installations for power generation, the closed cycle gas turbine is a viable proposition and indeed, several such units have been constructed. These can use low molecular mass gases, such as hydrogen, with high heat capacity and low pumping losses. For this type, external heat supply is necessary which normally implies external steady combustion. In addition, primary energy sources other than combustibles (nuclear, solar) are feasible. A combination of a closed cycle gas turbine with a steam plant where the waste heat from the former is used as a heat source for the latter is an important concept for high efficiency.

1.4.2 Future fuels for IC engines

While efficiency increases and reduced vehicle size together with further crude oil discoveries and non-oil fuel based energy strategies which were introduced mainly in nonvehicular areas since the 1970s have lengthened the time period at which depletion of petroleum resources were expected to become severe, there is little doubt that there will eventually be an increasing dependence of the transport industry on alternative fuels. Sometime during the 21st century, substitution is likely to increase rapidly. The first section of the market to be affected is likely to be the distillates (light diesel fuels) as these compete directly with aviation kerosene for the same section of the crude oil barrel. This is particularly so where the crude oil is high in light fractions. Possible alternative fuels need examining.

Alternative fuels for transport may be either liquid or gas with the former having an obvious advantage in that its much higher density allows a good quantity to be carried in the available volume. However, gaseous fuels are the most plentiful, readily available alternatives in many parts of the world. Liquid alternative fuels are derived from either coal, gas or biomass.

Gaseous fuels

These consist of natural gas (NG), petroleum gas usually liquefied by pressure (LPG), biogas from various sources, coal gas and hydrogen. Natural gas resources are large and various estimates put the supply at current usage rates as 40 to 80 years. As new discoveries are, compared with oil, relatively frequent, the latter seems most likely and may, in fact prove to be conservative. Natural gas is an excellent fuel for SI engines as it has a high octane number (Research Octane Number, RON approximately 130) giving it very good resistance to their end-gas autoignition type knock (pinking). It consists mainly of methane, generally around 90% by volume but its composition may differ noticeably from place to place. Nevertheless, it is always difficult to liquefy because of the methane and pressure alone is insufficient. The liquefaction temperature at atmospheric pressure is $-162\,°C$ and either excellent refrigeration is required or boil-off must be tolerated to maintain this temperature. While in large installations, such as ships, this can be accommodated with boil-off rates of less than 1% per day now being possible, it poses problems for smaller vehicles used for road transport. The solution usually proposed at present is compressed natural gas (CNG) with cylinders at 17 to 24 MPa. Unfortunately, their mass is some seven times that of the fuel and the quantity that can be carried is small both on a volume and mass basis, thereby limiting their range.

Petroleum gas is basically a byproduct of the oil industry, coming partially from wells and partially from the refining process. It therefore is produced regardless of demand and can, at times, be in glut proportions. Much of it may be exported from producing countries. While there are good supplies available, its long term potential is less than NG, estimates being of perhaps up to about 40 years supply. It consists mainly of propane and butane, their proportions varying in different parts of the world. Like natural gas, it also is an excellent fuel for SI engines with an octane number (RON) of about 110. It can be liquefied at about 10 atmospheres pressure at which it has an energy density of about 70% that of gasoline. Consequently the limitations of NG do not arise to anywhere near the same extent. As it is widely used already in automotive applications, a reasonable distribution network already exists in many countries.

The other gases are less likely to be viable candidates as general alternative fuels although they may have a niche under some local conditions, particularly for stationary engines. Biogas is also mainly methane and is derived from decaying animal or vegetable wastes. Sewerage gas, for example, is already used by the relevant authorities. Coal gas consists of carbon monoxide and hydrogen in varying proportions depending on the method of extraction and may also contain small quantities of methane. Nitrogen in substantial proportions is usually present. Thus it has a low heating value (about half that of NG). Modern methods of coal gas production are the Koffers Totzek, Winkler or Lurgi processes which use either high pressures to obtain more methane and/or oxygen rather than air to eliminate the nitrogen. Coal gas produced this way is noticeably more expensive than traditional methods. Hydrogen may, in the long term, become a transport fuel and its introduction may be hastened by the greenhouse effect. However, in the immediate future, its production, cost and storage problems count against it.

The use of either NG or LPG in SI engines presents few problems. This is because their octane numbers exceed those of petrol (RON 90 to 100 in most countries) and either higher compression ratios or larger diameter cylinders are possible. Fuel introduction and mixing is also simpler as the fuel does not have to be vaporized and the more even distribution allows better lean operation. Some power loss occurs due to the gas displacing air in the inlet manifold. Natural gas has a slow burning velocity and desirably, should have either greater turbulence or swirl designed into the head, a greater ignition energy and some spark timing advance. Liquid petroleum gas has similar burning characteristics to vapourized gasoline and the above aspects therefore need little modification.

For CI engines, greater problems occur with the use of gaseous fuels. One solution which allows substitution for diesel fuel while retaining fuel flexibility is a relatively simple conversion that can be carried out on existing CI engines. This is called dual-fuelling where the primary fuel, in this case gas, is introduced into the inlet manifold while distillate injection is retained at sufficient levels to provide pilot ignition and to overcome combustion problems. This solution may sometimes be attractive compared to conversion to large SI engines in transport applications because long routes and embryonic gaseous fuel outlet structure may sometimes necessitate a return to operation on diesel fuel alone. Also the low conversion cost renders it feasible as a retrofit option for existing vehicles. However, new engine problems now occur because the engine is being fuelled with two different fuels with opposing characteristics. The gas fuel has a poor (essentially zero) cetane number which is the index of a fuel's ability to resist 'diesel knock' while the distillate has a poor octane number. It is for these reasons that the gas cannot be directly injected into the cylinder as a complete substitute for distillate. Moreover, there may be some problems of interaction between the gaseous fuel with the pilot distillate which increases the ignition delay of the latter. Thus, over the speed/load range of the engine, substantially different fuel proportions are required to maximize the gas substitution and to ensure that knock of either variety is minimized. A diesel engine control adjusts only the fuel rather than the total mixture. Therefore, in addition to the knock problem, a high gas proportion at low load results in low distillate flows which can cause ignition failure. Alternatively, a very lean gas mixture can exhibit extensive flame quench away from the pilot spray and may prohibit flame propagation completely. Solutions such as partial throttling, use of substantial recirculated exhaust gas have been considered. Much applied research on the control of dual-fuelled engines is necessary as well as more fundamental research on the appropriate fluid mechanics of the spray, mixing and combustion.

Alcohol fuels

Of the possible alternative liquid fuels, alcohols the two common ones, methyl alcohol (methanol) and ethyl alcohol (ethanol) in particular, have received the most attention. This is because of their relatively easy production and the fact that they are both good SI engine fuels. Ethanol is derived from biomass sources and is the better fuel of the two as it has a higher heating value and blends better with petrol. However, except in rare circumstances, its labour intensive production makes it too expensive. Methanol, on the other hand, can be produced from natural gas or coal and is therefore the most likely possibility.

As with the gaseous fuels, alcohols are most suited to SI engines because of their high octane but low cetane numbers. For ethanol, their octane numbers are approximately 100 and 90 for ethanol on the Research Octane Number RON (for lightly loaded engines) and Motor Octane Number MON scales (for heavily loaded engines) respectively. For methanol, the equivalent values are 110 and 90. Typical automotive operation requirements may be approximated by the average of the two scales and hence the alcohols do not show up quite as well when compared to petrol (for example, at 97 and 98 approximately) as would be assumed by a simple RON comparison. Nevertheless, higher compression ratios are possible with them. They require less air in stoichiometric combustion than does petrol, air/fuel ratios being 6.5, 9.0 and 14.5:1 for methanol, ethanol and petrol respectively. Thus, for a given capacity engine, more alcohol fuel can be introduced per cycle thus compensating for their lower heating values (lower heating values are 20, 27 and 44 $MJkg^{-1}$ respectively for methanol, ethanol and petrol respectively). That is, higher power outputs are possible. The disadvantage is that volume fuel consumption is higher than for petrol giving either reduced range or requiring larger fuel tanks. Other disadvantages are their high latent heat giving difficult starting and requiring additional manifold heating during operation. In general, emissions are about the same for an optimized alcohol fuelled vehicle except for NO_x which tends to be lower and the production of high levels of formaldehydes. Most work in the 1970s concentrated on alcohols as a gasoline extender and octane booster with alcohol proportions of about 15% being considered. These allowed existing engines to be used without modification. It was found that the blend octane numbers of alcohol/petrol were lower than the average of the components and that, particularly with methanol, separation occurred in the presence of water. For alcohols to make a significant impact as an alternative fuel, high alcohol content fuels will be required. A substantial engine redesign will then be necessary to allow for the higher compression ratio, different fuel/air ratios, manifold heating changes and different timing. For use in CI engines, alcohols suffer from the same problems as gases and the solution has generally also been to dual-fuel with the alcohol being 'fumigated' into the inlet manifold.

Other alternative fuels

Several other possibilities exist, these basically producing synthetic fuels more akin to the current oil based products. They are, however, likely to be more expensive than the above due to the extra processing. The first are the modified alcohols, methyl (or ethyl) tertiary butyl ether, which eliminate the separation and heating problems of alcohols. The other alternatives are a synthetic gasoline or distillate produced either via methanol or by hydrogenation of coal. The technology for both exists and plants are or have been operational. Their commercial viability depends on the current world crude oil prices. Another alternative is shale oil with much the same composition as crude oil. The problems here lie in the economics of extraction, not on the operation of engines.

1.5 THE INTERNAL COMBUSTION ENGINE AS A GENERALIZED CONCEPT

The term **internal combustion engine** represents a range of different machines and each needs to be explored individually and in detail in order to improve its operating characteristics, efficiency and emission levels. However, some important basic ideas can be grasped by taking a macroscopic approach and regarding the engine as a steady state device which uses fuel and produces power using the combustion products as the working substance. With this concept, an inclusive picture of all internal combustion engines can be obtained. An appropriate arrangement is shown on Figure 1.1. In this generalized view, a compressor, combustor with attached reciprocator, a turbine to drive the compressor and a further power turbine is shown. Any of these components can be in single or multiple units.

Something of the type shown in Figure 1.1 would be normally termed a turbocompounded reciprocating engine. While such machines have been built around the CI engine, the many problems associated with the huge variation in output shaft speed have proven to be limiting. For example, for optimum operation, a typical CI engine might have an output shaft speed of $2\,000$ rev min^{-1} while that of the power turbine would be some 20 to 30 times higher. Coupling the two in an efficient compact and lightweight manner is a major problem.

The arrangement shown may be regarded as a generalized engine layout. If the reciprocator power goes to zero, this component becomes a combustor only and the engine is a gas turbine. This is one limit of the system. The other limit is when both turbine units do not produce power, hence there is no compressor pressure rise and the engine is a normally aspirated reciprocating engine. If power is produced only in the reciprocator and the first turbine, the layout now represents a turbocharged reciprocating engine. It

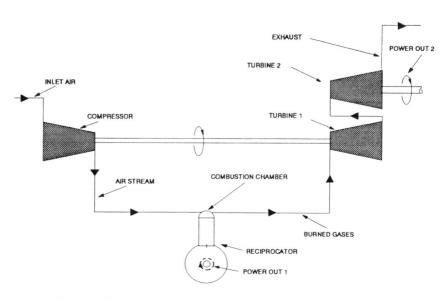

Figure 1.1 Generalized engine layout.

can therefore be seen that reciprocating engines and gas turbines need not be viewed as entirely different but, through a gradual transfer of the method of producing both the useful work and compression from a reciprocator to a power turbine and separate compressor, there is progression from one type to the other.

The arrangement shown in Figure 1.1 is not the only possible combination of components and a number of alternative arrangements are possible. For example, the reciprocator could be used to drive the compressor and all useful work could then be obtained from a power turbine. An interesting possibility is the differential compound engine. Here, by use of differential gearing, the compressor can be made to rotate faster as the output shaft speed falls. Generally with turbocharging, a fall in output shaft speed is accompanied by reduced gas flow through the system, lower turbocharger speed, lower compressor pressure ratio and hence a faster falloff in torque and power than in an equivalent normally aspirated engine. This produces peaky torque and power curves and gearboxes with many alternative ratios are required for heavy road vehicle operation where the load can vary considerably. The differential compound engine, by increasing the compressor speed as engine speed falls which is the reverse of normal turbocharging, provides better torque at the low speeds.

Engines are usually analysed by considering their processes one by one. Alternatively, if the flow entering and leaving the engine, the power output and the heat transfer rates are time averaged over a period, even reciprocating engines can be initially examined by use of a steady state approach. That is, the engines can be regarded as a 'black box' with only these mass and energy flows considered. Appropriate equations follow from a consideration of the basic thermal efficiency η_{TH} of an engine which relates the work W and heat addition Q.

That is,

$$W = \eta_{TH} Q \tag{1.1}$$

Power $\dot{W}$ is then

$$\dot{W} = \eta_{TH} \dot{m}_f H_f \tag{1.2}$$

where $\dot{m}_f$ and H_f are the mass flow rate and heating value respectively of the fuel.

Now the mass flow rate of fuel is dependent on the air flow rate $\dot{m}_a$ that can be provided for its combustion. That is, a primary concept in the design of an engine is to obtain the necessary air flow rate in a simple and efficient manner. Also there are limits on the ratio of fuel to air which can be used although, particulary in the case of lean mixtures, this may vary considerably. This ratio is best expressed as

$$\dot{m}_f / \dot{m}_a = \phi F_s \tag{1.3}$$

where ϕ is termed the equivalence ratio and is simply a term expressing the relativity of the actual fuel to air ratio to the 'chemically correct' or 'stoichiometric' ratio F_s. The stoichiometric ratio can be loosely defined as the mass of fuel to air that exists when there is just sufficient oxygen to burn all the carbon in the fuel to carbon dioxide and all the hydrogen to water. These terms will be discussed further in a later chapter.

From the above

$$\dot{W} = \eta_{TH}\phi F_s \dot{m}_a H_f \tag{1.4}$$

The term $\dot{m}_a$ will need defining in the most convenient manner for each particular engine type. For example, for a gas turbine, it is best expressed in terms of the engine volumetric flow rate $\dot{V}$ and the atmospheric air density ρ_a. Hence

$$\dot{m}_a = \dot{V}\rho_a \tag{1.5}$$

In a reciprocating engine, a better representation is in terms of the measured positive displacement (swept volume) V_s of the cylinder, a volumetric efficiency, η_V which designates how well they fill and the rate N (in revolutions per second, rev s^{-1}) at which the filling cycle repeats. The last itself depends on both the engine speed N_e (usually given in rev min^{-1} and must be converted to rev s^{-1}) and whether the engine is a two- or four-stroke variety. Here

$$\dot{m}_a = \eta_V V_s N \rho_a \tag{1.6}$$

The density ρ_a may be taken at any appropriate position and the volumetric efficiency expressed accordingly. However, an appropriate position is in the inlet manifold just upstream of the inlet port and this is usually chosen. The volumetric efficiency is then a relatively simple curve varying with engine speed for the reciprocator only and is not affected by such upstream devices as throttles, turbochargers etc.

The combination of (1.5) or (1.6) as appropriate with (1.4) will give an overview of the steady state performance. However, this will not provide the detail necessary to understand the way in which power output, efficiency and emission levels can be improved.

1.6 OPERATION OF CONVENTIONAL MODERN ENGINES

1.6.1 Reciprocating engines

Four-stroke engines

The basic principle of the operation of a four-stroke engine is shown on Figures 1.2 and 1.3. Figure 1.3 represents a spark ignition (SI) and Figure 1.3 a compression ignition (CI), diesel type. In the spark ignition engine, fuel and air are mixed at constant pressure by a spray basically controlled by

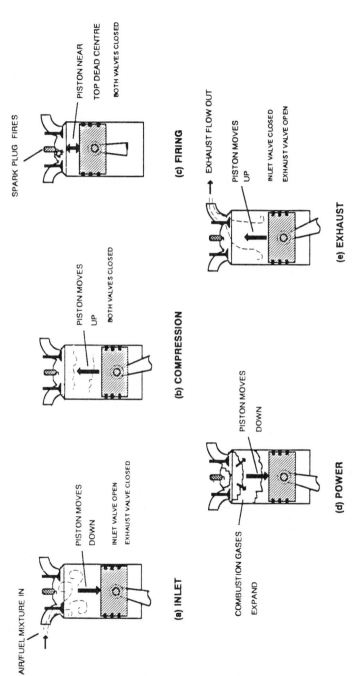

Figure 1.2 Four-stroke spark ignition engine operation. (a) inlet: inlet valve open, exhaust valve closed, (b) compression: both valves closed, (c) firing: both valves closed, (d) power, (e) exhaust: inlet valve closed, exhaust valve open.

AIR/FUEL MIXTURE IN

PISTON MOVES
DOWN

INLET VALVE OPEN
EXHAUST VALVE CLOSED

(a) INLET

PISTON MOVES
UP

BOTH VALVES CLOSED

(b) COMPRESSION

SPARK PLUG FIRES

PISTON NEAR
TOP DEAD CENTRE

BOTH VALVES CLOSED

(c) FIRING

COMBUSTION GASES
EXPAND

PISTON MOVES
DOWN

(d) POWER

EXHAUST FLOW OUT

PISTON MOVES
UP

INLET VALVE CLOSED
EXHAUST VALVE OPEN

(e) EXHAUST

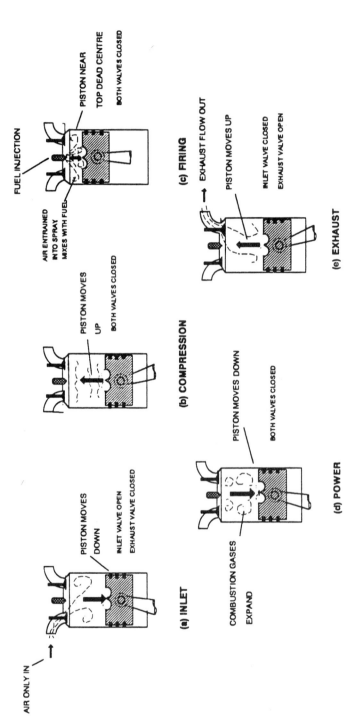

Figure 1.3 Four-stroke compression ignition engine operation, (a) inlet: inlet valve open, exhaust valve closed, (b) compression: both valves closed, (c) firing: both valves closed, (d) power: both valves closed, exhaust valve open.

pressure drop in a venturi with some means of compensation for different conditions such as full power or acceleration. The mixture may then be heated (using heat transfer from the exhaust manifold or engine coolant) to promote vaporization and passed to the cylinders. In the first stroke (a), the inlet valve is opened and the mixed gas flows into the cylinder as the piston moves down. The inlet value then closes and the mixture is compressed when the piston moves up, as in (b). The compression ratio, i.e. the volume ratio between the lowest and highest position of the piston, cannot be too great as the resulting temperature rise would then cause a spontaneous uncontrolled ignition of the fuel/air mixture. Just before the piston reaches the top of the stroke, ignition occurs at a controlled point by use of a high voltage spark. A flame front moves rapidly across the combustion chamber in a few degrees of crank angle movement (c) causing a considerable increase in pressure and temperature. The expansion stroke follows and as the piston moves down (d) work is extracted from the hot, high pressure gases. Towards the end of this stroke, the exhaust valve opens and the piston moves up (e) expelling the burnt gases. The engine is now back to its original state and the cycle can be repeated. In the compression ignition engine, the processes are essentially the same except that air only is drawn into the cylinder during the intake stroke. Compression can now be much greater as no spontaneous combustion can occur. Indeed, the compression ratio has to be very high to provide the necessary temperatures for ignition. The fuel is now sprayed directly into the combustion chamber by a high-pressure injector. Injector timing provides the controlled point to start combustion. Combustion occurs by means of a diffusion flame over a somewhat longer period than with the spark ignition engine. The pressure and temperature rises are thus somewhat more limited than with the SI engine. The remainder of the cycle is then identical to that of the SI engine.

Two-stroke engines

Figure 1.4 shows a two-stroke engine cycle. If, as shown in the figure, the piston starts at top dead centre (TDC) it moves downward in the power stroke. Towards the bottom of this stroke, exhaust ports are uncovered and, the exhaust gases 'blowdown' to the exhaust system. Simultaneously, inlet ports are opened and a pressurized charge, either a fuel/air mixture in the case of an SI engine or air alone for a CI one, is forced in through the ports. The geometry of the piston top is such that this scavenges the cylinder of burnt exhaust gases and leaves it charged with fresh air or mixture. The pressurized inlet charge is obtained either by compressing in the crankcase by use of the downward movement of the piston in the power stroke or by a separate blower. With CI engines in particular, the higher pressures obtained by a blower allow use of excess scavenging air to completely clean out the cylinders as no fuel can then be wasted. On rising in the second stroke, the piston seals off the ports and commences compression. The

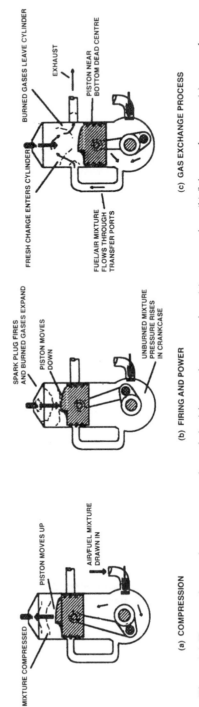

Figure 1.4 Two-stroke crankcase scavenged spark ignition engine operation. (a) compression, (b) firing and power, (c) gas exchange process.

remainder of the stroke is thus a similar compression to a four-stroke engine. The combustion processes are also similar. It should be noted that, although ports are widely used on two-stroke engines, they are sometimes replaced by poppet valves.

1.6.2 Gas turbines

In gas turbine operation the air is admitted continuously to a compressor which, in modern turbines, is generally axial flow. Figure 1.5 shows the various sections. Pressure is raised through two or three compressors each driven independently from its own turbine disc on coaxial shafts to allow each to turn at different speeds. With modern aircraft engines, the first of these, the low pressure compressor, is basically a fan which provides a large mass flow rate but only a small pressure rise. Only a proportion (as low as 25%) of the fan air actually passes through the engine core. The rest goes directly to the jet stream where the low velocity high mass flow augments jet efficiency. Most of the compression is carried out in the intermediate and high pressure compressors which themselves each have large numbers of rows of blades. Pressure ratios of the order of 20:1 to 30:1 are typical of modern engines.

In modern gas turbines, the turbomachinery nearly always consists of an axial flow compressor and an axial flow turbine. Very early engines used centrifugal compressors and occasionally these still have some application but they are not suited for the multistaging necessary for the very high

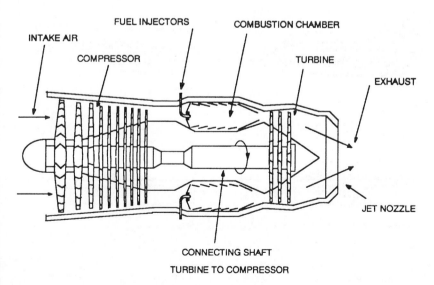

Figure 1.5 Arrangement of components in a gas turbine.

pressure ratios used on modern engines. The power for the compressor is taken from a turbine in the first stage of expansion and, as the compressor can consist of several stages, there may be two or even three turbine stages each driving a compressor stage via coaxial shafts. The remainder of the expansion is used for power in a jet nozzle (aircraft) or in further turbine sections to drive a power shaft. Figure 1.6 shows some typical gas turbine arrangements.

In modern aircraft engines, extra power is extracted in turbine sections to drive a fan, which may be regarded as a shrouded propeller. Although the fan becomes the first compressor stage, most of its air bypasses the engine and is used for thrust. Bypass ratios (mass flow rate around the core engine to mass flow rate through it) range from 1:1 to 5:1 but designs for unducted fans will, in the future, noticeably increase this value to around 7:1. This bypass air makes the jet thrust more efficient. The effect of bypass air may be simply calculated by use of the impulse–momentum principle. Assuming that the inlet and exhaust pressures are both atmospheric, the thrust force, T, for an engine is given by

$$T = \dot{m}(V_j - V_a) \qquad (1.7)$$

where V_j and V_a are the air velocities relative to the engine and $\dot{m}$ is the mass flow rate, neglecting the addition of the fuel, through it. Note that V_a is approximately the velocity of the aircraft and so the jet velocity must exceed this value to give a positive thrust.

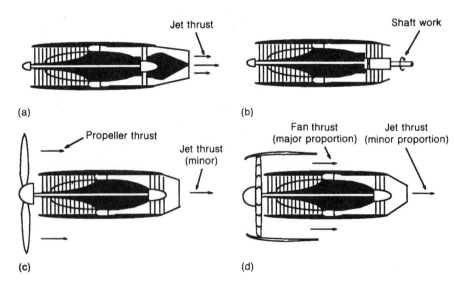

Figure 1.6 Some methods of power extraction from gas turbines. (a) axial flow turbojet, (b) industrial gas turbine, (c) turbopropeller, (d) turbofan.

The thrust can be increased by changing either V_j or $\dot{m}$ assuming that V_a is a fixed value. Now if the gas left behind in the atmosphere by the jet is considered, its absolute velocity is $V_j - V_a$.

Its kinetic energy is then

$$\text{Energy lost} = \tfrac{1}{2}\dot{m}(V_j - V_a)^2 \tag{1.8}$$

This is lost energy because once the jet is clear of the aircraft, it has no further value. The energy used can be obtained by multiplying the thrust by the aircraft velocity:

$$\text{Energy used} = TV_a = \dot{m}(V_j - V_a)V_a \tag{1.9}$$

Hence, the total energy is

$$\text{Total energy in jet} = \dot{m}(V_j - V_a)V_a + \tfrac{1}{2}\dot{m}(V_j - V_a)^2$$
$$= \tfrac{1}{2}\dot{m}(V_j^2 - V_a^2) \tag{1.10}$$

The jet propulsion efficiency can now be obtained by dividing the energy used by the total energy:

$$\eta_P = \frac{(V_j - V_a)V_a}{\tfrac{1}{2}(V_j^2 - V_a^2)} = \frac{2V_a}{V_j + V_a} \tag{1.11}$$

It can be seen that the maximum propulsion efficiency of 100% can only be obtained when V_a is equal to V_j at which stage the thrust is zero. As V_j is increased, the propulsion efficiency decreases but the thrust per unit of mass flow increases. Values are shown in Figure 1.7. For a given thrust and a high propulsion efficiency, it can be seen that it is best to increase the mass flow rate rather than the jet velocity. The high bypass ratio achieves this purpose.

A further benefit is that noise is proportional to, approximately, the seventh power of the jet velocity. Thus, a high bypass engine is also quieter.

Air is admitted to the combustion chamber which may be a series of 'cans' arranged around the engine or a fully annular device on modern

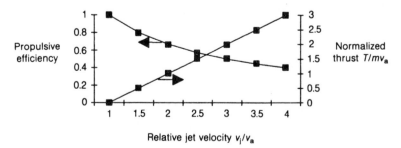

Figure 1.7 Effect of relative jet velocity on normalized thrust and propulsion efficiency of aircraft gas turbines.

engines. The fuel spray is admitted to the combustion chamber at its leading edge and the air flow is directed through a series of holes in the walls to maintain flame stability and promote cooling. Although some pressure loss occurs in the combustion chamber, it is essentially a constant pressure process. Expansion occurs initially through the turbine discs, the first high-pressure one driving the last high-pressure compressor and so on. Thus, two or three turbine stages are used, one for each compressor section. However, in many low-bypass engines, the fan and intermediate pressure compressor are connected to the one shaft and are essentially the same low-pressure compressor. Some air is also ducted directly from the compressor through the shaft to the turbine blades where it is used as a coolant. The remainder of the expansion may occur either in a jet nozzle for aircraft or through further turbine sections for stationary gas turbines in which shaft work is obtained. It should be noted that stationary gas turbines have no need for bypass air from a fan.

1.7 REFERENCES

Boyd, T. A. (1950) Pathfinding in fuels and engines, *Society of Automotive Engineers Trans.*, **4** (2).

Cummins, C. L. (1976) *Internal Fire*, Carnot Press, Lake Oswego, Oreg.

Eichelberg, G. (1939) Some new investigations on old combustion engine problems, *Engineering*, **148**, 463.

Goodenough, G. A. and Baker, J. B. (1927) A thermodynamic analysis of internal combustion engine cycles, Univ. Illinois Expt. Station Bull, **160**.

Hottell, H., Williams, G. and Satterfield, C. (1936) *Thermodynamic Charts for Combustion Processes, Vols. 1 and 2*, Wiley, New York.

Midgeley, T. Jr (1950) *The combustion of fuels in internal combustion engines, Society of Automotive Engineers Trans.*, **15**.

Mock, F. C. (1920) Design of intake manifolds for heavy fuels, *Society of Automotive Engineers Trans.*, Part 1.

Ricardo, H. (1933) *The High Speed Internal Combustion Engine*, Blackie, London.

Robins, R. (1986) *Following the Standards of Henry Royce*, Rolls Royce, Derby, England.

Walshaw, T. (ed.) (1950) *Modern Oil Engine Practice*, Newnes, London.

Wilson, D. (1984) *The Design of High-Efficiency Turbomachinery and Gas Turbines*, MIT Press, Cambridge, MA.

PROBLEMS

1.1 Estimate the amount of chemical energy (μMJ) in the tank of your car if the heating value of the fuel is 44 MJkg^{-1} and the fuel relative density (i.e. compared to water) is 0.74. Determine the time taken to fill the tank of your vehicle and hence estimate the energy filling rate in kW^{-1}. If the vehicle uses 12 kW to maintain a constant speed of 80 km h^{-1} and is 20% efficient (overall) at that speed, determine how long the fuel supply will last, the distance covered and the fuel consumption rate (either km l^{-1} or l/100 km). In practice, how would you expect normal operation (acceleration, deceleration, idling etc.) to alter this consumption rate?

1.2 The largest passenger jet aircraft require 40 kN of thrust/engine at cruise. Assuming the cruise speed to be 864 km h^{-1}, determine: (a) the power required by the aircraft; (b) the rate of energy supply in the fuel to all four engines assuming that they are 30% efficient under these conditions; (c) both the mass and volumetric fuel supply rate if the fuel relative density is 0.76 and its heating value 43 MJkg^{-1}. Compare this with the flow rate from the fuel bowser used for filling your car's tank in problem 1.1; (d) the total quantity (both volume in kl and mass in tonnes) of fuel which needs to be carried for a trans-Pacific crossing of 12.5 hours at this speed. For a first order estimate, neglect any differences due to take-off and landing fuel burn; (e) the saving in mass if the engine efficiency can be increased to 35%.

1.3 In the future, alternative energy sources are likely to be required for transport. For a car and aeroplane at constant speed as in 1.1 and 1.2, evaluate the additional volume, mass or surface area for energy collection for the same power requirement and range if the energy sources given below were used. The fuel tank mass for conventional fuel may be taken as 25% of the fuel mass for the car, 1% of the fuel mass for the aircraft. The heating value (lower) and density of conventional fuels and the engine efficiencies are given in Problems 1.1 an 1.2. The efficiencies may be assumed to be unchanged for conversion to an alternative by fuelled internal combustion engine while the other data is tabulated below. Comment on any other deficiencies of the alternative energy source.

Combustible fuels

Fuel	Heating value (MJkg^{-1})	Density (kgm^{-3})	Tank mass (% of fuel mass)
Methanol	20	706	25
Ethanol	26.7	794	25
CNG	46.7	133 (at 17MPa, 300K)	700 (steel), 170 (wrapped carbon fibre)
LNG	46.7	440	75

Other energy sources

Solar voltaic panels

Maximum energy collection with clear sunlight normal to the collecting
  surface:1 kW m^{-2} at sea level; 1.3 kW m^{-2}at 10 000 m.
Factor to allow for variation from the normal over 12 hours daylight: 0.6
Factor to allow for variation from the normal due to latitude: 0.90
  (35° latitude);   0.95 (average for journey 35° N to 35° S latitude)
Solar collection efficiency: 22%
Electric motor efficiency: 90%

Battery power

Battery discharge efficiency: 85%
Electric motor efficiency: 90%
Energy supply rate/lead acid battery: 120 W kg^{-1}
Total energy stored/lead acid battery: 144 kJ kg^{-1}
Energy supply rate/possible new technology: 300 W kg^{-1}
Total energy stored/possible new technology: 300 kJ kg^{-1}

1.4 The following equations will be remembered from a first course in
thermodynamics.

- first law of thermodynamics for a closed system

$$dq = du + dw$$

- work done by a moving boundary

$$dw = p dv$$

- ideal gas equation

$$pv = RT$$

- ratio of specific heats

$$\gamma = \frac{c_p}{c_v}$$

Show, for an adiabatic (i.e. $dq = 0$) process using an ideal gas in a closed
system, that

$$\frac{p_2}{p_1} = \left(\frac{v_1}{v_2}\right)^{\gamma}$$

$$\frac{T_2}{T_1} = \left(\frac{v_1}{v_2}\right)^{\gamma - 1}$$

$$\frac{T_2}{T_1} = \left(\frac{p_2}{p_1}\right)^{(\gamma - 1)/\gamma}$$

You may assume that $dq = c_v \, dT$ for *any* process in an ideal gas and that, also for an ideal gas, the gas constant is the difference in specific heats.

$$c_p - c_v = R$$

1.5 By taking a reversible process between two state points (p_1, v_1, T_1 and p_2, v_2, T_2), and using the first law together with the entropy relationship from the second law, i.e.

$$ds \geqslant \frac{dq}{T}$$

show that the entropy change depends only on the properties at the two state points and is described by the relationships

$$ds = c_v \frac{dT}{T} + R \frac{dv}{v}$$

$$ds = c_p \frac{dT}{T} - R \frac{dp}{p}$$

You may assume that the enthalpy, h, is given by

$$h = u + pv \quad \text{for any substance}$$

and

$$= u + RT \quad \text{for an ideal gas.}$$

Hence, for an isentropic (i.e. s constant) process, show that the relationships developed in question 1 also apply. Explain why both these derivations represent the same (isentropic) process.

1.6 Show, for an ideal gas expanding through any polytropic process which can be described by the equation (note that n can take different values to describe different processes)

$$pv^n = \text{const.}$$

that the boundary work done is given by

$$w_{1-2} = \frac{p_1 v_1 - p_2 v_2}{n - 1}$$

1.7 Air, which may be assumed to be an ideal gas, expands through an isentropic process (for which the value of $n = \gamma = 1.4$) from a pressure of either 0.4, 1 or 2 MPa to the atmospheric pressure of 100 kPa. Its initial temperature in each case may be assumed to be 3300 K. Determine the work (per kg) which can be produced by this expansion in each case and hence explain the importance of the compression process in an engine in relation to its potential work output. Also determine the temperature at the end of expansion in each case. What can you infer about the work transfer and the temperature in relation to these results?

1.8 A reciprocating engine is required to produce 150 kW of power at its design maximum power speed. Make an initial estimate of the required engine capacity (volumetric size) if it is: (a) a four-stroke SI engine, 25% thermal efficiency, 0.80% volumetric efficiency, $\phi = 1.0$, maximum power speed $= 5000$ rev min^{-1}; (b) a four-stroke CI engine, 35% thermal efficiency, 0.85% volumetric efficiency, $\phi = 0.7$, maximum power speed $= 3000$ rev min^{-1}; (c) a two-stroke SI engine, 20% thermal efficiency, 60% volumetric efficiency, $\phi = 1.0$, maximum power speed $= 6000$ rev min^{-1}; In each of these cases, the volumetric efficiency is related to an atmospheric air density which may be taken as 1.2 kg m^{-3}.

You may assume that for the fuel, $H_f = 44$ MJ kg^{-1}, and the air/fuel ratio $= 14.7{:}1$

1.9 For each single gas turbine of the aircraft operating as in Problem 1.2 determine the air mass flow rate through the core engine if the equivalence ratio ϕ fuel/air ratio at the time is 0.35. Determine the cross-sectional area and diameter (assuming it to be circular) required at inlet if the inlet velocity is assumed to be equal to the forward velocity of the aircraft and the local conditions are pressure, 30 kPa, temperature $-50\,^{\circ}$C. The gas constant R for air may be taken as 287 J kg^{-1}K^{-1}. Assuming that a bypass ratio of 4:1 is used, calculate the total mass flow rate of air, the cross-sectional area and diameter (assuming a circular section) required. For the fuel, $H_f = 43$ MJ kg, and the air/fuel ratio $= 14.7{:}1$

1.10 For the engine of Problem 1.9, compare the propulsive efficiency to that of one with equivalent thrust but no bypass air.

Air standard cycles

2.1 CYCLE ANALYSIS

A simple overview of an engine as a steady state device was discussed in
Chapter 1. Section 1.5 highlights the importance of the volumetric flow rate
of air, the fuel-to-air ratio, the heating value of the fuel and the thermal
efficiency on its power output. It does not, however, give any insight into
the fundamental effects, relationships and limitations of the various engine
parameters on these items or on others such as the composition of the
exhaust gases. To do this, the individual thermodynamic processes making
up the engine cycle need to be examined. Thus, cycle analysis is the basic
theoretical step through which the practical development of engines has
occurred. Without such a theoretical analysis, the levels of power output
and efficiency obtained by modern engines could not have been reached and
internal combustion engines would have remained clumsy devices more of
curiosity than of practical value.

From the beginning of internal combustion engine development, some
simple thermodynamics was used for evaluation. One notable example was
the realization, during some comparative evaluation tests of 1867, that the
Otto and Langen 'free piston' engine was more efficient (at about 11%
overall) than the Lenoir engine (about 5%) in spite of its more clumsy
mechanical arrangement because it allowed a much greater expansion ratio.
For many years, the major tool for analysis was the **air standard cycle** which
was first developed by Dugald Clerk towards the end of the nineteenth
century and essentially considers only the energy balance of the working
substance. In recent years, particularly since the advent of modern com-
puters, much more detailed theoretical cycles of this type have been available. These include more complex **thermodynamic** cycles (sometimes called
zero dimensional) which are essentially an extension of the air standard idea
but which may include such things as effects of the mixture of fuel and air
(rather than just the air) and the real gas properties of both the original
gas mixture and the dissociated combustion products which vary with temperature. They may be extended to study further detail by the addition of
submodels for such individual phenomena as burning rates and emission
production and are then known as **phenomenological** or for some specific
geometrical feature such as the fuel spray zone in CI engines, when the term
quasidimensional is used. If the full fluid mechanics formulation of continuity,
momentum and energy is considered, the cycles are called **multidimensional**

models. These have the advantage that they allow the full geometrical differences between engines to be considered but are much more complex.

While more detailed information about engine processes, emissions etc. can be obtained from a modern analysis, the air standard cycle is still of importance in that it allows simple mathematical equations to be developed which describe the basic trends associated with the major engine parameters thereby giving an easy guide to their effects. In addition, it provides the structure from which more complex engine cycle analysis can be later understood and developed. This chapter is therefore devoted to an examination of air standard cycle analysis for both steady flow (gas turbines) and intermittent flow (reciprocating) internal combustion engines.

2.2 AIR STANDARD CYCLES

Air standard anlaysis is, in fact, a familiar tool to those who have already dealt with basic thermodynamics in that it uses cycles made up of idealized processes, each with simple mathematical equations, represented on a state diagram as approximations for the very complex, real processes which occur in practice. The latter are, in fact, rapidly occurring non-equilibrium events for gas compositions which vary with time and in which a range of chemical, thermodynamic and flow processes are occurring. The air standard cycle assumes that the processes are **quasi-equilibrium**, i.e. ones in which equilibrium of the complete working substance is reached at each state point during any process, and that air alone exists as the working substance throughout the whole cycle. Mixtures of fuel and air and of burnt or partly burnt gases are therefore not considered. The combustion process is replaced by a heat transfer process to the air in the cycle using the equivalent amount of energy to that obtained in practice by burning the fuel. A further heat transfer from the gas is used in the air standard cycle to allow for the heat which is rejected to the atmosphere in the exhaust of most real engines. These heat transfers are an artificial mechanism in order to allow the cycle to be considered as a simple closed system. Usually, in air standard cycle analysis, the values of the specific heats, c_p and c_v, and their ratio, γ, are considered to be constant throughout. This is obviously inaccurate but it allows tractable equations to be developed. However, if greater accuracy is required, values of the internal energy, enthalpy etc. as functions of temperature can be obtained from appropriate air tables or complex equations of state but simple trends with variation in the basic parameters are then less obvious and it is more useful to save this addition for a more complex analysis. When constant specific heats are used, the temperature at which they are evaluated needs consideration. A **cold air standard** is often applied where c_p, c_v and γ are evaluated at atmospheric conditions but constant values at a higher temperature, for example the mean temperature of the cycle, are also in frequent use thereby improving the accuracy of the analysis. However, as trends rather than accuracy are the important element of air standard cycle analysis, this is more to illustrate the effect of a different isentropic index on the efficiency of the cycle than to try to obtain great precision. As a general comment, it should be emphasized that even the best results of air standard cycle analysis are not at all accurate. When compared with engine tests, the power output and efficiency could be in error by perhaps 30%. The value of air standard analysis lies in simply and rapidly assessing the relative effect of different parameters, processes, energy transfers and working substance properties on the cycle.

There are a number of well-known cycles which are designated by the names of some of the most significant inventors from the early period of engine development. Perhaps the best known to the reader (and the most important thermodynamic cycle from a theoretical point of view) is the Carnot cycle which does not represent any actual engine in common usage. Thermodynamically, it represents the most efficient heat engine cycle possible for operation between two temperature limits and, as it is considered in detail in most basic thermodynamics courses, it will not be discussed further here. Other well-known cycles are the Otto, Diesel and Brayton (sometimes called Joule) which are the air standard cycles to be considered here for internal combustion engines, the Rankine, a vapour cycle, and the Stirling and Ericsson air standard cycles which are important in external combustion engines. The Otto, Diesel or Brayton cycles are often taken as representing reciprocating spark ignition, reciprocating compression ignition and rotodynamic gas turbine operation respectively. However, particularly with reciprocating engines, the situation is more complex and a cycle somewhere between the Otto and Diesel actually more closely represents both engines. This goes under several designations, the Dual, the Mixed or the Limited Pressure cycle, and is of considerable importance. The Brayton, Otto, Diesel and dual cycles will all be discussed below and their relationship to the appropriate engines considered.

2.2.1 General equations for air standard cycle analysis

The most important trends required from an air standard cycle analysis are those of work output and efficiency. While each cycle will need to be considered on its own merits, some general principles apply. It is assumed that the reader has some basic familiarity with cycle analysis from a previous thermodynamics course for the following discussion.

Cycle analysis begins with the construction of a state diagram containing a series of processes which are the nearest depiction of the actual events, an evaluation of the thermodynamic state at each delineating cycle point and a determination of the energy balance for each process. Every process will have a work and heat transfer associated with it as, for example, w_{1-2}, q_{1-2} for that from state point 1 to state point 2. It should be noted that lower case letters designate intensive (values per kilogram) while upper case refer to extensive (total) quantities. The sign convention used here is the traditional one of work output and heat input being positive which is more suited to engines although it is noted that a number of modern texts apply the reverse sign to the work term. To obtain the net work output w_n of a cycle, the work for each process should be calculated and summed. That is

$$w_n = w_{1-2} + w_{2-3} + w_{3-4} + \cdots \tag{2.1}$$

Most processes will have some such work term, the exception being constant volume processes.

The total heat transfer for the cycle q_n may also be obtained by the summation of that for the individual processes. That is

$$q_n = q_{1-2} + q_{2-3} + q_{3-4} + \cdots \tag{2.2}$$

Adiabatic (which includes isentropic) processes have no heat transfer but all others do. In the cycles associated with engines, many of the processes will be approximated by an adiabatic. Thus the value q_n for a cycle is often easier to obtain than the value w_n.

Now for a cycle, the first law of thermodynamics is usually written as

$$\sum q = \sum w$$

In other words, the net cycle work, w_n (i.e. $\sum w$) can be obtained from the net heat transfer, q_n (i.e. $\sum q$). This is often expressed as

$$w_n = q_{\text{added}} - q_{\text{rejected}}$$

or, with a more simple nomenclature as

$$w_n = q_a - q_r \tag{2.3}$$

Here, the q_a and q_r represent the magnitude only of the heat transfer with its direction being accommodated by the signs within the equation itself.

The cycle thermal efficiency is expressed as the ratio of the net work (i.e. the **useful** output of the system) to the heat added (i.e. the input which somehow is a cost to the user). Thus, the thermal efficiency is

$$\eta_{\text{TH}} = \frac{w_n}{q_a} = 1 - \frac{q_r}{q_a} \tag{2.4}$$

The most useful form of the above equation will vary with individual cycles and both should be kept in mind for future usage.

Property values are now assumed as being those for air. Air at moderate to low temperatures closely follows the ideal gas assumption. While the temperatures reached in most cycles are high, they are generally still within the range where this is still reasonably valid. Therefore, the ideal gas relationships which also specify that the internal energy and enthalpy are functions of temperature only, hold. That is, the change in internal energy may be expressed, for constant specific heat values, as

$$\Delta u = c_v \Delta T \tag{2.5}$$

and the change in enthalpy as

$$\Delta h = c_p \Delta T \tag{2.6}$$

These relationships, it must be noted, hold for an ideal gas only, for any process.

2.3 EXAMINATION OF SPECIFIC CYCLES

2.3.1 Gas turbines: the air standard Brayton cycle and its derivatives

Gas turbines have become the dominant means of propulsion for aircraft during the last 20 years and are also used for land and sea-based operations. Their operation was described in Chapter 1 and it can be seen that they are normally steady-flow machines except during transients when, for example, they are warming up or changing speed or load and they essentially consist of a series of separate but interlinked components which themselves also operate predominantly under steady conditions. The basic components in order as the gas proceeds through the machine are:

- a rotodynamic compressor;
- a combustion chamber;
- a rotodynamic turbine section which occupies the initial stage of expansion and is used to drive the compressor;
- a further expansion stage where the useful work is extracted either in a nozzle as a jet for propulsion purposes or in extra rotodynamic stages connected to a power shaft.

For an air standard cycle analysis, each of these components must be given an idealized thermodynamic process which has an easily tractable equation. Thus many effects which are relatively minor are ignored. One of the keys to any successful cycle analysis is the exercising of appropriate judgement to determine not only which factors are of lower significance but also, at a later stage, to be able to assess how much of the difference between actual performance and the cycle analysis results is due to these simplifications. For the Brayton cycle which is the basic cycle representative of the gas turbine, the simplification process may be described as follows.

1 **Compression.** During compression, the air moves rapidly through the compressor and, while some heat transfer will occur, probably to the air in the initial stages and from it in the later ones, the actual quantity of heat transferred per unit of mass flow is very small compared to the work input. Thus the process can be designated approximately as adiabatic.

However, while an adiabatic process can be handled directly by the first law of thermodynamics, the establishment of the end point properties requires a process equation in terms of the working substance properties. A compressor rotating at very high speeds will produce a reasonable degree of disturbances to the fluid which render the process irreversible. However, as a first approximation, the process may be taken as isentropic (reversible). Later, an improved assumption of irreversible but still adiabatic compression can be included by use of a comparison, expressed as a compressor efficiency, between isentropic and irreversible, adiabatic processes.

2 **Heat addition.** Following compression, the fuel and air are introduced in practice as two separate streams into the combustion chamber where there is sufficient temperature for vaporization of the fuel and ignition to occur. The energy release is replaced by a simple heat addition of equivalent magnitude. Some pressure drop is necessary to promote the flow but, as this is a small percentage of the absolute pressure of the chamber, the heat addition process is usually regarded as constant pressure.

3 **Expansion.** The expansion process may be viewed as isentropic following a similar rationale to compression although heat transfer here is likely to be more significant because the gas temperatures are now very high. Nevertheless, it is still small in comparison to the work output and so, again, an isentropic process is chosen for the Brayton cycle. Again, a later comparison of the effect of the adiabatic irreversible process is possible by use of a turbine efficiency.

4 **Heat rejection.** The work-producing expansion process continues, in an open cycle machine, until atmospheric pressure is reached. This is identical to the compressor intake pressure but the exhaust temperature is considerably higher. Although the gas in the exhaust is different from that at intake, the air standard cycle assumption is that heat transfer cools it to atmospheric temperature giving it identical working substance conditions to intake. Thus, a final constant pressure determines the heat transfer from the working substance. It should be noted that, in a closed cycle gas turbine which does in fact use a heat exchanger for this purpose, there is an actual link between the exhaust and intake gas and some small pressure drop is necessary to maintain the flow through it.

Gas turbines are well suited for an initial examination of the usefulness of air standard analysis because there are many variations in their layout which can be highlighted in relation to a few parameters, the overall engine pressure ratio r_p and the maximum and minimum cycle temperatures. The cycle most relevant directly to them is the Brayton cycle and it will now be examined for the basic machine and for more complex variations. Additional cycles which are all therefore modifications of the Brayton cycle will be studied and these are called the **regenerative cycle**, the **intercooled** and the **reheat cycle**. Combinations of these are possible.

2.3.2 Analysis of the Brayton cycle

The basic Brayton cycle as drawn on Figure 2.1 can be analysed with individual processes being either a closed or an open system and it should be emphasized that the results are the same in both cases. As the gas turbine is, in practice, essentially composed of components operating as steady flow open systems, this will be considered first.

Open system analysis

For a steady flow system, the first law of thermodynamics can be written as

$$q + h_x + \mathrm{ke}_x + \mathrm{pe}_x = w + h_y + \mathrm{ke}_y + \mathrm{pe}_y \qquad (2.7)$$

where ke and pe are the kinetic and potential energies per unit mass respectively. Here the subscripts x and y represent respectively the beginning and end of any process. Assuming, as is generally the case, that the potential and kinetic energy per unit mass are small relative to the other terms and can therefore be ignored, the equation reduces to

$$q + h_x = w + h_y \qquad (2.8)$$

This equation can now be applied to each of the processes in turn.

The equation developed in section 2.1.2 can also be used. That is

$$w_n = q_a - q_r \qquad (2.9)$$

Heat transfer does not occur during either the compression or expansion processes as these have been assumed to be isentropic, and is added only in process 2–3 (Figure 2.1) which represents the combustion chamber and

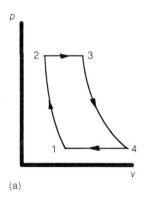

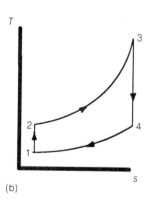

(a)

(b)

Figure 2.1 Air standard Brayton cycle for gas turbine cycle analysis. (a) pressure–specific volume diagram, (b) temperature–entropy diagram. 1–2, isentropic compression; 2–3, constant pressure heat addition; 3–4, isentropic expansion; 4–1, constant pressure heat rejection.

rejected only in process 4–1 (Figure 2.1) which represents cooling of the exhaust gases. In neither of these processes is work done during the steady flow as they have no devices such as turbomachinery by which work transfer can occur. Hence, applying (2.8) gives

$$q_a = h_3 - h_2 \tag{2.10}$$

Considering the magnitude of the heat transfer only, the heat rejection is

$$q_r = h_4 - h_1 \tag{2.11}$$

Hence, the net work becomes

$$w_n = q_a - q_r = (h_3 - h_2) - (h_4 - h_1) \tag{2.12}$$

or

$$w_n = (h_3 - h_4) - (h_2 - h_1) \tag{2.13}$$

From (2.8) it can be seen that (2.13) is also the expansion work minus the turbine work as would be expected

$$w_n = w_t - w_c \tag{2.14}$$

Closed system analysis

Now, a closed system analysis may be used instead. Here the basic first law of thermodynamics is

$$q - w = \Delta u \tag{2.15}$$

But for both constant pressure heat addition and rejection processes,

$$w = \int p \Delta v = p \Delta v \tag{2.16}$$

Hence

$$q = \Delta u + p \Delta v = \Delta(u + pv) = \Delta h \tag{2.17}$$

Therefore

$$w_n = q_a - q_r = (h_3 - h_2) - (h_4 - h_1) \tag{2.18}$$

which is the same result as for the open system. It should be noted, however that a closed system analysis does not strictly represent actual gas turbine processes. In particular, it assumes that boundary work exists during the heat addition and rejection processes.

Trends in the thermal efficiency

In either case, using the ideal gas relations for enthalpy,

$$w_n = c_p(T_3 - T_2) - c_p(T_4 - T_1) \tag{2.19}$$

and thermal efficiency becomes

$$\eta_{TH} = 1 - \frac{T_4 - T_1}{T_3 - T_2} \tag{2.20}$$

Now if r_P is used to designate the ratio of the maximum to minimum pressure in the gas turbine (i.e. the combustion pressure to the intake pressure) which can be called the engine pressure ratio, and remembering that it applies over the isentropic processes of both the compressor and turbine for the simple Brayton cycle, the thermal efficiency can be modified.

$$\frac{T_2}{T_1} = (r_P)^{(\gamma - 1)/\gamma} \tag{2.21}$$

$$\frac{T_3}{T_4} = (r_P)^{(\gamma - 1)/\gamma} \tag{2.22}$$

Therefore

$$\eta_{TH} = 1 - \frac{1}{(r_P)^{(\gamma - 1)/\gamma}}\left[\frac{T_4 - T_1}{(T_4 - T_1)}\right] = 1 - \frac{1}{(r_P)^{(\gamma - 1)/\gamma}} \tag{2.23}$$

Thus the thermal efficiency of the Brayton cycle is dependent only on the engine pressure ratio and increases with increase in r_P. This should also be a fundamental factor in the variation that occurs in the gas turbine. The increases are not linear but are of decreasing rate as the pressure ratio becomes higher as shown on Figure 2.2. It would therefore seem that all that is necessary for gas turbine development is to use as high a pressure ratio as is compatible with the structural integrity of the engine. In fact this is an oversimplification and needs further air standard cycle investigation.

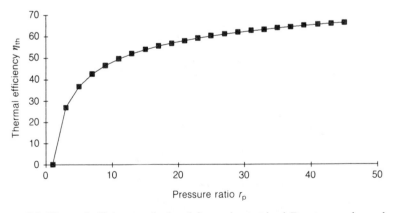

Figure 2.2 Thermal efficiency calculated from air standard Brayton cycle analysis. Isentropic index, $\gamma = 1.4$.

2.3.3 Maximum work from the Brayton cycle

While the above formula give the efficiency and work output of the cycle, let us now investigate some practical limitations on their application. In this section, the maximum work is to be considered.

Maximum engine temperature

Consider the cycle shown on Figure 2.3 on a temperature–entropy diagram. From (2.23) it is obvious that increase in the engine pressure ratio increases the efficiency. However, from the figure it is also obvious that increase in pressure must also increase maximum engine temperature if the same quantity of heat is added per unit mass of working substance. This is simply because the higher pressure ratio increases the end of compression temperature and feeds through from there to the remaining state points. The effect of high temperature is particularly critical in practice when the gas enters the first turbine stage. Here it has just left the combustion chamber and is passing over high-speed machinery which should be of low mass and which must be carefully designed, as will be seen later, to retain high engine efficiency. Thus the temperature at point 3 of the cycle, called the **turbine entry temperature** (TET), must be limited to a value which depends on the current state of the art in both high-temperature metallurgy and in blade cooling design. A maximum temperature is shown on Figure 2.3 as the horizontal line on the diagram.

The question must be, what happens to the work output if this temperature, T_3, is fixed and the pressure ratio varied from that of the original cycle. In order that the limiting maximum temperature not be exceeded,

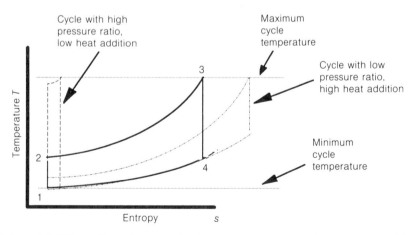

Figure 2.3 Effect of maximum and minimum cycle temperature on gas turbine performance.

two approaches are possible. These are that the pressure ratio is increased with a corresponding cutback in the heat addition or that it is reduced allowing greater heat addition to take place. It is obvious that the first should increase the cycle thermal efficiency while the second should reduce it. Two alternative cycles are shown on the figure, these approaching the extreme cases. That is, the first has a pressure ratio which is such that the maximum temperature is almost achieved with compression alone allowing very little heat addition while the second has a pressure ratio approaching unity (i.e. virtually no compression) and a very large heat input. It can be seen that, in both these cases, the cycle area is extremely small and in fact approaches zero. That is, there is little work output, and a maximum work output must exist for a cycle with a pressure ratio in between the two cases.

Now for the isentropic compression and expansion processes

$$(r_P)^{(\gamma - 1)/\gamma} = \frac{T_2}{T_1} = \frac{T_3}{T_4} \tag{2.24}$$

which on rearranging, gives

$$T_4 = \frac{T_3 T_1}{T_2} \tag{2.25}$$

Eliminating T_4 from (2.19) by use of (2.25) gives

$$w_n = c_P \left(T_3 - T_2 - \frac{T_3 T_1}{T_2} + T_1 \right) \tag{2.26}$$

But T_1 is the ambient temperature which is fixed and T_3 is the maximum temperature which, as previously discussed, is also fixed. Therefore, the work must vary only with temperature T_2 and will have a maximum value at a particular end of compression temperature (and pressure). Differentiating and equating to zero to find this maximum gives

$$\frac{dw_n}{dT_2} = -1 + \frac{T_3 T_1}{T_2^2} \tag{2.27}$$

Therefore

$$\frac{T_3 T_1}{T_2^2} = 1$$

and

$$T_2 = (T_3 T_1)^{1/2} \tag{2.28}$$

for maximum work output to occur.

Resubstituting for T_2 into (2.23) and (2.24) gives

$$w_{n,max} = c_p(T_3^{1/2} - T_1^{1/2})^2 \tag{2.29}$$

at a pressure ratio of

$$(r_p)^{(\gamma - 1)/\gamma} = \left(\frac{T_3}{T_1}\right)^{1/2} \tag{2.30}$$

If the maximum temperature, T_3, is low, the maximum possible work and hence power output is also low. This explains why early gas turbines were not practical because engine temperatures were severely limited. It also shows why the gas turbine is suited to high altitudes where the cold ambient temperature allows an increase in the maximum work.

2.3.4 The effect of component efficiency

In the simple Brayton cycle analysis, the compressor and turbine processes were assumed to be isentropic. This is untrue as, although heat transfer may approach zero, there is always some friction with the flow through high-speed turbine blades which increases the entropy. That is, there is more work required for the compression and less obtained from the expansion than the simple analysis indicates. We will now see how much the isentropic assumption affects the final thermal efficiency value.

A modified Brayton cycle with a more realistic temperature–entropy $(T-s)$ diagram is shown on Figure 2.4 with the dashed lines showing the

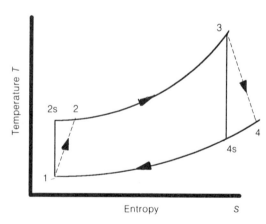

Figure 2.4 Air standard Brayton cycle modified for adiabatic but non-isentropic compression and expansion. 1–2s, isentropic compression; 1–2, actual adiabatic compression; 2–3, constant pressure heat addition; 3–4s, isentropic expansion; 3–4 actual adiabatic expansion; 4–1, constant pressure heat rejection.

irreversible adiabatic processes. That is, although they have no heat transfer, their entropy must increase and it can be seen in both cases that the entropy at the end of the process is greater than that at the beginning.

To account for the increase in entropy, process efficiencies can be defined as follows.

1. For the compressor, the adiabatic efficiency is

$$\eta_T = \frac{w_{\text{isentropic}}}{w_{\text{actual}}} = \frac{h_{2s} - h_1}{h_2 - h_1}$$

and, for an ideal gas, this becomes

$$\eta_c = \frac{T_{2s} - T_1}{T_2 - T_1} \tag{2.31}$$

2. For the turbine, the adiabatic efficiency is

$$\eta_T = \frac{w_{\text{actual}}}{w_{\text{isentropic}}} = \frac{h_3 - h_4}{h_3 - h_{4s}}$$

Again for an ideal gas

$$\eta_T = \frac{T_3 - T_4}{T_3 - T_{4s}} \tag{2.32}$$

Note that the definitions are inverted between the compression (work input) and expansion (work output) cases. This is because the frictional effects always result in a less efficient process. The values of η_c and η_T are always less than unity. Returning now to a conventional approach to the cycle analysis, the work output is as before although the location of the state points are different as can be seen on Figure 2.4 and a thermal efficiency can be obtained. That is

$$\eta_{\text{TH}} = \frac{w_n}{q_a}$$

$$= \frac{\eta_T(T_3 - T_{4s}) - 1/\eta_c(T_{2s} - T_1)}{T_3 - T_1 - 1/\eta_c(T_{2s} - T_1)} \tag{2.33}$$

Note that the numerator is the actual turbine work minus the actual compressor work and that, while the work output is reduced, the heat addition is also reduced because the extra compression work raises the end of compression temperature.

Assuming a constant pressure combustion as before still with a pressure ratio of r_P, the above equation can be modified to

$$\eta_{TH} = \frac{\eta_T\left(T_3 - \dfrac{T_3}{r_P^{(\gamma-1)/\gamma}}\right) - \dfrac{1}{\eta_c}(T_1 r_P^{(\gamma-1)/\gamma} - T_1)}{T_3 - T_1 - \dfrac{1}{\eta_c}(T_1 r_P^{(\gamma-1)/\gamma} - T_1)}$$

$$= \left[1 - \frac{1}{r_P^{(\gamma-1)/\gamma}}\right]\left[\frac{\eta_T T_3 - \dfrac{T_1 r_P^{(\gamma-1)/\gamma}}{\eta_c}}{T_3 - T_1 - \dfrac{T_1}{\eta_c}(r_P^{(\gamma-1)/\gamma} - 1)}\right] \qquad (2.34)$$

The term in the first bracket is identical to that for the thermal efficiency of the Brayton cycle while that in the second bracket expresses the effect of the modification due to the component efficiencies. Note that this term reduces to unity for η_c and η_T equal to one (i.e. 100%). A typical plot of the thermal efficiency against pressure ratio obtained from this equation is shown on Figure 2.5.

For η_c and η_T of 100%, the thermal efficiency is identical to that of the Brayton cycle and continues to rise as pressure ratio increases. However, as either η_c, η_T or both are reduced, the thermal efficiency now peaks at particular values of r_P. The maximum thermal efficiency and the pressure ratio at which it occurs both become less as the component efficiencies reduce. Of the two, it can be seen by comparing the lines on Figure 2.5 that the turbine losses have a more deleterious effect than those for the compressor.

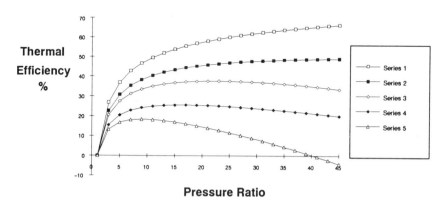

Figure 2.5 Thermal efficiency of gas turbine with different turbine and compressor efficiencies. Minimum cycle temperature, $T_1 = 300\,\text{K}$; maximum cycle temperature, $T_3 = 1760\,\text{K}$; $\gamma = 1.4$. Series 1: $\eta_T = 100\%$, $\eta_c = 100\%$; series 2: $\eta_T = 90\%$, $\eta_c = 90\%$; series 3: $\eta_T = 90\%$, $\eta_c = 70\%$; series 4: $\eta_T = 70\%$, $\eta_c = 90\%$; series 5: $\eta_T = 70\%$, $\eta_c = 70\%$.

In spite of this, while both compressor and turbine have received considerable attention in relation to their development since gas turbines were introduced, the former has received the most. This is because the heating effect of the entropy increase at each stage (or element of the process) as the process progresses makes compression a little harder and expansion a little easier to obtain and high overall isentropic compressor efficiencies are more difficult to achieve than turbine ones. This stage reheating effect will be dealt with in a later chapter. It can be seen that, when both η_c and η_T fall to a low value of, say, 70%, the maximum thermal efficiency has reduced to a low 18.4% at a pressure ratio of just over 9:1. High turbomachinery component efficiencies are therefore of extreme importance but are difficult to obtain. Again, this was another factor which inhibited the development of the gas turbine.

2.3.5 The regenerative gas turbine cycle

Even with high component efficiencies, a very high pressure ratio is required for a conventional gas turbine in order to obtain a good thermal efficiency. This has disadvantages in the size and complexity of the turbomachinery, particularly the compressor and the mass of the engine which must provide a strong structure to withstand the very high pressures. Also, engine durability is likely to suffer. However, the regenerative gas turbine provides a potential alternative although, to date, it has only partially fulfilled its potential in practice due to the difficulty of obtaining the necessary high regenerator efficiency.

A conventional Brayton cycle for a low pressure ratio engine is shown on Figure 2.6. It can be seen that the temperature of the gas at the end of the expansion process is still very high, substantially above that of the gas emerging from the compressor. The use of the exhaust to preheat the compressed gas to a temperature T_x at state point x part way along process 2–3 is therefore a possibility and, as this would reduce the heat input required from the fuel without altering either the turbine or compressor work, it can be seen that the thermal efficiency of the cycle could be improved. This is the basic concept of the regenerative gas turbine.

The effectiveness of the regenerator is extremely important and the best possible result would be for T_x to be identical to the temperature T_4. This value cannot be exceeded and is impossible to completely achieve in practice but a lower value is possible. The effectiveness of the regenerator can be defined as

$$\varepsilon_r = \frac{\text{Actual temperature rise of the gas being heated}}{\text{Maximum possible temperature rise}}$$

$$= \frac{T_x - T_2}{T_4 - T_2} \tag{2.35}$$

Obviously an effectiveness of 1 gives T_x equal to T_4 and it is only this best

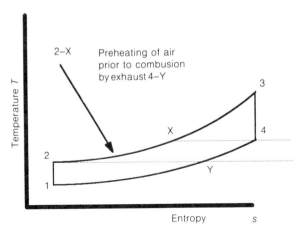

Figure 2.6 Preheating of air prior to combustion by exhaust gas.

possible case that will be considered in the present analysis. Needless to say, any case with an inferior effectiveness will result in a lower thermal efficiency. The cycle analysis proceeds very easily by evaluating the work from the turbine minus that from the compressor as previously and dividing by the now reduced heat input:

$$\eta_{TH} = \frac{(T_3 - T_4) - (T_2 - T_1)}{(T_3 - T_x)} \tag{2.36}$$

If $T_x = T_4$, this becomes

$$\eta_{TH} = 1 - (T_2 - T_1)/(T_3 - T_4) \tag{2.37}$$

Substituting for T_2 and T_4 in terms of the pressure ratio r_p then gives

$$\eta_{TH} = 1 - \frac{T_1(r_P^{(\gamma-1)/\gamma} - 1)}{T_3(1 - 1/r_P^{(\gamma-1)/\gamma})} = 1 - \frac{T_1 r_P^{(\gamma-1)/\gamma}}{T_3} \tag{2.38}$$

This is plotted on Figure 2.7. It can be seen that the efficiency is a maximum for r_P equal to one and reduces as r_P increases. This is the opposite trend to that of the conventional Brayton cycle.

At r_P equal to one

$$\eta_{TH} = 1 - \frac{T_1}{T_3} = \frac{T_3 - T_1}{T_3} \tag{2.39}$$

That is, it is equal to the efficiency of the Carnot cycle which operates between the same maximum and minimum temperatures. This is also the value reached by the conventional Brayton cycle with a pressure ratio such

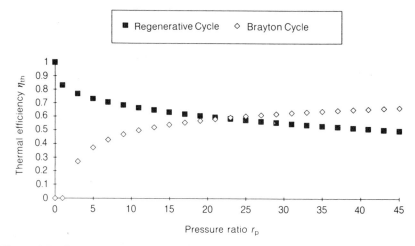

Figure 2.7 Thermal efficiency of regenerative gas turbine. Isentropic index $\gamma = 1.4$; effectiveness of heat exchanger $= 100\%$; minimum cycle temperature $T_1 = 300\,\mathrm{K}$; maximum cycle temperature $T_3 = 1760\,\mathrm{K}$.

that T_2 reaches the maximum cycle temperature T_3. In neither case does the cycle produce any work output. If the thermal efficiencies of the two cycles are plotted on the one graph, it can be seen that they cross at a particular pressure ratio, the regenerative gas turbine having the better efficiency below this point while the conventional gas turbine (Brayton) cycle is superior above it. The intersection point can be determined by equating the thermal efficiencies of the two cycles. That is

$$1 - \frac{T_1 r_{\mathrm{P}}^{(\gamma-1)/\gamma}}{T_3} = 1 - \frac{1}{r_{\mathrm{P}}^{(\gamma-1)/\gamma}} \tag{2.40}$$

Hence, the pressure ratio at the intersection can be determined as

$$(r_{\mathrm{P}})^{(\gamma-1)/\gamma} = \left(\frac{T_3}{T_1}\right)^{1/2} \tag{2.41}$$

This is identical to the value for maximum Brayton cycle work determined previously. Hence the regenerative cycle has no advantage over the Brayton cycle here and its greater complexity in practice make it an unlikely option. However, when structural lightness together with a greater thermal efficiency is required and some sacrifice of specific work output can be tolerated, the regenerative cycle has potential. For example, it has been investigated as a power source for automobiles with a pressure ratio as low as about 4:1. The major problem lies in producing robust, lightweight and effective regenerators, rotating ceramic types showing the greatest potential.

2.3.6 Intercooling and reheating for gas turbine cycles

Intercooling

As a gas is compressed, its temperature rises together with its pressure reducing the gain in density and requiring more work input to achieve a given incremental pressure ratio late in compression compared to the beginning. If, however, the compression is carried out in stages with cooling, normally called intercooling, between them, the total work requirement is reduced. This can be utilized to increase the net work output of the gas turbine. The cooling is normally close to a constant pressure process and it is possible to have several intercooled stages. An example of the cycle is shown on Figure 2.8 of two compression stages with a single intercooling

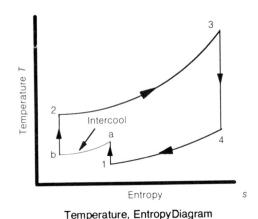

Temperature, Entropy Diagram

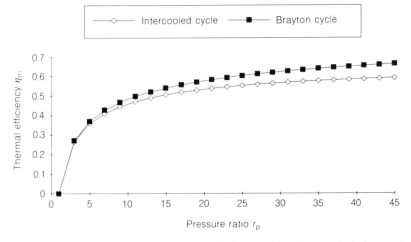

Figure 2.8 Thermal efficiency of an intercooled gas turbine. Isentropic index $\gamma = 1.4$; minimum cycle temperature $T_1 = 300\,\text{K}$; maximum cycle temperature $T_3 = 1760\,\text{K}$.

process between. Because the ambient is the normal restriction on the cooling, this process can extend down to that temperature but is limited to it.

Assuming all compression processes are isentropic and that the specific heats are constant, the compressor work for the original single stage is

$$w_c = c_P(T_2 - T_1) = c_P T_1(r_P^{(\gamma-1)/\gamma} - 1) \quad (2.42)$$

whereas, for the two stages, it is

$$w_c = c_P(T_a - T_1) + c_P(T_2 - T_b)$$
$$= c_P T_1(r_{P1}^{(\gamma-1)/\gamma} - 1) + c_P T_b(r_{P2}^{(\gamma-1)/\gamma} - 1) \quad (2.43)$$

Here r_{P1} and r_{P2} are the pressure ratios of the first and second stages respectively. It should be noted that

$$r_P = r_{P1} \cdot r_{P2} \quad (2.44)$$

When the limiting case is reached where T_b is equal to T_1, the compressor work becomes

$$w_c = c_P T_1(r_{P1}^{(\gamma-1)/\gamma} + r_{P2}^{(\gamma-1)/\gamma} - 2) \quad (2.45)$$

To determine the value of the stage pressure ratios to minimise the total compressor work for a given overall pressure ratio, either r_{P1} or r_{P2} can be eliminated. Hence

$$w_c = c_P T_1[r_{P1}^{(\gamma-1)/\gamma} + (r_P/r_{P1})^{(\gamma-1)/\gamma} - 2] \quad (2.46)$$

To obtain the minimum value

$$\frac{dw_c}{dr_{P1}} = c_P T_1\left[\frac{\gamma-1}{\gamma}r_{P1}^{((\gamma-1)/\gamma)-1} - \frac{\gamma-1}{\gamma}r_P^{(\gamma-1)/\gamma}\bigg/r_{P1}^{((\gamma-1)/\gamma)+1}\right] = 0 \quad (2.47)$$

This gives the value of r_{P1} as

$$r_{P1} = r_P^{1/2} \quad (2.48)$$

with r_{P2} having the same value.

The amount of work input that has been saved can be calculated by substitution into (2.46) and its subtraction from (2.42) giving

$$w_{saved} = c_P T_1(r_P^{(\gamma-1)/2\gamma} - 1)^2 \quad (2.49)$$

which can be quite substantial.

Although less work compared to a single stage compression process has now been used to reach the desired final pressure, it does not follow that the thermal efficiency has increased. It can be seen by inspection of the $T-s$ diagram of Figure 2.8 that the cooling has resulted in a lower end of compression temperature than would have otherwise occurred requiring more heat input to regain the original value. This heat is added at a low average temperature which is detrimental to high efficiency. The overall thermal efficiency of the intercooled cycle will now be explored using the minimum compressor work case examined above as an example. It must be remembered that the following equations apply only to this case and are not general for all intercooling applications.

The thermal efficiency is:

$$\eta_{TH} = \frac{w_t - w_c}{q_a}$$

$$= \frac{c_P(T_3 - T_4) - c_P(T_2 - T_b) - c_P(T_a - T_1)}{c_P(T_3 - T_2)}$$

$$= \frac{T_3(1 - 1/r_P^{(\gamma-1)/\gamma}) - 2T_1(r_P^{(\gamma-1)/2\gamma} - 1)}{T_3 - T_1 r_P^{(\gamma-1)/2\gamma}}$$

$$= \frac{(1 - 1/r_P^{(\gamma-1)/\gamma}) - \dfrac{2T_1}{T_3}(r_P^{(\gamma-1)/2\gamma} - 1)}{\left(1 - \dfrac{T_1}{T_3} r_P^{(\gamma-1)/2\gamma}\right)} \qquad (2.50)$$

This is plotted on Figure 2.8 together with the Brayton cycle thermal efficiency using a temperature ratio T_3/T_1 of 5.87 (i.e. 1760/300) and an isentropic index, γ of 1.4 ($\gamma - 1/\gamma = 0.286$). The intercooled cycle can be seen to have a lower thermal efficiency throughout the pressure ratio range. Even if the total heat input had been kept constant and the final temperature lowered correspondingly, the result would have been similar. Its advantage lies, therefore, not in increasing the efficiency but in giving a greater power output for a given size machine. Some increase in mechanical complexity is, however, required.

Reheating

Reheat is best known in gas turbines as a means of obtaining additional thrust in high-performance supersonic aircraft by burning more fuel just upstream of the exhaust nozzle but downstream of the turbine unit which drives the compressors. It can be seen from a cursory examination of the cycle on a T–s diagram (Figure 2.9) that this extends the cycle area and hence results in a greater work output. It should be noted, however, that the increased work is obtained at the expense of additional fuel and does not increase the thermal efficiency; in fact, it will reduce it because the pressure ratio over which the gas can be further expanded in order to obtain work from the additional energy input is now smaller.

The concept is not limited to the above application and may be used between turbine stages of a gas turbine. In theory, as with intercooling, a large number of reheats are possible but, in practice, these are limited because of the mechanical complexity. The maximum temperature which can be tolerated during reheat is identical to the normal end-of-combustion value because the next stage turbine blades have the same limiting design conditions as the first stage. In the case of reheat in the exhaust pipe for nozzle expansion only, the conditions are different and some increased temperature may be possible.

The analysis follows a similar pattern to that for intercooling. Assuming all expansion processes are isentropic, as before, the turbine work for the original single stage is

$$w_c = c_P(T_3 - T_4) = c_P T_3 (1 - 1/r_P^{(\gamma-1)/\gamma}) \qquad (2.51)$$

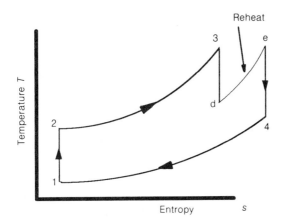

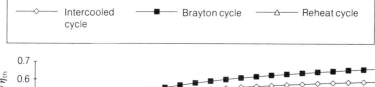

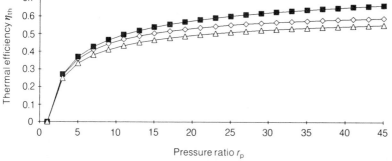

Figure 2.9 Thermal efficiency of a reheated gas turbine. Isentropic index $\gamma = 1.4$; minimum cycle temperature $T_1 = 300\,\text{K}$; maximum cycle temperature $T_3 = 1760\,\text{K}$.

whereas, for the two stages, it is

$$w_t = c_P(T_3 - T_d) + c_P(T_e - T_4)$$

$$= c_P T_3 (1 - 1/r_{P1}^{(\gamma-1)/\gamma}) + c_P T_e (1 - 1/r_{P2}^{(\gamma-1)/\gamma}) \tag{2.52}$$

For T_e equal to T_3, the work output in the latter case becomes

$$w_t = c_P T_3 (2 - 1/r_{P1}^{(\gamma-1)/\gamma} - 1/r_{P2}^{(\gamma-1)/\gamma}) \tag{2.53}$$

Substituting for either r_{P1} and r_{P2} in terms of the overall pressure ratio, r_P and maximizing the value of the expression by differentiation as in the intercooled case, gives

$$\frac{dw_t}{dr_{P1}} = c_P T_3 \left[\frac{\gamma-1}{\gamma} \middle/ r_{P1}^{((\gamma-1)/\gamma)+1} - \frac{\gamma-1}{\gamma} r_{P1}^{((\gamma-1)/\gamma)-1} \middle/ r_P^{(\gamma-1)/\gamma} \right] = 0 \tag{2.54}$$

Hence, a value of r_{P1} (and r_{P2}) is obtained as before

$$r_{P1} = r_P^{1/2} \tag{2.55}$$

The additional work output for this maximum case follows as

$$w_{additional} = c_P T_3 (1 - 1/r_P^{(\gamma-1)/2\gamma})^2 \tag{2.56}$$

which again can be quite substantial.

In this case, an additional heat input is required which is an operating cost unlike the intercooled case where the extra heat rejection required only greater mechanical complexity. The additional heat is added at constant pressure (in the real situation, some small pressure drop would exist) and so is

$$q_{additional} = c_P(T_e - T_d)$$

$$= c_P(T_3 - T_3/r_{P1}^{(\gamma-1)/\gamma})$$

$$= c_P T_3 (1 - 1/r_{P1}^{(\gamma-1)/\gamma})$$

$$= c_P T_3 (1 - 1/r_P^{(\gamma-1)/2\gamma}) \tag{2.57}$$

The ratio

$$\frac{w_{additional}}{q_{additional}} = 1 - 1/r_P^{(\gamma-1)/2\gamma} \tag{2.58}$$

which is less than one. That is, as would be expected, all the additional heat input cannot be turned into useful work.

The efficiency of the reheat gas turbine cycle is reduced below that of the original Brayton cycle. For the maximum work pressure ratio explored above, the reheat cycle efficiency is

$$\eta_{TH} = \frac{w_t - w_c}{q_a}$$

$$= \frac{c_P(T_3 - T_d) + c_P(T_e - T_4) - c_P(T_2 - T_1)}{c_P(T_3 - T_2) + (T_e - T_d)}$$

$$= \frac{T_3(2 - 2/r_P^{(\gamma-1)/2\gamma}) - T_1(r_P^{(\gamma-1)/\gamma} - 1)}{T_3(2 - r_P^{(\gamma-1)/\gamma} T_1/T_3 - 1/r_P^{(\gamma-1)/2\gamma})}$$

$$= \frac{(2 - 2/r_P^{(\gamma-1)/2\gamma}) - T_1/T_3(r_P^{(\gamma-1)/\gamma} - 1)}{(2 - r_P^{(\gamma-1)/\gamma} T_1/T_3 - 1/r_P^{(\gamma-1)/2\gamma})} \qquad (2.59)$$

This, together with the intercooled and Brayton cycle efficiencies is plotted on Figure 2.9 for a T_3/T_1 value of 5.87. It can be seen that there is a noticeable drop in thermal efficiency throughout the pressure ratio range and it is, in fact, greater than that of the equivalent intercooled case.

Combined reheat and intercooling

Reheat and intercooling are commonly proposed for use together in order to obtain a machine with a high specific power output (i.e. power per unit mass). Here the cycle is a direct combination of the two as shown on Figure 2.10 with the separate advantages accruing in the additional area of the two parts of the cycle. As in the above cases, the minimum compressor and maximum turbine work occur (for a two-stage process in each) when the stage pressure ratios are each the square root of the overall pressure ratio. Although this is again a specific case, it will now be examined for its overall effect on thermal efficiency.

The net work output is given by the sum of the two turbine minus the two compressor stages and the heat added is that during the initial combustion plus that during reheat. That is

$$w_n = c_P[(T_3 - T_d) + (T_e - T_4) - (T_2 - T_b) - (T_a - T_1)] \qquad (2.60)$$

$$q_a = c_P[(T_3 - T_2) + (T_e - T_d)] \qquad (2.61)$$

Using the same procedure as before to introduce the pressure ratio, the thermal efficiency becomes

$$\eta_{TH} = \frac{(2 - 2/r_P^{(\gamma-1)/2\gamma}) - T_1/T_3(2r_P^{(\gamma-1)/2\gamma} - 2)}{2 - r_P^{(\gamma-1)/2\gamma} T_1/T_3 - 1/r_P^{(\gamma-1)/2\gamma}} \qquad (2.62)$$

This, also plotted on Figure 2.10 for T_3/T_1 equal to 5.87, shows that the combination of the two is, under these conditions, less efficient than either

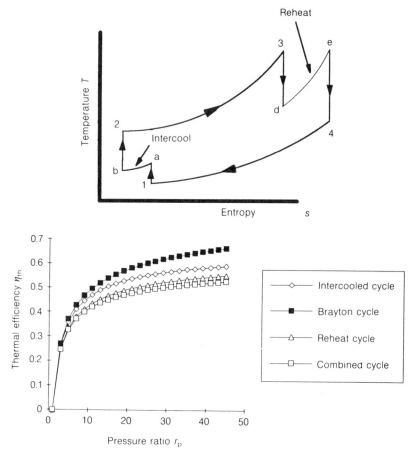

Figure 2.10 Thermal efficiency of a combined reheat-intercooled gas turbine. Isentropic index $\gamma = 1.4$; minimum cycle temperature $T_1 = 300\,\text{K}$; maximum cycle temperature $T_3 = 1760\,\text{K}$.

separately and considerably less efficient than the simple Brayton cycle. However, it can be seen from the T–s diagram that the low end of compression and high end of expansion temperatures improve the prospects of this cycle for regeneration. When this occurs, the cycle efficiency improves over the basic Brayton cycle. By introducing more stages for both compression and expansion, the limiting case can be reached where an infinite number of extremely small intercooled and reheated stages are used. Here the cycle approaches one with constant temperature compression and expansion with full regenerative heat transfer between the constant pressure processes. This is the Ericsson cycle which can be shown to have a thermal efficiency equal to that of the Carnot cycle operating between the same temperature limits.

The thermal efficiency for the reheat intercooled cycle with n stages for each of the compression and expansion with the same pressure ratio is

$$\eta_{TH} = \frac{(n - n/r_P^{(\gamma - 1)/n\gamma}) - T_1/T_3(nr_P^{(\gamma - 1)/n\gamma} - n)}{n - r_P^{(\gamma - 1)/n\gamma} T_1/T_3 - (n - 1)/r_P^{(\gamma - 1)/n\gamma}} \tag{2.63}$$

It should be noted that there is no reason why the number of stages in compression and expansion must be identical and the reader may establish appropriate equations for say n stages of the former and m of the latter by following the same techniques as above.

Regeneration with the intercooled reheated cycle

For any of the intercooled reheated cycles, the thermal efficiency falls although the work output increases. Regeneration can be added to any of these in order to improve efficiency. Taking the same situation as described by (2.63) and assuming that with 100% regeneration is available, the equation for thermal efficiency requires only a modification of the denominator. The result is

$$\eta_{TH} = \frac{(n - n/r_P^{(\gamma - 1)/n\gamma}) - T_1/T_3(nr_P^{(\gamma - 1)/n\gamma} - n)}{n - \dfrac{n}{r_P^{(\gamma - 1)/n\gamma}}}$$

$$= 1 - \frac{T_1}{T_3} r_P^{(\gamma - 1)/n\gamma} \tag{2.64}$$

The variation of the thermal efficiency with the number of stages can now be calculated and is shown for a pressure ratio of 20 and a temperature ratio, T_3/T_1, of 5.87 on Figure 2.11. The Brayton cycle, the regenerative cycle and the Carnot cycle efficiencies for this case are 57.5%, 59.9% and 83.0% respectively. The relationships should be noted for $n = 1$ and for large values of n.

2.3.7 Assessment of air standard analysis for gas turbines

As pointed out previously, cycle analysis is particularly useful for gas turbines. This is because they consist of a series of linked but separate mechanical components with continuous steady flow, each then being able to follow a single process equation more closely than, for example, happens in reciprocating machines where everything occurs in the one unit. Also, the separate components allow considerable flexibility in the arrangement of the machine as a whole with many simple trends to be explored. These have been discussed in the previous sections. While this has provided a good insight into the changes in efficiency and power possible with the various arrangements, the quantitative answers will not accurately reflect

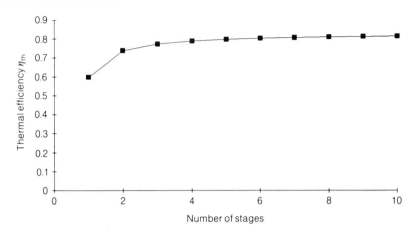

Figure 2.11 Thermal efficiency of multistage combined cycle gas turbine. Isentropic index $\gamma = 1.4$; minimum cycle temperature $T_1 = 300\,\text{K}$; maximum cycle temperature $T_3 = 1760\,\text{K}$.

actual conditions. Some of the reasons for this are fluid friction, mechanical losses, the properties of the working substance, combustion, and heat transfer.

Fluid friction

While the fluid friction can be taken into account by use of isentropic compression and turbine efficiencies as demonstrated, these are essentially empirical and experimental data must be obtained to evaluate them. This needs to be quite reliable. Other important effects of fluid friction are the pressure drops which exist as the fluid passes from one component to another and particularly in the combustion chamber itself. This loss is partially due to the swirling pattern which must be imposed on the air entering the primary combustion zone so that good mixing with the fuel will take place and partially on the large quantity of air which must be made to pass through narrow slots or holes along the periphery of the chamber in the secondary region for cooling purposes.

Mechanical losses

While mechanical friction is not a large loss, it is nevertheless significant and may be of the order of 10% of the work output. Losses occur in the bearings of the shaft or shafts joining the turbine and compressor units and, in an industrial type, on the power output turbine. Note that the turbine work output for the compressor will therefore always be greater than the work absorbed. Different configurations will have different losses.

Peripheral machinery such as fuel and oil pumps also use energy and this also needs consideration. Each specific gas turbine must then be studied separately.

Properties of the working substance

The gas which is the working substance in the engine is only air for part of the cycle. After compression, it consists of the burnt products of the combustion and the remnants of the air. Because gas turbines operate on relatively lean fuel-to-air ratios, the deviation from air properties is not as marked as in reciprocating engines, particularly SI engines. Temperature also noticeably modifies the gas properties and the simple air standard analysis does not include this although it is possible without much greater complexity. Typically, variations in c_p, c_v and γ between say $250\,K$ and $1700\,K$ are 10% to 20% or more. When taken as constant, even if average values are used, some considerable error is possible.

Combustion

Gas turbine combustion is an efficient process with few direct losses. Typically, a 98% combustion efficiency can be assumed. Nevertheless, if the addition of heat is taken as the lower heating value of the fuel, a considerable error results. This is because, at the high combustion temperatures, some of the products do not form completely and remain dissociated into other species releasing less energy than anticipated to the working substance. Recombination only occurs during expansion and much of this may be so delayed that very little of the energy is recoverable. Typical dissociations of this type are the CO and O_2 which eventually combine to form CO_2 and H_2, OH and O_2 which form H_2O. A full combustion calculation of the heat release is therefore necessary for an accurate cycle analysis.

Heat transfer

In any combustion engine, heat transfer losses are inevitable. In gas turbines, these occur through the outer casing of the engine to the atmosphere. In aircraft engines, some of this is recoverable in the bypass air stream which is ducted around the core but, nevertheless, a considerable loss occurs. A full cycle analysis should include an assessment of heat transfer.

Other losses

Various leakage losses are possible around the tips of the turbomachinery blades. They are of particular importance in compressors where they induce backflow, making additional work necessary for the same mass flow rate but they also reduce the work output of the turbine where they occur in the flow direction. In modern gas turbines, active tip clearance control has helped reduce them. Other losses may be deliberate and include air which is bled from the compressor and ducted to the turbine blades for cooling purposes. Also, under some conditions, air is bled from the compressor for control purposes. This is particularly the case during transients. While all these factors can be modelled, they require knowledge of the particular unit and are difficult to include at an early stage of the analysis.

2.4 RECIPROCATING ENGINE CYCLES: THE OTTO, DIESEL AND DUAL CYCLES

Reciprocating engines provide by far the most common types of prime movers in use throughout the world. They exist as both petrol (gasoline) and diesel oil engines which will be designated as spark ignition (SI) and compression ignition (CI) respectively because of their respective ignition systems. In both cases, the cycles are complex because of the intermittent processes which make the analysis less accurate than that for the steady-flow gas turbines. Nevertheless, cycle analysis is the only way that an accurate understanding and quantitative evaluation of engines can be carried out and it has been instrumental in their rapid improvement in both efficiency and emission levels in recent years. While considerable improvements over the air standard approach are extremely important here, the initial understanding can best be obtained by simple considerations as with the gas turbines.

2.4.1 The air standard Otto cycle

The Otto cycle is usually regarded as the most representative cycle for the study of SI engines, although modern high speed CI engines are often closer to this cycle than to the air standard Diesel cycle. Although Otto invented the four-stroke engine, the gas exchange processes of induction and exhaust are not involved in the basic Otto cycle which considers only compression, heat addition expansion and heat rejection. Thus it is identical for a four-stroke and a two-stroke cycle although additional processes can be added to represent the gas exchange (i.e the inlet and exhaust) processes if necessary.

The air standard Otto processes in detail are:

1. isentropic compression from the initial condition representing the air in the cylinder at the end of intake;
2. constant volume heat addition;
3. isentropic expansion of the heated gas to produce work;
4. constant volume heat rejection to the atmosphere.

These can be related to the operation of an SI engine by considering the following simplifications of the real processes. For the time being, the gas exchange processes, intake and exhaust, will be ignored.

The compression, as an isentropic process, is a reasonable approximation even though some turbulence must occur due both to a carry-over from the intake disturbances which are likely to be reinforced by compression and to additional disturbances generated during the process itself. However, the net heat transfer is likely to be small, being to the gas during the start of the process and from the gas towards the end. This is because a typical compression temperature range for the gas would be 40°C to less than 400°C while the wall temperatures are in excess of 100°C. Hence, the total entropy change is not likely to be great. The heat addition is probably the most problematical process to idealize in relation to the Otto cycle. The combustion process is rapid as the fuel and air are pre-mixed and a flame front moves rapidly through the mixture after ignition. In practice, the ignition takes place well before the compression stroke is completed so that the slow initial flame development is of little relevance. This allows the peak pressure to be reached about 15 crank angle degrees after top dead centre which represents only a small volume increase. Hence, a constant volume process is, if not a good approximation, one that is justifiable as a starting point to give simple equations. The expansion process as isentropic is again a relatively poor assumption as wall temperatures are well below the gas temperature. A typical gas temperature range is likely to be from over 2000°C to about 800°C and much of the heat transfer occurs here. Finally, the heat rejection in practice takes place as the exhaust gas emerges rapidly with a pressure wave on exhaust valve opening (called *blowdown*) and with the remainder of the gas as it is forced out in the following exhaust stroke. In the Otto cycle, heat rejection takes place as a single constant volume process which is close to the blowdown process although the actual process starts a little before bottom dead centre due to valve timing and ends a little after because of the time the rarefaction wave takes to clear the cylinder. The proportion of heat rejected in the exhaust stroke is ignored in the Otto cycle.

Thus the Otto cycle is a rough approximation only of the processes in a real SI engine. In spite of this, it has proven to be a very valuable aid in understanding and developing this type of engine. It is most usefully shown on a pressure–volume diagram where it can be drawn using either p–v or p–V ordinates. A temperature–entropy (T–s) diagram is also often used. Figure 2.12 shows these three diagrams.

The difference between using specific or total volume appears in the intake and exhaust processes. In the former case, these processes occur only as a point because the state is the same at their beginning, throughout and at their end. However, it must be remembered that they are not closed system processes as are those for the Otto cycle proper and the mass in the cylinder varies as the piston moves up or down drawing in or expelling gas. That is, while the specific volume remains constant, the total volume is proportional to the mass in the cylinder. Consequently, the intake and exhaust processes, idealized as constant pressure processes for an initial

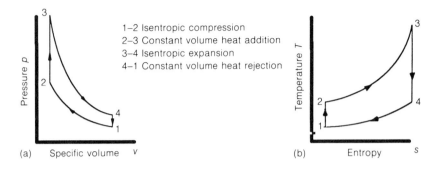

1–2 Isentropic compression
2–3 Constant volume heat addition
3–4 Isentropic expansion
4–1 Constant volume heat rejection

(a)   Specific volume   v

(b)   Entropy   s

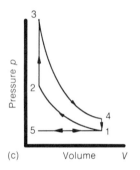

5–1 Constant pressure intake
1–2 Isentropic compression
2–3 Constant volume heat addition
3–4 Isentropic expansion
4–1 Constant volume heat rejection
1–5 Constant pressure exhaust

(c)   Volume   V

Figure 2.12 Air standard Otto cycle for reciprocating (SI) engine cycle analysis. (a) pressure–specific volume diagram, (b) temperature–entropy diagram, (c) pressure–volume diagram.

investigation, show up, in a four-stroke application, as horizontal lines on the p–V diagram but as points only on the p–v one. Even for the p–V application, if they are both assumed to be at the same pressure, they overlap completely and have no net effect on the cycle the remainder of which can be analysed as a closed system only.

2.4.2 Analysis of the Otto cycle

As the Otto cycle is a closed system, quantities such as work, heat addition, internal energy and specific volume may be expressed as total (extensive) quantities or on a per unit mass (intensive) basis. It is generally more usual thermodynamic practice to use the latter and that will be followed here except in cases where the gas exchange processes are being considered. An engine consuming one kilogram per cycle would be very large, around 700 to 800 litre capacity and so the actual quantity of air to be dealt with is normally very much less than a kilogram. This should be kept in mind where an analysis based on intensive quantities is used. Also, the gas exchanged in any engine cycle is not equal to the mass at any of the state points of the cycle. The exact quantity of gas exchanged needs further discussion at a later stage but a convenient, if not completely accurate, assumption is that the incoming air will be sufficient at the initial density to just occupy the volume swept out by the piston motion. This is the gas which brings in the new fuel charge and hence provides the heat to be added to the cycle. Moreover, it is often the most desirable base for work output. It is

$$m_i = \rho_i(V_1 - V_2) \tag{2.65}$$

where the incoming density, ρ_i, is assumed to be equal to ρ_1. The total mass in the cylinder is, however, the incoming mass plus the residual mass m_r at the end of exhaust and it is this which should be used in the intensive analysis. That is, a conversion of specific work and heat transfer quantities to a different mass base may be necessary depending upon how information is given or required. It will be assumed in the following analysis that they are both based on total mass unless otherwise noted. The total cycle mass is

$$m = m_i + m_r \tag{2.66}$$

The compression ratio, r, of the engine is defined as the volume ratio between bottom dead centre (BDC) and top dead centre (TDC) of the piston motion. A volume ratio is used rather than a pressure ratio because it is a direct result of the engine geometry and can be precisely measured. It can be seen from the p–v diagram that

$$r = (\text{swept volume} + \text{clearance volume})/\text{clearance volume}$$

$$r = V_1/V_2 = v_1/v_2 \tag{2.67}$$

(i.e. total volume or specific volume ratio).
    For a basic cycle analysis

$$\text{Net work, } w_n = \text{Heat added} - \text{Heat rejected}$$

Now the processes are all closed system and the heat is added and rejected at constant volume. Therefore, applying the first law of thermodynamics for a closed system gives, for these two processes

$$q = \Delta u + \int p\,dv$$

$$= \Delta u \text{ for a constant volume process}$$

$$= c_v \Delta T \text{ for an ideal gas} \tag{2.68}$$

Hence, the work for the Otto cycle is

$$w_n = c_v(T_3 - T_2) - c_v(T_4 - T_1) \tag{2.69}$$

The heat added is

$$q_a = c_v(T_3 - T_2) \tag{2.70}$$

Therefore, thermal efficiency becomes

$$\eta_{TH} = w_n/q_a = 1 - \frac{(T_4 - T_1)}{(T_3 - T_2)} \tag{2.71}$$

But the isentropic compression and expansion both occur over a volume ratio of r (i.e. the compression ratio). Hence

$$T_2 = T_1 r^{\gamma - 1} \tag{2.72}$$

and

$$T_3 = T_4 r^{\gamma - 1} \tag{2.73}$$

Substitution of these into (2.71) allows the thermal efficiency to be expressed in terms of the compression ratio only as

$$\eta_{TH} = 1 - \frac{1}{r^{\gamma - 1}} \tag{2.74}$$

It can be seen that, according to this equation, increasing the compression ratio must always effect an increase in the engine thermal efficiency. This can best the understood in terms of the expansion ratio which is directly related to the compression ratio; identical in this case but a little less in other cycles. For given conditions at the commencement of expansion, the work which can be obtained during the power stroke of the engine is governed by how much it can be expanded. Therefore a large expansion ratio is a major factor in practice and the only parameter in the Otto cycle which affects efficiency. The variation of the Otto cycle efficiency with r is shown on Figure 2.13 and will be further discussed in the next section.

While a number of other Otto cycle equations can be developed, one which is useful is the **mean effective pressure**. As discussed previously, this is essentially a parameter for comparison of engines of different sizes and types and although having units, is used in a similar manner to a dimensionless group. It is defined as the total work per cycle W_n divided by the swept volume. That is

$$\text{MEP} = \frac{W_n}{V_s}$$

$$= \frac{W_n}{V_1 - V_2} = \frac{W_n}{V_1(1 - 1/r)} \tag{2.75}$$

Also, from the definition, above, it can be seen that

$$W_n = (\text{MEP})V_s = (\text{MEP})AL \tag{2.76}$$

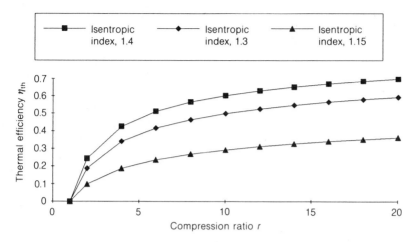

Figure 2.13 Thermal efficiency calculated from air standard Otto cycle analysis.

where A and L are the piston area and length of stroke respectively. That is MEP is equivalent to a constant pressure which, if acting on the area of the pistons over a single stroke, would produce the same output as the engine. While this is not its definition and no such constant pressure exists, it is helpful in visualizing its effect. It needs to be reiterated that the mean effective pressure is not a real pressure but merely a means of normalizing engines of different capacities so that reasonable comparisons of their outputs can be made.

Engine power is a more usually required parameter than work. To obtain the power from the net work output, all that is required is to multiply the total work per cycle by the number of cycles per second. This gives

$$\dot{W} = W_n \frac{N}{60} \tag{2.77}$$

where N is the number of cycles per minute and is equal to the revolutions per minute or to half this number for two-stroke and four-stroke engines respectively. Alternatively

$$\dot{W} = w_n m \frac{N}{60} = w_n(\rho_1 V_1)\frac{N}{60} \tag{2.78}$$

where w_n is the work per unit mass of working substance in the cycle, ρ_1 is the density of the air and V_1 is the total volume, swept plus clearance, at point 1 of the cycle.

2.4.3 Results of the air standard Otto cycle analysis

Typical results for the thermal efficiency obtained by using (2.74) are shown on Figure 2.13. As the compression ratio, r, is increased, the thermal efficiency, η_{TH}, also increases although at an increasingly slower rate at higher values of r. This tendency is also shown by tests on engines although, for practical reasons the range of useable compression ratios is quite small. The constant value for the isentropic index, γ, which is assumed for calculation purposes can be seen to have marked affect on thermal efficiency. A value of about $\gamma = 1.15$ may be seen to give a reasonable agreement with a range of experimental results at moderate compression ratios but this is not the correct way to go. It should be noted that the experimental values cease to rise at a compression ratio of around 15:1 to 16:1 and there are therefore other reasons for the difference. Air has a value of γ of about 1.29 even at a temperature of 2500 K and the effect of the fuel in the mixture in lowering it further makes only a small difference. A typical average cycle temperature is about 1600–1700 K at which γ for air is about 1.31. It has been estimated (Caris and Nelson, 1959) that reasonable compression and expansion indices are 1.31 but decreasing slightly with compression ratio and 1.29 respectively and it is therefore reasonable to assume that the overall value for the cycle should be similar. Thus, a constant value of around $\gamma = 1.3$ is often used as an approximation. The gas specific heat properties are therefore not the main reason for the discrepancy noted with experiment which must be sought elsewhere. The air standard cycle itself must be at fault for the high efficiency values and variations of the real cycle from the Otto cycle must be examined.

2.4.4 Comparison of air standard Otto and real spark ignition cycles

Figures 2.14 and 2.15 show typical comparisons between air standard Otto cycles and pressure–volume diagrams taken for real SI engines with fully opened throttles. These need to be compared process by process.

First, it can be seen that the four-stroke cycle (Figure 2.14) has an intake and exhaust process line which deviates from the constant volume process of the air standard cycle. During intake, the pressure in the cylinder is below atmospheric pressure, particularly in the middle of the stroke where the piston velocity is at its highest, while during exhaust cylinder pressure is above atmospheric pressure. Some small pressure difference between the cylinder and ambient is essential for flow to take place but the restrictions of the manifold, ports and valves can create significant pressure drops particularly at high engine speeds. This is because pressure loss is roughly proportional to gas velocity squared. Towards the end of the stroke, the pressure is regained as the piston comes to a halt and the gas continues to flow but overall, considerable work is required to move the gas into and out of the engine. The intake and exhaust stroke may be considered as a separate loop from the remaining processes of the Otto cycle and can therefore be used for approximate calculations of the pumping work in gas exchange. That is, the anticlockwise loop formed by them represents negative work

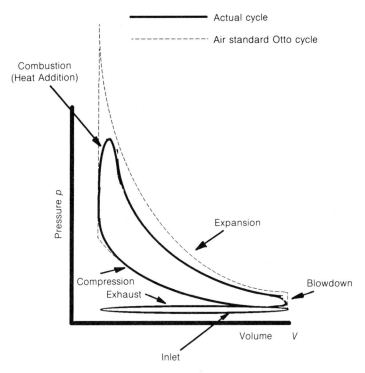

Figure 2.14 Comparison of an actual and air standard Otto cycle for a four-stroke reciprocating (SI) engine.

which must be subtracted from the positive work obtained from the compression, combustion and expansion processes.

The compression stroke of the real cycle follows the air standard cycle reasonably well as anticipated in the assumptions, particularly in the first half of the stroke. Some deviation occurs from isentropic due to the heat transfer from the walls and right at the end of the stroke because combustion has already begun. The combustion process itself differs quite markedly from the ideal cycle. Not only is it not constant volume, but the total pressure rise is well below that calculated if the heating value of the fuel is used to determine the heat addition for the air standard cycle. This is because dissociation of the combustion products prevents complete combustion and hence limits heat input, a problem which will be dealt with more fully later. During expansion, some deviation from the ideal isentropic process occurs partially due to entropy increase in the highly turbulent gas, partly due to heat transfer to the walls but mainly because of the additional heat energy becoming available as the dissociated combustion products recombine as the temperature reduces. Finally, towards the end of expansion, the exhaust valve opens. The blowdown part of the exhaust process is not constant volume as the air standard cycle assumes.

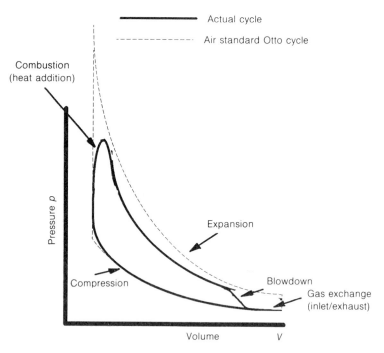

Figure 2.15 Comparison of an actual and air standard Otto cycle for a two-stroke reciprocating (SI) engine.

The two-stroke cycle of Figure 2.15 is similar in most respects to the four-stroke cycle. However, the additional two strokes giving the negative pumping work are missing. Although not shown, the additional work from the pressure on the back of the piston for crankcase scavenging or the work for the blower must be subtracted from the cycle work shown. The other major difference is that because exhaust and inlet occur during the two strokes normally used for positive work, blowdown must occur earlier than the four-stroke engine and starts when the exhaust ports are uncovered. Blowdown in a two-stroke engine represents a work loss from the ideal cycle of about 5% compared with 2% for a four-stroke cycle. It should also be noted that the pressure does not start to rise during compression until after the ports have been covered. Thus the real compression ratio is somewhat less than the usual definition of r indicates. In the example shown, the real compression ratio is perhaps 6.5:1 compared with 9:1 obtained by calculation from V_1/V_2. However, as the initial pressure in the cylinder is a little higher than in the four-stroke due to the forced induction of gas, the actual compression ratio is slightly indeterminate.

For all spark ignition engines, fuel to air ratios lie between about 1:10 and 1:22. Engines do not run well at the extremes and ignition will not occur outside these limits. With normal fuels, the stoichiometric ratio is about 1:14.5–this is the point where there are exactly sufficient oxygen molecules to combine with all the carbon and hydrogen. Usual mixture extremes are 1:12 (maximum power) to 1:16 (maximum economy). They have only a small effect on the heat addition on the p–v diagram.

2.4.5 The air standard Diesel cycle

Diesel's original engine concept assumed constant temperature heat addition to make it close to a Carnot cycle and therefore achieve that efficiency. However, it was soon realized that this was impractical and that very little heat input would be achieved per cycle. The modified concept used constant pressure heat addition and it is for this reason that a theoretical cycle of this type now carries Diesel's name. The original CI engines were pressure limited and this still applies on very large engines which can be, to some extent, represented by the Diesel cycle. Most small to medium diesel engines, however have fairly rapid mixing and combustion with some parts of the combustion process, to be discussed later, occuring very rapidly. Although some of the burning is controlled by the spray mixing processes, the overall rate is fast enough for the constant pressure heat addition to be a poor assumption as there is, in practice, a considerable pressure rise. Nevertheless, a study of the Diesel cycle is important as it shows the importance of the combustion rate on engine efficiency.

The Diesel cycle, as with the Otto cycle does not necessarily include the gas exhange processes. If these are excluded, it also consists of four processes which are:

- isentropic compression of the intake air;
- constant pressure heat addition;
- isentropic expansion of the heated gas to produce work;
- constant volume heat rejection to the atmosphere.

In other words, the only difference to the Otto cycle is the heat addition process. Figure 2.16 shows the cycle.

2.4.6 Analysis of the Diesel cycle

Most of the basic comments made in relation to the Otto cycle apply also to the Diesel cycle. These include the use of extensive or intensive properties and the mass of intake charge. The equations developed for MEP and for power in relation to cycle work are also equally valid. The analysis of the cycle is similar with some additional assumptions.

As before, the compression ratio, r, is defined as before as a volume ratio. That is

$$r = V_1/V_2 = v_1/v_2 \tag{2.79}$$

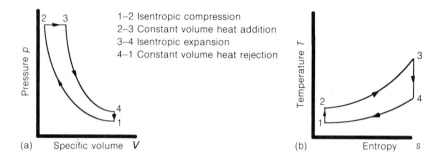

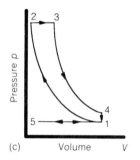

Figure 2.16 Air standard Diesel cycle for reciprocating compression ignition engine cycle analysis. (a) pressure–specific volume diagram, (b) temperature–entropy diagram, (c) pressure–volume diagram.

As can be seen from the figure, the compression and expansion ratios for this cycle, unlike the Otto cycle, are not identical. Hence, another ratio is required for analysis. The most convenient, and the one that can be related to the actual engine operation, is the ratio of the volume at the end of fuel injection to that at the beginning and it is assumed that this is when heat addition occurs. It is assumed for this purpose that injection commences at TDC. This is reasonable because the injection pump in a diesel engine controls the fuel addition and the pump itself is normally directly coupled mechanically to the engine. Both start of injection and end of injection control are used with the latter being the most common and it is this type of operation that the cycle represents. That is, the time during the expansion stroke in which fuel is injected is measured as a distance down the stroke or, effectively, a volume ratio. It is called the cutoff (i.e. fuel cutoff) ratio, r_c. Thus, cutoff ratio

$$r_c = V_3/V_2 = v_3/v_2 \qquad (2.80)$$

The heat addition maintains the pressure until volume v_3 in spite of the piston motion away from TDC but it then starts to drop when no more fuel is delivered.

For the closed system, heat addition follows from the first law

$$q_a = \Delta u + \int p\,dv = \Delta u + p\int dv$$

$$= \Delta u + p\Delta v \text{ for a constant pressure process} \qquad (2.81)$$

The cycle analysis now follows

$$\text{net work } w_n = \text{heat added} - \text{heat rejected}$$

$$= c_p(T_3 - T_2) - c_v(T_4 - T_1)$$

The heat added is

$$q_a = c_p(T_3 - T_2)$$

$$= \gamma c_v(T_3 - T_2) \qquad (2.82)$$

Therefore thermal efficiency becomes

$$\eta_{TH} = \frac{w_n}{q_a}$$

$$= 1 - \frac{(T_4 - T_1)}{\gamma(T_3 - T_2)} \qquad (2.83)$$

Now

$$\frac{v_1}{v_2} = r, \quad \frac{v_3}{v_2} = r_c, \quad \text{and} \quad \frac{v_4}{v_3} = \frac{r}{r_c}$$

Hence

$$\frac{T_2}{T_1} = r^{\gamma-1}, \quad \frac{T_3}{T_2} = r_c, \quad \text{and} \quad \frac{T_4}{T_3} = \left(\frac{r_c}{r}\right)^{\gamma-1}$$

as these are isentropic, constant pressure and isentropic processes respectively. Substituting for T_4, T_3 and T_2 in (2.83) in terms of T_1 and cancelling T_1 gives

$$\eta_{TH} = 1 - \frac{(r_c/r)^{\gamma-1}r_c r^{\gamma-1} - 1}{\gamma(r^{\gamma-1}r_c - r^{\gamma-1})}$$

$$= 1 - \frac{1}{r^{\gamma-1}}\left[\frac{r_c^{\gamma} - 1}{\gamma(r_c - 1)}\right] \qquad (2.84)$$

It should be noted that should the term in the square bracket, which varies with the cutoff ratio but not the compression ratio, equal unity, the thermal efficiency is identical to that of the Otto cycle. That is, it is this term which provides information on the way the rate of heat addition affects the efficiency.

2.4.7 Results of the air standard Diesel analysis

The thermal efficiency of the Diesel cycle from (2.84), depends on the compression ratio, the cutoff ratio and the isentropic index. Typical results

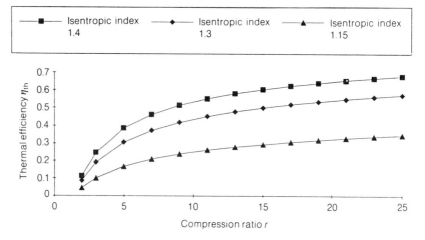

Figure 2.17 Thermal efficiency calculated from air standard Diesel cycle analysis, showing the effect of compression ratio and isentropic index γ for a cutoff ratio of 2:1.

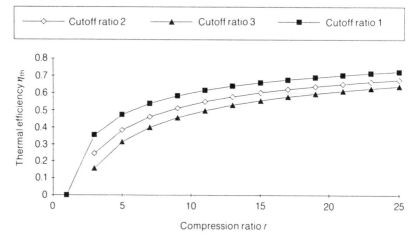

Figure 2.18 Thermal efficiency calculated from air standard Diesel cycle analysis, showing the effect of compression and cutoff ratios for an isentropic index γ of 1.4.

are plotted on Figures 2.17 and 2.18, the first being for a fixed cutoff ratio, taken here as 2:1 and the second for a fixed isentropic index taken as 1.4.

Figure 2.18 shows the effect of changing the value of γ and r. These provide a similiar variation to those found for the Otto cycle and no further comment is required. Figure 2.20 shows the effect of the cutoff ratio and it can be seen that there is a decrease in efficiency as cutoff ratio is increased. This is fundamentally due to the fact that the expansion ratio for the last gas to be heated has been reduced. However, increasing the cutoff ratio increases the work output of the cycle. This can be seen from the increased area of the cycle on the p–v diagram of Figure 2.19. It is the way in which speed control is obtained with a diesel engine and has advantages over the throttling arrangement used for an SI engine. The different control mechanisms will be discussed later. Obviously, the greater the cutoff ratio, the greater is the area of the cycle which represents work output. However, in practice, the maximum cutoff ratio is limited to about $r_c = 2.4$ as, past this point, excessive fuel with limited time left to burn causes '*smoking*' from the exhaust – i.e. the black smoke emission common to diesel engines. Normal full load fuel/air ratios for diesel engines are thus about 1:20. These are much lower than stoichiometric.

2.4.8 Comparison of air standard Otto and Diesel cycles

The difference between the Otto and Diesel cycle thermal efficiencies is the term expressing the effect of the cutoff ratio. For convenience, this will be

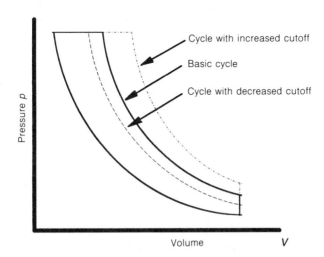

Figure 2.19 Diesel cycle showing the effect of changing the cutoff ratio on work output.

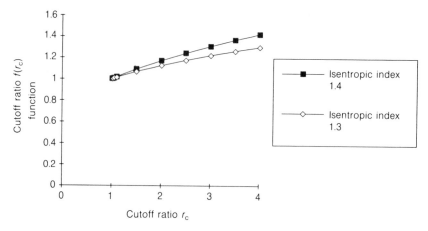

Figure 2.20 Behaviour of the cutoff ratio function for Diesel cycle analysis.

termed a function of r_c or simply $f(r_c)$. That is

$$f(r_c) = \left[\frac{r_c^\gamma - 1}{\gamma(r_c - 1)} \right] \qquad (2.85)$$

An Otto cycle has a cutoff ratio of one but a Diesel cycle with that value produces no work as can be seen from the p–v diagram where the expansion process then merely overlaps the compression. Nevertheless, it is worthwhile examining the Diesel cycle efficiency as r_c is reduced. For this comparison, $f(r_c)$ is plotted against r_c on Figure 2.20 for values of r_c ranging from 1 to 4. It can be seen that $f(r_c)$ reduces in value and approaches one as r_c approaches one. The larger the value of $f(r_c)$, the lower the thermal efficiency of the Diesel cycle. When $f(r_c)$ equals one, the thermal efficiency of the Diesel cycle becomes

$$\eta_{TH} = 1 - \frac{1}{r^{\gamma-1}} \qquad (2.86)$$

which is identical to the Otto cycle value. That is, the Diesel cycle must normally have a lower thermal efficiency than an Otto cycle with the same compression ratio but its efficiency increases towards the Otto cycle value as the cutoff ratio reduces to one. The difference between the two cycles is that the addition of heat in the Otto cycle occurs in the most rapid way possible which is when the piston is exactly at its top dead centre postion or, in other words, instantaneously. This allows a greater expansion ratio for the completely combusted gas and hence more work must be obtainable. Fast heat addition is therefore an important factor in obtaining high thermal efficiencies. The loss in efficiency from the Otto cycle values due to a slower heat addition rate is sometimes referred to as a **time loss**. The effect

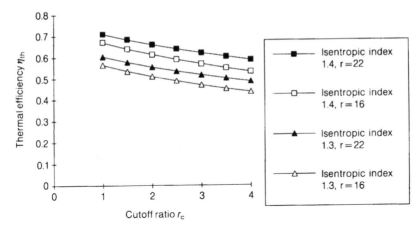

Figure 2.21 Effect of changing the cutoff ratio on the thermal efficiency of a Diesel cycle.

of cutoff ratio increase on the thermal efficiency is shown on Figure 2.21 for several different combinations of r and γ.

This illustrates the rising efficiency with falling cutoff ratio and it can be seen that a cutoff ratio of, say 2:1 gives a loss of about 5% from the equivalent Otto cycle value which is appreciable. This is surprising to some who think that the Diesel cycle is more efficient than the Otto, it being a generally held belief that these cycles represent CI and SI engines respectively and it is widely known that CI engines have a superior fuel economy. There are, however, reasons for this: both engines in reality have cycles intermediate between the Otto and Diesel and a modern high-speed CI engine has a combustion time which is not much longer than an SI engine. Also, factors other than the cycle play a role, these being the different energy per unit volume of the different fuels used, the method of part load control and the compression ratios in normal use. While these points will be dealt with later, it should be noted that the compression ratio of a CI engine is much higher than, perhaps double, that of an SI engine because, having no fuel in the intake charge, knock caused by premature ignition of the end gas (last part of the charge to combust) cannot occur. This tendency is exacerbated in SI engines by high compression ratios.

2.4.9 Comparison of air standard Diesel and real compression ignition cycles

Figure 2.22 shows a typical pressure–volume diagram for a compression ignition engine compared with both air standard Diesel and Otto cycles. As noted above, it can be seen that the real diagram lies somewhere in between the theoretical cycles and is only a little less peaked than that for a real spark ignition engine (Figure 2.14). This is because the flame front in the

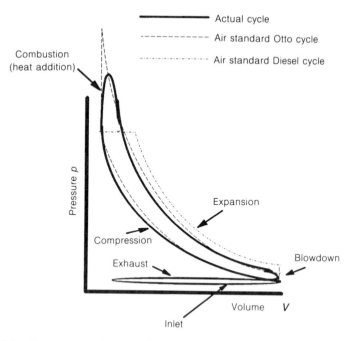

Figure 2.22 Comparison of an actual and air standard Otto and Diesel cycles for a four-stroke reciprocating compression ignition engine.

spark ignition engine takes a finite time to cross the combustion chamber to complete the burning process. While this is normally more rapid than that for the fuel injection, mixing and diffusion controlled burning in the compression ignition engine, the difference is not nearly as great as the theoretical cycles suggest. This is particularly noticeable with modern small, high speed diesels.

Other comments on the cycle are similar to those made in section 2.1 for the spark-ignition engine. Again, a most important factor is the dissociation of the burnt products of combustion at high temperature which limits the temperature and pressure rise at the end of combustion to less than would be expected from a simple combustion analysis. Non-isentropic compression effects are likely to be a little more apparent in the CI engine because of its higher compression ratio giving very high end of compression temperatures. The expansion and the variation from constant volume in the blowdown process produce similar deviations to those discussed with SI engines.

2.4.10 The air standard dual cycle

It is apparent that a cycle falling somewhere between the Otto and Diesel cycles is likely to provide a more realistic assessment of actual engine

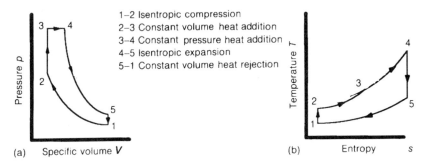

Figure 2.23 Air standard dual cycle for general reciprocating engine cycle analysis. (a) pressure–specific volume diagram, (b) temperature–entropy diagram.

operation for both SI and CI types. A number of possibilities exist but, in order to retain the simple air standard approach where equations can be developed, it is best to use some combination of processes which can be described by the polytropic expression $pv^n = $ constant. A cycle that comes readily to mind has the heat addition made up of a constant volume followed by a constant pressure process. By varying the proportion of heat input between these two processes, it is possible to obtain a reasonable approximation for a range of actual combustion types. Those of this type are usually referred to as the dual cycle because of this two-stage combustion but other names are common, these being the mixed cycle or the limited pressure cycle, the latter because it is essentially a type of operation where as much heat as possible is added at constant volume subject to pressure limitations, the rest then being added at constant pressure. The pressure limitation may be due to either structural or combustion reasons. A diagram of the cycle is shown on Figure 2.23. All processes are as on the Otto cycle except that combustion has been replaced by the combination of a constant volume and a constant pressure process.

For analysis, three parameters are now required, these being:

- the compression ratio $$r = \frac{V_1}{V_2} = \frac{v_1}{v_2} \qquad (2.87)$$

- the cutoff ratio $$r_c = \frac{V_4}{V_3} = \frac{v_4}{v_3} \qquad (2.88)$$

- the pressure ratio $$r_P = \frac{p_3}{p_2} \qquad (2.89)$$

The first two ratios are as described before while the third is based on the limited maximum pressure of the cycle.

Analysis follows a similar method to that of the previous sections

$$w_n = c_p(T_4 - T_3) + c_v(T_3 - T_2) - c_v(T_5 - T_1)$$

$$= c_v[\gamma(T_4 - T_3) + (T_3 - T_2) - (T_5 - T_1)] \qquad (2.90)$$

while heat input is

$$q_a = c_p(T_4 - T_3) + c_v(T_3 - T_2)$$

$$= c_v[\gamma(T_4 - T_3) + (T_3 - T_2)] \qquad (2.91)$$

Therefore, the cycle thermal efficiency is

$$\eta_{TH} = 1 - \frac{(T_5 - T_1)}{\gamma(T_4 - T_3) + (T_3 - T_2)} \qquad (2.92)$$

As before, all temperatures T_2 to T_5 can be written in terms of T_1 by use of the defined ratios. That is, using

$$\frac{v_1}{v_2} = r, \quad \frac{p_3}{p_2} = r_p, \quad \frac{v_4}{v_3} = r_c, \quad \frac{v_5}{v_5} = \frac{r}{r_c}$$

gives

$$\frac{T_2}{T_1} = r^{\gamma-1}, \quad \frac{T_3}{T_2} = r_p, \quad \frac{T_4}{T_3} = r_c, \quad \frac{T_5}{T_4} = \left(\frac{r_c}{r}\right)^{\gamma-1}$$

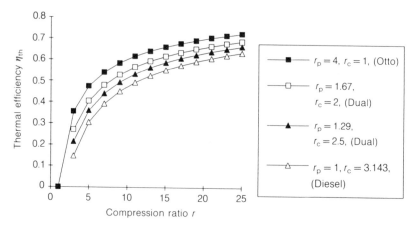

Figure 2.24 Dual cycle thermal efficiency for constant heat addition calculated from air standard analysis, showing effect of increasing constant pressure component of heat addition from zero (Otto cycle) to maximum (Diesel cycle) for an isentropic index γ of 1.4.

as these are isentropic, constant volume, constant pressure and isentropic processes respectively. Substituting in (2.92) in terms of T_1 gives

$$\eta_{TH} = 1 - \frac{(r_c/r)^{\gamma-1} r_c r_p r^{\gamma-1} - 1}{(r^{\gamma-1} r_c r_p - r^{\gamma-1} r_p) + (r^{\gamma-1} r_p - r^{\gamma-1})}$$

$$= 1 - \frac{1}{r^{\gamma-1}} \left[\frac{r_c^\gamma r_p - 1}{\gamma r_p (r_c - 1) + (r_p - 1)} \right] \qquad (2.93)$$

Note that, for $r_p = 1$, this reverts to the Diesel cycle equation while $r_c = 1$, it becomes that of the Otto cycle. Heat addition can now be proportioned between the two processes for values of r_p up to a maximum of

$$r_p = \frac{q_a}{c_v T_1 r^{\gamma-1}} + 1 \qquad (2.94)$$

which is the Otto cycle value. For a fixed heat addition, as r_p increases, the cutoff ratio r_c must decrease and so the cycle thermal efficiency increases. Typical plots of the efficiency are shown on Figure 2.24.

2.4.11 Inlet pressure effects from air standard cycle analysis

While air standard cycles are fundamentally used to study the work output and efficiency trends of the closed system cycle made up of the compression, combustion, expansion and blowdown processes, they can also be used to assess other variations. An important example is the study of the effect of inlet pressure variation. While it should be noted that the thermal efficiency in all the reciprocating engine cycles is independent of inlet pressure, the gas exchange processes are not. Two effects will be examined here: governing and supercharging. The effects on any of the air standard cycles described previously will be similar but the basic Otto and Diesel cycles will be used here for illustrative purposes.

Governing of engines

Governing is the term used to describe the control of engine power output in order to adjust to changed conditions which may occur either without a change of speed due to a different load requirement such as a change in gradient for a road vehicle or at a new speed. As pointed out previously, a compression ignition engine is governed by controlling directly only the rate of fuel addition. No throttling device exists in the intake manifold, the air enters at, or close to, atmospheric pressure and the variation in air/fuel ratio alters the heat added to the cycle. Hence the work output per cycle changes allowing the engine to match the new load conditions. This is often referred to as **quality governing** because it changes the fuel/air ratio. The effect on the cycle is shown on Figure 2.19.

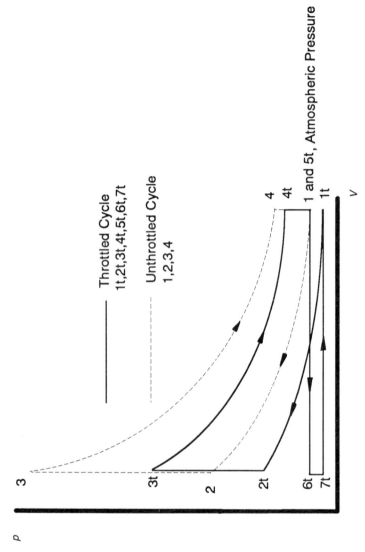

Figure 2.25 Throttled and unthrottled Otto cycles.

With a spark ignition engine, the same type of control is not possible because of the narrow flammability limits of the fuel when ignited by a point source. A different type of control is therefore used and this is termed **quantity governing**. The term is derived from the fact that a changed quantity of gas (in this case, fuel/air mixture) is drawn into the engine as the means of control reducing the quantity of fuel, as in CI engine, but also reducing the quantity of air in proportion. This is achieved by use of a throttling valve in the intake system, which was traditionally the butterfly valve in the carburettor just upstream of the fuel jet.

Now consider the air standard Otto cycle of Figure 2.25. Point 1 represents intake conditions with a wide-open throttle giving a manifold pressure at low to moderate speeds of only a little less than one atmosphere.

A throttling process, neglecting changes in kinetic and potential energies which are usually very small, is thermodynamically a constant enthalpy process during which the pressure drops. For ideal gases, this therefore takes place at constant temperature although it should be noted that for non-ideal gases this may not be so. The butterfly valve then reduces the pressure of the charge entering the cylinder although, naturally the swept volume of the cylinder remains the same. The charge density is reduced roughly in proportion to the pressure and so the engine experiences a smaller mass charge per cycle and hence a lower mass flowrate which reduces its power output. Looked at in more detail from the point of view of the engine thermodynamic cycle, the pressure at the start of the cycle falls from point 1 to a new point designated here as 1_t.

The work output for the cycle, $1_t, 2_t, 3_t, 4_t$ can now be calculated in relation to the original cycle 1, 2, 3, 4. Starting from point 1_t, it can be seen that p_{1t}, is less than p_1 in the proportion which will be designated here as the throttling ratio, r_t, which has a value less than one. That is

$$p_{1t} = p_1 r_t \qquad (2.95)$$

Assuming that the working substance is an ideal gas (the air standard assumption), the inlet temperature due to the throttling is unchanged.

$$T_{1t} = T_1 \qquad (2.96)$$

Now for both the wide-open and part-throttle cases, the pressure increases during compression as given by

$$p_2 = p_1 r^\gamma \quad \text{and} \quad T_2 = T_1 r^{\gamma - 1} \qquad (2.97)$$

Therefore

$$p_{2t} = p_2 r_t \quad \text{and} \quad T_{2t} = T_2 \qquad (2.98)$$

Because the quantity of fuel is reduced in the same ratio as the mass inflow of air into the cylinder, so is the heat output. That is, the heat addition q per unit mass of charge remains constant. For constant volume heating

$$T_{3t} - T_{2t} = T_3 - T_2 = \frac{q}{c_v} \qquad (2.99)$$

Remembering that $T_{2t} = T_2$, this gives

$$\frac{T_{3t}}{T_{2t}} = \frac{T_3}{T_2} = \frac{q}{c_v T_2} + 1 \qquad (2.100)$$

Therefore

$$\frac{p_{3t}}{p_3} = \frac{p_{2t}}{p_2} = r_t \qquad (2.101)$$

The expansion process follows a similar pattern to the compression giving

$$\frac{p_{4t}}{p_4} = r_t \qquad (2.102)$$

That is, the whole cycle is therefore reduced proportionally to the initial pressure and the work output is lower. The efficiency, however, depends only on the compression ratio and is the same. This is essentially because both the net work output and heat addition reduce in the same proportion. This is true as long as the intake and exhaust strokes are ignored. However, in the real engine, the efficiency is reduced because the pumping work increases during throttling. This can be seen on Figure 2.25 where the

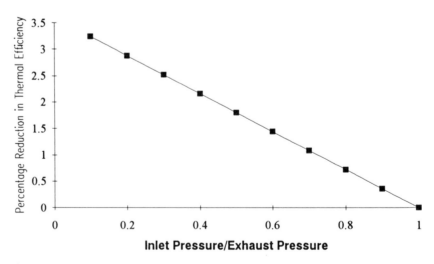

Figure 2.26 Reduction in thermal efficiency from Otto cycle value due to pumping. Compression ratio $r = 10:1$; $p_1 = 100\,\text{kPa}$; $v_1 = 1\,\text{m}^3\,\text{kg}^{-1}$; $q_a = 2.5\,\text{MJ}\,\text{kg}^{-1}$ of air plus fuel.

intake and exhaust strokes make an anticlockwise loop on the diagram indicating negative work. At low loads, a substantial lowering of efficiency occurs. Assuming both these to be constant pressure processes at an exhaust pressure p_e (i.e. p_{5t} to p_{6t}) and an inlet pressure p_i (i.e. p_{7t} to p_{1t}), the pumping work is

$$W_P = p_e(V_{5t} - V_{6t}) - p_i(V_{1t} - V_{7t})$$
$$= p_{5t}(V_{1t} - V_{2t}) - p_{1t}(V_{1t} - V_{2t})$$
$$= (p_{5t} - p_{1t})(V_{1t} - V_{2t}) \tag{2.103}$$

As noted previously, these must be expressed as total, not specific values. The net cycle work becomes

$$W_n = mc_v[(T_3 - T_2) - (T_4 - T_1)] - W_P \tag{2.104}$$

and the heat added is, as usual

$$Q_a = mc_v(T_{3t} - T_{2t}) = mq_a \tag{2.105}$$

The thermal efficiency can therefore be calculated and the amount by which the Otto cycle value is reduced is given by

$$\eta_{TH_{(pumping)}} = -\frac{p_{it}v_{it}}{q_a}\left(1 - \frac{p_i}{p_e}\right)\left(1 - \frac{1}{r}\right) \tag{2.106}$$

A plot showing its reduction from the Otto cycle value with variation in the ratio p_i/p_e below unity is shown on Figure 2.26. It should be remembered in comparison that the quality control of the compression ignition engine raises the efficiency as the speed or power is reduced by reducing the cutoff ratio.

Supercharging (or turbocharging)

Supercharging and, in its alternative form, turbocharging has been in use for a long period, particularly with high-performance SI aircraft engines used before the advent of the gas turbine. Essentially, as far as the basic thermodynamic cycle is concerned, both supercharging and turbocharging are identical in that they increase the mass of charge in the cylinder at intake. They are not identical, however, when the intake and exhaust processes are included. The difference is that the former uses a mechanically driven blower (low pressure compressor) and does not suffer from a slow response to load and speed changes while the latter obtains its power from a turbine in the exhaust which utilizes some of the waste energy from the engine. However, it does increase the engine backpressure and so the work obtained is substantially, but not entirely, 'free'.

Figure 2.27 shows air standard cycles which approximate both these arrangements so that a simple analysis can be considered. This is a consideration of the changed initial conditions only using a simple assessment of the

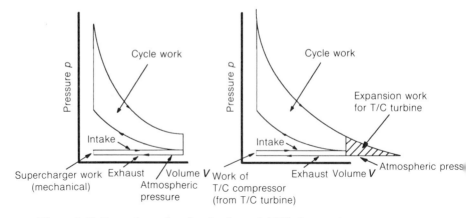

Figure 2.27 Supercharged and turbocharged (T/C) Otto cycles.

intake and exhaust strokes again as constant pressure processes. Turbocharging, in particular, is much more complex and will be dealt with in more detail later although specialist texts are recommended for a complete treatment.

The compressor in a supercharged engine raises both the pressure and temperature during compression although intercooling may be used between the compressor and engine, as in the gas turbine, to further increase the density of the inlet charge. The effect of raising the inlet pressure is the exact opposite of throttling. The pressures rise throughout the cycle, the diagram increases in size and more work is produced. Efficiency would remain the same (for a theoretical Otto cycle analysis) if the compression ratio was kept constant. In an SI engine, in order to reduce peak pressures in the cylinder and to prevent knock (detonation) due to the higher temperature and more dense charge, the compression ratio is usually in practice lowered. This will lower efficiency, although this may be partially offset by the more nearly constant volume operation with a faster flame front at the higher pressure. Alternatively, higher octane fuels may allow the previous compression ratio to be retained.

Diesel engines may be supercharged with better results. Peak pressures are increased but, in this case no extra demands are made on the fuel quality. In fact, supercharging allows use of a wider range of fuels because the delay period for ignition is reduced and the rate of pressure rise is more controlled. Even better results may be obtained by supercharging two-stroke diesels. With two-stroke spark ignition engines, too much intake pressure results in excessive scavenging and hence wasted fuel. In diesels, only air is wasted. Thus supercharging of two-stroke diesels is common practice with perhaps 40% of the air being used for scavenging and passing directly to the exhaust. This has the additional benefit of cooling the engine.

It is often used in large marine installations with horizontally opposed pistons.

2.5 SUMMARY OF AIR STANDARD CYCLE ANALYSIS

It can be seen that an air standard cycle analysis can quickly provide useful insights into engine operation. With gas turbines, the effect of pressure ratio, the limitations imposed by materials to withstand the high continuous temperatures after combustion and the effect of inefficient turbomachinery can be clearly seen. With reciprocating engines, the trends towards improved efficiency with compression ratio and the losses due to pumping during gas exchange and to volume increase during heat addition are clearly evident. However, the results of the analysis cannot be relied upon to provide accurate values and in most cases are incorrect. An example of this is that experimental work on high-compression engines shows that efficiency peaks at a value of around 16:1 to 17:1 rather than continuing to increase as the analysis indicates. It is somewhat dependent on specifics of the engine design. Also, as a thermodynamic model only, the effect of geometry (such as, for example, combustion chamber shape) in general on engine performance cannot be determined although it is known to be significant in practice. It is necessary to improve on the air standard cycle both as a thermodynamic model and as a dimensional model where the fluid mechanics is considered. Also heat transfer to the walls and friction losses need assessment.

In summary, therefore, the following factors need much more detailed attention before accurate modelling is possible. In general they require theory outside the scope of a normal first course in thermodynamics, fluid mechanics and heat transfer and so the fundamentals will be given in the following chapters. The reader should then be in a position to comprehend more fully the material in specialist internal combustion engine texts.

Consideration of the air standard cycle indicates that a number of factors need further examination. The most important are listed below and include factors that are directly related to the thermodynamic cycle of compression, heat addition, expansion and heat rejection and those affecting the gas exchange processes which determine the initial conditions in the cycle itself. An improved understanding is required of:

1. thermodynamic properties, including mixtures of ideal gases and non-ideal gas vapours for the induction process and, more particularly, of gases at high temperatures for most of the thermodynamic cycle;
2. the basic details of compressible fluid flow which are important in all multidimensional models but are particularly relevant to the gas exchange processes;

3. engine turbomachinery which also modifies the gas exchange and is of great importance in relation to gas turbine flows but also for those of turbocharged reciprocating engines;
4. combustion, both in relation to the basic chemical formulations which occur and the assessment of the energy release and in the interaction of the combustion with the fluid dynamics;
5. heat transfer both steady and unsteady because of its modification to the individual processes within the cycle.

2.6 REFERENCES

Caris D. F. and Nelson, E. E. (1959) A new look at high compression engines, *Society of Automotive Engineers, USA, Trans*, **67**.

PROBLEMS

2.1 A simple air standard Brayton cycle representing gas turbine operation has inlet pressure and temperature conditions of 100 kPa and 300 K respectively. If its pressure ratio is 10:1 and the heat addition is 850 kJ kg^{-1} of air in the cycle, determine the pressure, temperature and gas density at each cycle point. Also calculate the work input for the compressor, the work output from the turbine and the net work output and efficiency of the cycle. It may be assumed that the isentropic index, $\gamma = 1.4$ and the gas constant, R, for air is 0.287 kJ kg K^{-1}.

2.2 For the cycle on Problem 2.1, it may be assumed that the first expansion stages in the turbine supply the work for the compressor. Calculate the pressure and temperature at the exit from these stages and the pressure ratio over the remaining turbine (or jet nozzle) expansion which will produce the useful work output. Check this value by comparing the work from that final expansion with that calculated from the heat addition and heat rejection processes.

2.3 The cycle of Problem 2.1 is to be modified for a jet aircraft to suit initial conditions of $-53\,°C$ and 25 kPa. Assuming that the same maximum cycle temperature will be achieved with the same heat addition as before, calculate the pressure ratio now required, the temperatures and pressures throughout the cycle, the work output and thermal efficiency. Also determine the pressure ratio for the jet nozzle.

2.4 Repeat Problem 2.3 but assume now that the maximum cycle temperature can be raised to 1550 K.

2.5 For Brayton cycles with the following conditions, determine the pressure ratio which will give the greatest work output. Hence evaluate their respective work outputs and thermal efficiencies.

Cycle	Maximum cycle temperature (K)	Minimum cycle temperature (K)
1	1425 K	300 K
2	1425 K	220 K
3	1750 K	220 K

2.6 Gas turbines operating under conditions specified in the table of Problem 2.5 are represented for evaluation purposes by an air-standard Brayton cycle modified to include isentropic compressor and turbine efficiencies. Using this, determine for each case the work output and cycle thermal efficiency if the compressor isentropic efficiency is 88% and the turbine isentropic efficiency 92%.

2.7 For the first of the cases tabulated in Problem 2.5 and assuming that the compressor and turbine isentropic efficiencies are identical, determine their minimum value at which the gas turbine unit just starts to produce a net power output.

2.8 For the third case only of Problem 2.6, the pressure ratio is now modified but all other parameters kept constant. Determine the pressure ratio at which the maximum thermal efficiency now occurs and compare its value and the net work output with those calculated for the basic Brayton cycle at maximum work output in Problem 2.5 and the modified cycle of Problem 2.6.

2.9 In the gas turbine of Problem 2.6 (third case only), the power producing turbine is replaced with a jet nozzle of 97% isentropic efficiency. Treating this in the same way as a turbine expansion, determine the pressure ratio across the nozzle and the net work output and thermal efficiency now available.

2.10 A regenerative gas turbine in which the effectiveness of the regeneration can be assumed to be 100% is to be developed for the maximum and minimum cycle temperatures specified in the first case of Problem 2.5. If its pressure ratio is to be 4:1, compare its work output and thermal efficiency with that at the maximum work output condition for the basic Brayton cycle as calculated both in Problem 2.5 and the initial case of pressure ratio 10:1 in Problem 2.1.

   If the efficiency of the regeneration process falls to 90%, determine how much the thermal efficiency and work output are effected.

2.11 The gas turbine of Problem 2.1 is to be intercooled at the mid pressure ratio point (i.e. the pressure ratios of stages 1 and 2 during compression are identical at 3.16:1) and the intercooling is back to the initial intake temperature. The maximum cycle temperature remains the same but the heat addition to the cycle is changed. Determine the total compressor work, the heat input, net work output and cycle thermal efficiency and compare these with the Problem 2.1 results.

2.12 Determine, for the gas turbine of Problem 2.11 with 100% regeneration and with reheat to the maximum cycle temperature also at the mid point of expansion (two equal stages, each of pressure ratio 3.16:1), the net work output and thermal efficiency. If the regeneration efficiency falls to 90%, and compressor and turbine stages have an isentropic efficiency of 88% and 92% respectively, determine now the net work output and thermal efficiency.

For the first part of this question only, calculate the work output and thermal efficiency if the overall compressor and turbine stages are each divided into five equal, fully intercooled (i.e. to the inlet temperature) and fully reheated (i.e. to the maximum cycle temperature) stages.

2.13 For an initial analysis of an SI engine, the basic Otto cycle can be used. The following data applies:

Ambient pressure and temperature, 101 kPa, 310 K;
Swept volume 2 l (2000 cm^3);
Design, 4 cylinder, 4 stroke, compression ratio 9.5:1;
Design speed range, 1000 to 5000 rev min^{-1};
Pressure in the cylinder at the start of compression (at wide open throttle), 1000 rev min^{-1} 85 kPa, 3000 rev min^{-1} 90 kPa, 5000 rev min^{-1} 80 kPa;
Temperature at start of compression: ambient;
Air to fuel ratio 15:1;
Heating value of the fuel 44 MJ kg^{-1}.

Calculate the heat input per kg of mixture for the cycle.

Using an isentropic index of 1.4 for cold air and a value of R of 0.287 kJ kg^{-1} K^{-1} determine the pressure and temperature conditions at each point in the cycle and estimate the work and power output, the mean effective pressure and the thermal efficiency at each speed.

Repeat the above for an isentropic index of 1.3 noting the variation from the previous calculation.

2.14 Repeat Problem 2.13 (using the isentropic index of 1.3) but assume that pumping work has now to be subtracted from each cycle. For this calculation, assume that the inlet and exhaust strokes are both constant pressure, the inlet being at an average pressure 5 kPa below the final in-cylinder value at the start of compression while the exhaust is 5 kPa above ambient. Calculate the approximate quantity of work and power required for pumping at these full throttle conditions and the revised work, power and thermal efficiency estimates for the cycle. What has the most significant effect in increasing the power output?

2.15 The engine of Problem 2.13 is to be turbocharged, raising the inlet pressure to 50% more than the normally aspirated precompression value at 3000 rev min^{-1}. Assume that this turbocharging compression process is isentropic with an index of 1.4 while the cycle index remains at 1.3. Calculate the maximum cycle pressure and temperature, work and power output and thermal efficiency at this speed. Neglect pumping effects.

If it is found that the compression ratio must be reduced to eliminate engine knock to a value which gives a peak cycle pressure halfway between that of the original normally aspirated engine and the turbocharged one above, determine this new compression ratio and its effect on the work output and thermal efficiency.

Estimate how much improvement in work and power output could be gained by intercooling the air following the compressor back to the ambient temperature with the compression ratio at the new, lower value.

2.16 For an initial analysis of a CI engine, the basic Diesel cycle can be used. The following data applies:

Ambient pressure and temperature, 101 kPa, 310 K;
Swept volume, 12 l (12 000 cm^3);
Design, 6 cylinder, 4 stroke, compression ratio 16:1;
Design speed range, 1000 to 3000 rev min^{-1};
Pressure in the cylinder at the start of compression, 3000 rev min^{-1} 85 kPa;
Temperature at start of compression: ambient;
Air to fuel ratio at maximum power, 21:1;
Heating value of the fuel, 44 MJ kg^{-1}.

Using an isentropic index of 1.3 for air and a value of R of 0.287 kJ kg^{-1} K^{-1}, determine the pressure and temperature conditions at each point in the cycle and estimate the cutoff ratio at the operating point specified. Calculate the work and power output, the mean effective pressure and the thermal efficiency at each speed.

If the cutoff ratio is reduced to half the value calculated above, determine the new thermal efficiency. Compare the expansion ratios for each case from the end of combustion and the final temperatures and pressures at the end of expansion.

2.17 The engine of Problem 2.16 is to be turbocharged, raising the inlet pressure to 90% more than the normally aspirated precompression value at 3000 rev min^{-1}. Assume that this turbocharging compression process is isentropic with an index of 1.4. and the air is intercooled following the compressor back towards the ambient temperature with a 90% effectiveness. Calculate the maximum cycle pressure and temperature, work and power output and thermal efficiency at this speed.

2.18 Show that, for a fixed heat addition per unit mass in a dual cycle, that the pressure ratio r_p is related to the cutoff ratio, r_c by

$$r_p = \frac{r_{po}}{1 + \gamma(r_c - 1)}$$

where r_{po} is the pressure ratio of the Otto cycle with the same heat addition.

Also show that

$$r_p = \frac{r_{co} - \left(\frac{\gamma - 1}{\gamma}\right)}{r_c - \left(\frac{\gamma - 1}{\gamma}\right)}$$

where r_{co} is the cutoff ratio of the Diesel cycle with the same heat addition.

Also show for the dual cycle that the work, w is given by

$$w = c_v r^{\gamma - 1} T_1 \left[\gamma r_p (r_c - 1) + (r_p - 1) - (r_p r_c^\gamma - 1)/r^{\gamma - 1} \right]$$

2.19 The Otto cycle analysis of the engines of Problem 2.13 and Problem 2.16 (isentropic index 1.3, 3000 rev min^{-1} condition) are now to be examined by use of the dual cycle. Calculate the work output and thermal efficiency in each case if the heat addition remains as in the original question but in case 1 (the Otto cycle), the pressure difference $(p_3 - p_2)$ is 100%, 75%, 50%, 25%, 0% of its original value to allow an increase in cutoff ratio for the dual cycle while in case 2 (the Diesel cycle), the cutoff volumes are modified in the same way by the same percentages to provide an increase in pressure ratio.

Thermodynamic properties | 3

3.1 THERMODYNAMIC PROPERTIES FOR ENGINE ANALYSIS

The basic concepts of thermodynamic properties will have been covered by the reader in a first course in thermodynamics and it is not the intention in this text to recover that ground. In summary, the normal first course approach deals with two groups of substances: ideal gases where simple formulas are used and fluids undergoing liquid–vapour transformations where tabulated properties are available. It is suggested that the reader briefly revises these topics by examining one of the many excellent thermodynamic texts available. Several useful texts are listed in the bibliography at the end of the book. However, in order to obtain a detailed understanding of engine cycles and other thermodynamic processes, particularly those that occur at high temperatures, more advanced concepts are necessary. These, when covered in detail are extensive and some require texts to be devoted completely to them. A full coverage is not possible here. That is, the material that will be presented in this chapter is not intended to be comprehensive but should give the reader sufficient insight to be able to obtain the necessary property values for the processes under consideration in engine cycles and related subjects. It is therefore aimed at indicating where non-ideal gas properties might apply at moderate temperatures and how ideal gas ones vary at very high temperatures. Other properties, such as the **Helmholtz** and **Gibbs** functions, which are necessary in the general development of property formulations, and **availability** which is useful in a second law analysis of heat engines, will also be discussed.

3.2 SOME ADDITIONAL PROPERTIES

3.2.1 The Helmholtz and Gibbs functions

The development of general thermodynamic property equations requires two further thermodynamic properties to be defined, these being the Helmholtz function, A or a, and the Gibbs function, G or g, in their extensive and intensive forms respectively. As with enthalpy, these are combinations of other thermodynamic properties which provide useful rather than necessary groups. They are most conveniently defined in their intensive form although the extensive form is identical and is obtained by appropriate replacement of any intensive properties in the group by the equivalent extensive. These definitions are

$$a = u - Ts \qquad (3.1)$$

$$g = h - Ts \qquad (3.2)$$

While, as with enthalpy, the Helmholtz and Gibbs functions have a range of applications, the examination of a particular usage is sometimes important in order for the reader to obtain a feel for their relevance. This is carried out below. Here, they are examined in relation to the maximum possible work which can be obtained from constant temperature systems such as those existing in reacting flows. Typical examples in practice would be gas turbine combustion in engines, while other well-known devices include batteries and fuel cells, these last two operating at or near a pressure and temperature which is in equilibrium with their surroundings. In these cases, their efficiency is not governed by the well-known Carnot efficiency formula which applies only to heat engines, thereby requiring energy to be transferred in to the working substance as heat transfer. This is fortunate, as human beings and other living creatures are also essentially fuel cell devices and it can easily be shown that at typical blood and ambient temperatures, their efficiency would be very low if they were a heat engine. Strictly, internal combustion engines are also not heat engines because their energy is obtained directly from a reaction (combustion) but because of their high temperature operation during the energy addition mode, a heat engine approach is quite useful in their analysis.

Maximum work in closed systems

Assume that the constant temperature reacting system to be considered is closed and that the total work output, mechanical or electrical, is dw. The appropriate first law statement for this case is

$$dq = du + dw \qquad (3.3)$$

Now from the definition of the Helmholtz function, it can be seen that a simple differentiation gives

$$da = du - Tds - sdT \qquad (3.4)$$

For maximum work output, the second law states that the process must be reversible and hence

$$dq = Tds \qquad (3.5)$$

Therefore, on substitution and rearrangement, (3.4) becomes

$$du = da + dq + sdT \qquad (3.6)$$

which on substitution into the first law statement (3.3), becomes

$$dq = da + dq + s\,dT + dw \tag{3.7}$$

Now the work term can be isolated as

$$dw = -da - s\,dT \tag{3.8}$$

which is the general expression for the maximum work. If, however, the temperature remains constant as in some of the applications mentioned above

$$dw = -da \tag{3.9}$$

which can be easily integrated to obtain the maximum work. The maximum possible work obtainable from the working substance in the above case is then

$$w_{1-2} = (a_1 - a_2)_T \tag{3.10}$$

where the subscript T indicates that it applies to an isothermal process. That is, the maximum possible work for a constant temperature process in a closed system is equal to the change in the value of the Helmholtz function for the working substance. It should be noted that this equation, in fact, gives the work done by the fluid or working substance inside the device which may therefore be referred to as internal work. Now, any expansion in a closed system produces movement of the boundaries and therefore causes work to be done against the constant pressure atmosphere which must be subtracted from the above in order to obtain the net (i.e. usable) work. The reader is invited to demonstrate by this simple modification of (3.10) that, under these circumstances, the maximum possible net work is equal to the change in the Gibbs function.

Maximum work in steady flow, open systems

A similar relation to the above can be derived for a steady flow open system. The first law, neglecting kinetic and potential energy terms, can then be expressed as

$$dq = dw + dh \tag{3.11}$$

Following the same procedure and logic as before gives

$$T\,ds = dw + dh \tag{3.12}$$

$$dg = dh - T\,ds - s\,dT \tag{3.13}$$

which, on substituting for dh and rearranging gives the maximum work as

$$dw = -dg - s\,dT \tag{3.14}$$

Again, at constant temperature

$$\mathrm{d}w = -\,\mathrm{d}g \tag{3.15}$$

which, on integration, becomes

$$w_{1-2} = (g_1 - g_2)_{\mathrm{T}} \tag{3.16}$$

That is, the maximum possible net work for an isothermal process in steady flow is equal to the change in the Gibbs function when changes in kinetic and potential energies are neglected. If these last terms are important, they can be simply added in to (3.16).

The Gibbs and Helmholtz functions are therefore of great importance in all chemically reacting systems. However, as pointed out previously, they are of more general use in the development of ideas related to properties.

3.2.2 Availability

All substances possess energy but it is not always available to do work. For example, sea water is a massive source of easily transportable energy but it cannot normally be used as an energy supply because it is at the temperature of the surroundings. The energy can only be used if it is in a high-grade form. For energy that will be transferred as heat, this means that it must be above the temperature of the surroundings. What is of great interest is the maximum possible value of the work from a given source of energy when it is obtained via a heat transfer process as occurs in heat engine analysis. This is called the **available energy** and shall be termed here, e_{a}. In fact, while it derives from the heat transfer and determines the maximum possible work output both of which are path functions, when the concept is applied to the working substance of the system, it can be shown that it is a property of this working substance and is then called simply the **availability** or sometimes, the **energy**. In fact, two properties are actually used these being called the closed system availability and the open system (or stream) availability designated by the symbols ϕ, Φ and ψ, Ψ. The lower and upper case symbols designate the intensive and extensive value respectively.

Closed system availability

Availability refers to the maximum possible energy which can be transferred for a given process and the work and heat transfer therefore need to be considered. First, examine a closed system. If work alone is done on a substance in the absence of dissipative effects such as friction, all of it is potentially returnable for later use. For example, consider a piston compressing gas in a cylinder. A certain quantity of work is required for that compression and, if there is no heat transfer or friction, the process can be reversed with the same magnitude for the work being returned from the system but with the opposite sign. Hence, the increase in the available energy $\Delta e_{a,w}$ of a substance due only to the work being done on it is equal to that quantity of the work, $-w$:

$$\Delta e_{a,w} = -w \qquad (3.17)$$

This is not so with heat transfer because, from Carnot's principle, it is known that some of the energy received must be rejected. That is, some of the heat transfer is available as useful work while some is unavailable. To determine the maximum work that can be achieved from the heat transfer, let us examine a simple Carnot cycle. Here, in order to obtain the maximum possible quantity of work from the heat added during a process, the heat rejection must take place to the lowest possible temperature sink. This is the temperature of the surroundings which may be regarded, for explanatory purposes, as constant. Any other lower temperature will require additional machines to maintain it and ultimately the final heat flow must be to the surroundings. In a Carnot cycle which rejects heat to the atmosphere, the energy being added per unit mass of the working substance of the system in the form of heat is q_a and the heat rejected is q_r. The work done is then the available energy which can be called e_a. The unavailable energy, e_u, which cannot be used to produce work is the heat rejected. This is depicted in Figure 3.1.(a).

A Carnot cycle consists of four particular processes, specifically two isothermal and two isentropic ones. In the above example, the energy available from the heat addition during the high-temperature isothermal process is e_a. A more general consideration for any process where the temperature varies during the heat addition is as shown on Figure 3.1.(b). For an incremental change of state which takes place during this process between points, say b and c, the temperature may be considered constant forming an infinitesimal Carnot cycle. Here the heat dq is added at temperature T while that rejected at the ambient temperature T_0 is dq_r. Hence

$$de_{a,q} = dq - dq_r \qquad (3.18)$$

where $de_{a,q}$ is the change in the available energy of the system due to the heat transfer only.

But from the Carnot relationship

$$dq_r = dq(T_0/T) \qquad (3.19)$$

from which

$$de_{a,q} = dq - dq(T_0/T) = dq - T_0(dq/T) \qquad (3.20)$$

Integrating for T_0 constant gives

$$\Delta e_{a,q} = q - T_0 \Delta s \qquad (3.21)$$

It should be noted that all the property values in (3.21) apply to the beginning and end of the process except for T_0 which is the ambient temperature. Hence, the change in what can be called the internal availability ϕ_i may now be determined for the original process and is, for a closed system

$$\phi_{i,2} - \phi_{i,1} = \Delta e_{a,w} + \Delta e_{a,q} = -w + q - T_0\Delta s = \Delta u - T_0\Delta s \qquad (3.22)$$

This may be positive, zero or negative depending on the relative values of q, $T_0\Delta s$ and w. It will be noted that all the terms on the right-hand side of (3.22) are properties and hence the availability must also be a property.

The total availability ϕ can now be evaluated and is the maximum possible work that can be obtained in expanding to an appropriate datum which is the atmospheric conditions subscripted 0. Here, the internal availability needs to be considered but, in addition the external effects of the moving boundary during expansion or compression process, are to be included to give the total or net effect, That is, the additional work which is done against the atmosphere must be subtracted and this is given by $p_0\Delta v$.

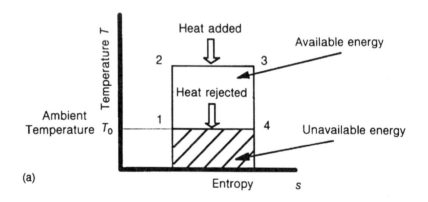

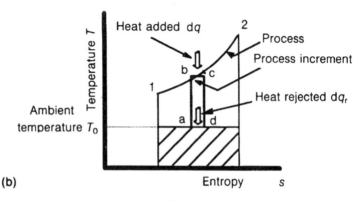

Figure 3.1 (a) Available and unavailable energy in the Carnot cycle. (b) constant temperature representation of an incremental part of a process.

Therefore the net closed system availability, ϕ, of the substance is given by

$$\phi = (u + p_0 v - T_0 s) - (u_0 + p_0 v_0 - T_0 s_0) \qquad (3.23)$$

Steady flow system (stream) availability

For a control volume (called cv), expansion from a general (unsubscripted) initial condition to that of the surrounding state, during steady flow, the first law can be written as follows:

$$q_{cv} + h + \tfrac{1}{2}v^2 + gz = w_{cv} + h_0 + \tfrac{1}{2}v_0^2 + gz_0 \qquad (3.24)$$

where q_{cv} and w_{cv} are the heat and work transfers with the control volume. For maximum work, reversible processes which expand the working substance completely to the conditions of the surroundings are required. However, the control volume itself and the surroundings are often at different temperatures even though the control volume outlet may be at state 0 which is that of the surroundings. Heat transfer from or to the control volume therefore represents either a source of additional useful energy if it is at a higher temperature than the surroundings or a consumer of energy if it is at a lower temperature. It should therefore be taken into account in the equation for the total availability which expresses the maximum work possible for the process. This heat transfer can only be used to its maximum by placing a reversible heat engine (or heat pump) between the control volume and surroundings. This is shown on Figure 3.2. This additional work will be called w_h (from the heat transfer) as distinct from the normal control volume work w_{cv} the sum of the two then making up the total work, called here w_t. The heat transfer from any infinitesimal section of the control volume at temperature T is dq_{cv} while that received by the heat engine is an equal quantity dq_h but is of opposite sign. Also, over the whole control volume, the total heat transfer is q_{cv} which is equal and opposite to that received by the engine, q_h.

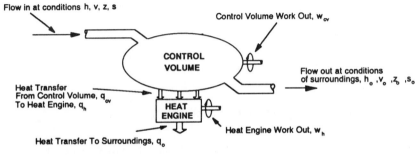

Figure 3.2 Availablity from a steady flow system.

Now

$$w_t = w_{cv} + w_h \tag{3.25}$$

But for the heat engine

$$w_h = q_h - q_0 \tag{3.26}$$

and

$$\frac{q_0}{T_0} = \int \left(\frac{dq_h}{T}\right) = \int \left(-\frac{dq_{cv}}{T}\right) \tag{3.27}$$

where T, the control volume temperature, is a variable. But, if heat is transferred reversibly from the control volume for the flow as it expands to conditions 0, the latter term in (3.27) can be evaluated as the entropy change of the flow between exit and entry of the control volume. That is

$$s_0 - s = \int \left(\frac{dq_{cv}}{T}\right) \tag{3.28}$$

Therefore

$$q_0 = -T_0(s_0 - s)$$

and hence

$$w_h = q_h + T_0(s_0 - s) = -q_{cv} + T_0(s_0 - s) \tag{3.29}$$

This gives

$$w_{tot} = w_{cv} - q_{cv} + T_0(s_0 - s) \tag{3.30}$$

Substituting for $(w_{cv} - q_{cv})$ into the first law equation above, now gives

$$h + \tfrac{1}{2}v^2 + gz = w_{tot} - T_0(s_0 - s) + h_0 + \tfrac{1}{2}v_0^2 + gz_0 \tag{3.31}$$

Rearranging and calling w_t the stream availability ψ gives

$$\Psi = (h - T_0 s + \tfrac{1}{2}v^2 + gz) - (h_0 - T_0 s_0 + \tfrac{1}{2}v_0^2 + gz_0) \tag{3.32}$$

Note that this is a simplified version of the derivation applying to the steady flow situation. For maximum cycle work, the mass leaving the control volume at 0 is in equilibrium with its surroundings. As well as the thermodynamic properties T_0, h_0, and s_0 being at such a minimum possible state, the physical properties z_0 and v_0 should also be minimized. That is, v_0 should be zero and z_0 should represent the datum position, also zero. Hence, these two variables are often excluded from the equation.

3.3 GENERAL THERMODYNAMIC RELATIONS

It will no doubt have been observed by the reader that there are no instruments to measure the energy quantities such as the internal energy, enthalpy and entropy of a substance directly. Only four of the thermodynamic properties, pressure, volume, temperature and mass, are easily measured and, in intensive terms, these reduce to three by dividing the volume by the mass to form specific volume. The gradients of pressure, specific volume and temperature each with respect to one another are also measurable as long as the third property is fixed as, for example, with the change in volume with respect to pressure at constant temperature. Other combinations of the gradients of pressure, temperature and volume can also be obtained. However, the above is very limited information when considered in light of the precise values required for all the other properties and gradients over an extremely wide range of different thermodynamic states which are necessary in order to encompass all possible thermodynamic applications. These must be able to be determined from a combination of theory and those measurements which are available. A powerful tool in accomplishing this is the group of general partial differential equations derived from simple basic relationships which are called **general thermodynamic relations** (GTRs). These are used to relate all thermodynamic properties to those which can be either simply measured or otherwise determined from fundamental theory (e.g. by the methods of statistical thermodynamics).

3.3.1 Some general mathematical relationships

Before examining the GTRs and the thermodynamic property equations which can be developed from them, several mathematical relationships which are required in the derivations need highlighting.

The total differential equation

Any property of a single phase substance may be expressed thermodynamically as a function of two other properties. If these properties are designated as x, y and z, they can be written, for example, as

$$z = f(x, y) \tag{3.33}$$

Then the total differential equation expressing the change in z with changes in both x and y is

$$dz = \left(\frac{\partial z}{\partial x}\right)_y dx + \left(\frac{\partial z}{\partial y}\right)_x dy \tag{3.34}$$

The exact differential rule

If properties x, y and z are related by an equation of the form

$$dz = M dx + N dy \tag{3.35}$$

then

$$\left(\frac{\partial M}{\partial y}\right)_x = \left(\frac{\partial N}{\partial x}\right)_y \tag{3.36}$$

The cyclic rule

Here, the gradients of the three properties x, y and z are related by

$$\left(\frac{\partial x}{\partial y}\right)_z \left(\frac{\partial y}{\partial z}\right)_x \left(\frac{\partial z}{\partial x}\right)_y = -1 \tag{3.37}$$

Proofs of the above will not be given here but can be found in most mathematical textbooks.

3.3.2 Development of the GTR equations

The GTR equations are simply developed from the fundamental general equations of the thermodynamic properties in differential form by obtaining partial derivatives from them. The basic equations required are those for internal energy, enthalpy, Helmholtz and Gibbs functions in terms of pressure, volume, temperature and entropy. The first two will be well known from a first course in thermodynamics and are obtained by choosing a reversible process between two state points and applying the first and second laws of thermodynamics to it. These, sometimes called the Gibbs equations, are

$$T\mathrm{d}s = \mathrm{d}u + p\mathrm{d}v \tag{3.38}$$

$$T\mathrm{d}s = \mathrm{d}h - v\mathrm{d}p \tag{3.39}$$

Note that, as these relate properties only, the final result is independent of the path or type of process, the reversible process being specified for evaluation purposes only. Rearranging gives the two required equations

$$\mathrm{d}u = T\mathrm{d}s - p\mathrm{d}v \tag{3.40}$$

$$\mathrm{d}h = T\mathrm{d}s + v\mathrm{d}p \tag{3.41}$$

The second two equations are obtained by differentiating the Helmholtz and Gibbs function relationships which are available from their definitions. This gives

$$\mathrm{d}a = \mathrm{d}u - T\mathrm{d}s - s\mathrm{d}T \tag{3.42}$$

$$\mathrm{d}g = \mathrm{d}h - T\mathrm{d}s - s\mathrm{d}T \tag{3.43}$$

The appropriate equations are then obtained by eliminating $du - T\,ds$ and $dh - T\,ds$ by use of (3.40) and (3.41).

$$da = -p\,dv - s\,dT \qquad (3.44)$$

$$dg = v\,dp - s\,dT \qquad (3.45)$$

When the exact differential rule is applied to (3.40), (3.41), (3.44) and (3.45), the following set of partial derivatives are then obtained. These relate p, T, v and s and are normally designated as Maxwell's relations.

$$\left(\frac{\partial T}{\partial v}\right)_s = -\left(\frac{\partial p}{\partial s}\right)_v \qquad (3.46)$$

$$\left(\frac{\partial T}{\partial p}\right)_s = \left(\frac{\partial v}{\partial s}\right)_p \qquad (3.47)$$

$$\left(\frac{\partial p}{\partial T}\right)_v = \left(\frac{\partial s}{\partial v}\right)_T \qquad (3.48)$$

$$\left(\frac{\partial v}{\partial T}\right)_p = -\left(\frac{\partial s}{\partial p}\right)_T \qquad (3.49)$$

These can be seen to be immediately useful. For example, to obtain the entropy change with volume during an isothermal process, the third of the Maxwell relations shows that all that is necessary is to measure the pressure change with temperature in a constant volume process. This is at least possible.

By appropriate differentiation of (3.38), (3.39), (3.44) and (3.45) and the equating of the results, a further set of partial derivative equations can be obtained:

$$\left(\frac{\partial u}{\partial s}\right)_v = \left(\frac{\partial h}{\partial s}\right)_p = T \qquad (3.50)$$

$$\left(\frac{\partial g}{\partial p}\right)_T = \left(\frac{\partial h}{\partial p}\right)_s = v \qquad (3.51)$$

$$\left(\frac{\partial u}{\partial v}\right)_s = \left(\frac{\partial a}{\partial v}\right)_T = -p \qquad (3.52)$$

$$\left(\frac{\partial a}{\partial T}\right)_v = \left(\frac{\partial g}{\partial T}\right)_p = -s \qquad (3.53)$$

Finally, a further two equations which will be well known from a first course in thermodynamics come from the definitions of specific heats. These make up the final set of ten GTR equations:

$$c_v = \left(\frac{\partial u}{\partial T}\right)_v \qquad (3.54)$$

$$c_p = \left(\frac{\partial h}{\partial T}\right)_p \qquad (3.55)$$

3.3.3 GTR property equations

For use in engines and most other thermodynamic applications, general differential equations in quantifiable terms are specifically required for the energy properties, these being the internal energy, enthalpy and entropy. They must be applicable for all substances, not only to ideal gases. It is most convenient to develop that for entropy first from the GTRs and to then obtain the others from it by further substitution.

Entropy

Entropy, as with any other property of a single-phase, single-component substance, may be defined in terms of any other two properties. It is most conveniently taken as being related to the temperature and specific volume or the temperature and pressure which are all measurable quantities. Using the first of these

$$s = f(T, v) \qquad (3.56)$$

Hence, the total derivative of entropy may be expressed as

$$ds = \left(\frac{\partial s}{\partial T}\right)_v dT + \left(\frac{\partial s}{\partial v}\right)_T dv \qquad (3.57)$$

But

$$\left(\frac{\partial s}{\partial T}\right)_v = \left(\frac{\partial u}{\partial T}\right)_v \bigg/ \left(\frac{\partial u}{\partial s}\right)_v = c_v/T \qquad (3.58)$$

Then, by substitution of (3.58) and (3.48) into (3.57)

$$ds = \frac{c_v}{T}dT + \left(\frac{\partial p}{\partial T}\right)_v dv \qquad (3.59)$$

Alternatively, writing the entropy as a function of temperature and pressure and repeating the procedure, a second equally valid equation for entropy results. This is

$$s = f(T, p) \tag{3.60}$$

giving

$$ds = \frac{c_p}{T} dT - \left(\frac{\partial v}{\partial T}\right)_p dp \tag{3.61}$$

Internal energy and enthalpy

A differential equation for the change in internal energy, du, can now be easily determined by substitution of (3.59) for ds into (3.40):

$$du = c_v dT + \left\{ T\left(\frac{\partial p}{\partial T}\right)_v - p \right\} dv \tag{3.62}$$

Similarly, using (3.61) and (3.41), dh can be obtained:

$$dh = c_p dT + \left\{ v - \left(\frac{\partial v}{\partial T}\right)_p \right\} dp \tag{3.63}$$

The equations (3.62), (3.63), (3.59) and (3.61) are quite general and can be used with measured values of specific heat and the relationships between the properties p, v and T determined from appropriate state equations to establish quantitative values for the change of internal energy, enthalpy and entropy. Substituting from the ideal gas equation reduces them to the frequently used equations for the internal energy and enthalpy where the second term disappears and internal energy and enthalpy are then functions of temperature only, as expected. The entropy equation reduces to the well-known formula for ideal gases as a function of temperature and pressure or volume. With other, more complex state equations, the second term remains and must be fully evaluated. That is, the internal energy and enthalpy are now functions of both temperature and either pressure or volume while the entropy equation is more complex than in the ideal gas case.

Many other equations can be developed using GTRs. Of particular interest in studies of engines and combustion are those for the specific heats themselves, the difference in specific heats and the ratio of specific heat. These will be examined in the exercises at the end of this chapter.

3.4 EQUATIONS OF STATE FOR REAL GASES

The majority of thermodynamic processes which are of interest to a combustion engineer or engine technologist occur in the gas phase and many substances, such as air and the combustion products, can be approximated with reasonable accuracy by the ideal gas equation throughout the temperature ranges of relevance. However, others such as the fuel vapour may be too near to the saturated vapour condition for this assumption to be reasonable and their properties need to be evaluated by other means. These are sometimes termed **real gases** to distinguish them from the **ideal gases** which are only a useful approximation to reality although a very close one in many instances. Values near to the saturation conditions for the commonly used substances, for example water vapour and a range of refrigerants, are readily obtainable from tabulations and similar data for other substances are available but are more difficult to obtain. For all substances, tabulations are often not convenient, particularly for use in computer programs, and alternative procedures which can be used in conjuction with the GTR equations to establish the appropriate property values are then necessary. Suitable equations of state exist to cover the non-ideal gas range and are substantially empirical in derivation although usually with some logical base from which their formulation developed. Nevertheless, they are very useful and, in addition, help in developing in the reader a more complete understanding of the thermodynamic properties of matter. While many such equations have been formulated, some of the most widely used will be discussed below.

3.4.1 Van der Waals equation

The earliest and perhaps most descriptive equation of this type is called the van der Waals equation after its originator, Dutch scientist Johannes Diderck van der Waals. It is based on the idea that gases at higher pressures start to deviate from the ideal gas equation for two reasons, both of which stem from the space occupied by the molecule itself in relation to the mean free path between the molecules. In these non-ideal gases, the mean free path is sufficiently small so that the molecules can no longer be regarded as isolated except for the collisions which occur between them, this being the ideal gas assumption. That is, interactive effects must now be considered. There are two effects. The available volume for compression is reduced by the size which the molecule must occupy and also the individual molecules can influence each other by attractive forces. These are essentially corrections from the ideal gas concept and the van der Waals equation is therefore expressed as a deviation from the ideal gas equation.

In more detail, the modifications required for the ideal gas equation are that the volume and pressure terms both need correction to allow for the relatively closer packing of the molecules. If the volume in the equation can be taken as actually representing the '*free*' space which is available for compression rather than the total volume occupied, then the term should be modified by deducting from the total volume, v, an amount, say b, which accounts for the '*solid*' space occupied by the molecule. In the ideal gas case, this latter volume is negligible but, for other gases, it may be significant.

Therefore the volume available for compression, v', can be expressed as

$$v' = v - b \qquad (3.64)$$

where v is the total volume and b a constant representing that volume occupied by the molecule itself.

The modification to the pressure term occurs because that which should be used in the equation is an internal parameter which exists between the molecules and which differs from that experienced on the walls or on a measuring device. It may be termed an **effective pressure**. The difference occurs due to a substantial number of other molecules being sufficiently close so that their attractive forces are significant. For molecules which impact on a solid surface, this attractive force is exerted on their gas side only which reduces their momentum towards the surface. All other molecules will have the attractive force totally surrounding them and hence it is cancelled with no net effect. Thus the measured pressure, p, at the surface is a distortion of what prevails throughout the gas and it is less than the effective pressure, which can be designated as p'. In an ideal gas, the wide molecular spacing ensures that the attractive force from any other molecule on that approaching the surface is negligible and hence the difference between p and p' approaches zero. In real gases, the effective pressure term for the equation must be calculated by the addition of an attractive term to the measured value. The effective pressure is then given by

$$p' = p + c \qquad (3.65)$$

Now c in (3.65) is due to two factors. These are the molecular density (a greater number of molecules providing a larger attraction) and the spacing of the molecules (each individual molecule providing a greater force when it is closer). Van der Waals postulated that it should be proportional to the density squared which in thermodynamic terms is the inverse square of the specific volume:

$$c \propto \frac{1}{v^2} = \frac{a}{v^2} \qquad (3.66)$$

where a is a constant.

Writing the gas equation in an identical manner to the ideal gas law but in terms of the effective pressure p' and the volume available for compression v' rather than the conventional p and v and substituting from (3.64), (3.65) and (3.66) gives:

$$\left(p + \frac{a}{v^2}\right)(v - b) = RT \qquad (3.67)$$

This is the van der Waals equation. Note the following points.

- As b approaches v, p approaches infinity. Here, a small change in v produces a large change in p which is the case for an incompressible substance. That is, it is the case for a liquid or a solid.

- If $p \gg a/v^2$ and $v \gg b$, the equation becomes the ideal gas equation $pv = RT$.
- For gases, the first term in (3.67) is usually the more significant as deviation from the ideal gas law occurs
- If p and T are fixed, the equation is a cubic in v and there are three roots.

It is now necessary to determine the constants a and b which will be different for every gas and this can be most easily achieved by considering the isotherms which pass through the critical point. These have a point of inflexion in p–v coordinates and hence both the first and second derivatives can be equated to zero:

$$\left(\frac{\partial p}{\partial v}\right)_T = 0 \quad \text{and} \quad \left(\frac{\partial^2 p}{\partial v^2}\right)_T = 0 \tag{3.68}$$

The van der Waals equation can be rewritten in terms of p and differentiated to give

$$\left(\frac{\partial p}{\partial v}\right)_T = -\frac{RT}{(v-b)^2} + \frac{2a}{v^3} \tag{3.69}$$

and

$$\left(\frac{\partial^2 p}{\partial v^2}\right)_T = -\frac{2RT}{(v-b)^3} + \frac{6a}{v^4} \tag{3.70}$$

Substituting now the critical point values of $v = v_c, T = T_c$ and $p = p_c$, equating to zero gives the following solution for a and b

$$a = \frac{9RT_c}{8}v_c = 3p_c v_c^2 \tag{3.71}$$

$$b = \frac{v_c}{3} \tag{3.72}$$

Now state equations and, indeed properties in general, may be expressed in terms of what are called 'reduced properties' which are the values relative to those at the critical point. This is often a very useful form because different substance are effectively normalized by this technique. These are defined as

Reduced pressure $p_r = p/p_c$ (3.73)

Reduced temperature $T_r = T/T_c$ (3.74)

Reduced specific volume $v_r = v/v_c$ (3.75)

By substitution and rearrangement of (3.73) into (3.69) to (3.72) a reduced form of the van der Waals equation is available. This is

$$\left(p_r + \frac{3}{v_r^2}\right)\left(v_r - \frac{1}{3}\right) = \frac{8\,T_r}{3} \tag{3.76}$$

3.4.2 Some other state equations

Other state equations have been developed using similar logic with modifications to fit them to the known data. They have generally been designated by the names of their originators. While it is not possible to give a full explanation and an exhaustive list here, some of the most common are

Clausius
$$p = \frac{RT}{(v-b)} - \frac{c}{T(v+d)^2} \tag{3.77}$$

Dieterici
$$p = \frac{RT}{(v-b)} e^{-a/RT_v} \tag{3.78}$$

Berthelot
$$p = \frac{RT}{(v-b)} - \frac{a}{Tv^2} \tag{3.79}$$

Redlich–Kwong
$$p = \frac{RT}{(v-b)} - \frac{c}{T^{1/2}v(v+b)} \tag{3.80}$$

Beattie–Bridgeman
$$p = \frac{RT(1-\varepsilon)}{v^2}(v+b) - \frac{a}{v^2} \tag{3.81}$$

Virial Equation
$$Z = \frac{pv}{RT} = 1 + \frac{B}{v} + \frac{C}{v^2} + \frac{D}{v^3} + \cdots \tag{3.82}$$

Equations (3.77) to (3.80) follow a modified van der Waals approach. Equations (3.77) and (3.79) are based on the assumption that intermolecular attractive forces vary with temperature while (3.78) and (3.80) are modifications to give better agreement with measured values. Equations (3.81) and (3.82) are empirical fits to the measured data, the latter being a power law relationship in terms of v (as given here) or p. It should be noted that (3.81) was used by Hottel, Williams and Satterfield (1936) in the development of the thermodynamic charts for internal combustion engines which were widely used for cycle analysis before the advent of computer codes for that purpose.

3.5 IDEAL GASES AT HIGH TEMPERATURES

The non-ideal gas equations apply to relatively dense gases, i.e. ones at high reduced pressures and moderate-to-low reduced temperatures. In many combustion and engine problems, the gases are well removed from the saturation conditions. This includes the air and the products of the combustion process. The ideal gas equation of state is then a close approximation to reality and the properties' values required can be obtained from it. The reader is reminded that the internal energy and enthalpy are the functions of temperature only, that is

$$du = c_v dT \tag{3.83}$$

$$dh = c_p dT \tag{3.84}$$

and entropy may be obtained from one of several equations of the form

$$ds = c_v dT/T + Rdv/v \tag{3.85}$$

These can easily be integrated if R, c_v and c_p (and hence γ) are assumed to be constant. In fact, apart from R which is a constant, c_v, c_p and γ are not constant and may be quite complex functions of temperature and, to a lesser extent, pressure at elevated conditions as shown for the example of hydrogen on Figure 3.3. The complex behaviour is due to changes in the molecule energy storage and to the breakup in the subparticles, ions and electrons as the temperature becomes extreme. Empirical relationships are available in the low to moderate temperature range and a power law series in terms of temperature for c_v or c_p is often convenient as

$$c_v = A + BT + CT^2 + DT^3 \ldots \tag{3.86}$$

However, a theoretical approach describing the way in which these properties are obtained is required. This is best obtained by use of *statistical thermodynamics* which is a branch of the subject that approaches it from the statistical analysis of the motion of many particles and the probability of certain states existing. It is a study in its own right and is too complex to be considered other than cursorily here. However, some comprehension of the molecular basis of thermodynamics is essential and a suitable approach, which may be regarded as a precursor of statistical thermodynamics, called the **kinetic theory**, needs examination. It relies on the laws of mechanics for a simple assessment of the linear and rotational motion of particles which can be related via the ideal gas law to the temperature for an evaluation of the energy properties. Under appropriate circumstances, it is surprisingly accurate. When temperatures become very high or molecules excessively large, additional effects from particle vibration need to be included to provide a reasonable account of the known variations. These will be included here as an arbitrary modification although the reader is advised to consult specialist texts such as Sonntag and Van Wylen for a full account of the basic theory which determines the additional contribution to the thermodynamic properties from vibration and other modes. It should be noted that at even higher temperatures, dissociation and ionization play a role in the gas properties and these will be examined in a later chapter.

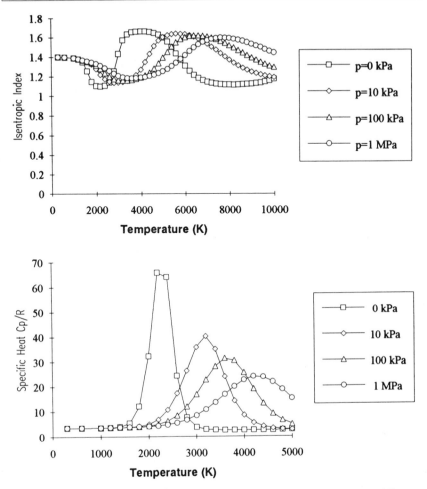

Figure 3.3 Variation with temperature of isentropic index and specific heat (dimensionless) at constant pressure for hydrogen (source: Kubin, R. F. and Presley, L. L., 1964, NASA SP-3002).

3.5.1 The kinetic theory of gases

Although developed in the middle of the nineteenth century, the kinetic theory of gases provides a good starting point for the understanding of the particle approach to the properties of gases. It suggests that the thermodynamic properties of matter derive from the kinetic energy of the particles either as linear (translational) kinetic energy or as rotational kinetic energy when the moment of inertia of the molecule is significant. It does not consider the potential energy of molecular attraction which is reasonable when applied to ideal gases because of the wide spacing of the molecules.

Neither does it consider the vibrational energy, electronic energy etc. which are a part of more modern theories. The vibrational energy, in particular is an essential component once temperatures become moderately high and so will be included here by a simple addition to the kinetic theory values. This provides a brief, workable although only partial model of the thermo-dynamic behaviour of ideal gases.

3.5.2 Basic postulates of the kinetic theory

The kinetic theory makes none of the assumptions of modern physics re-garding *quantized* energy and in fact predates the quantum theory. It as-sumes that matter is composed of a very large number of molecules which, for fluids, are in a state of continuous motion, the thermodynamics then being simply related to the well-known laws of mechanics. The following assumptions apply.

- Molecules are hard spheres or combinations of spheres which are perfect-ly elastic.
- Molecules travel in random directions but with a uniform velocity in a straight line between the many collisions which occur between them or with the walls in a normal gas.
- The time of encounter on impact is negligible as are frictional effects between molecules.
- The size of molecules is negligible with respect to the mean free path.
- The pressure on a wall is solely due to the molecular collisions with it.

3.5.3 Basic equations of the kinetic theory

Now consider a molecule of mass m in a Cartesian coordinate system x, y, z which has a velocity $\boldsymbol{v}$ with components $\boldsymbol{v}_x, \boldsymbol{v}_y$ and $\boldsymbol{v}_z$, the x direction being towards a wall as shown on Figure 3.4. The change in x momentum of one molecule on impact is equal to $2m\boldsymbol{v}_x$ if it is assumed to rebound with an equal and opposite velocity as required by the perfectly elastic assumption. Because of the very large number of molecules in the gas, assume now that n molecules having an identical value of the component $\boldsymbol{v}_x$ exist per unit volume, these being but some of the many molecules in that region. The volume of a slab of gas δx thick near the wall therefore contains $nA\delta x$ of the above molecules where A is the wall area. Again, because of the large numbers, it may be assumed that, on average, half of these move towards the wall and half away from it with, of course, only the former impacting with it. Therefore the total momentum change in the above volume due to these molecules is

$$\delta M = \frac{A}{2}\delta x n\, 2m\boldsymbol{v}_x = A\delta x n m\boldsymbol{v}_x \tag{3.87}$$

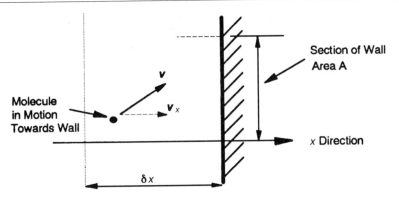

Figure 3.4 Molecular interaction with a wall in the kinetic theory.

All these molecules must impact with the wall within a time span equalling that taken for the furthest from the wall to reach it, this time being given by

$$\delta t = \frac{\delta x}{v_x} \tag{3.88}$$

The force on the wall is the time rate of change of momentum giving

$$F = \frac{\delta M}{\delta t} = Anm v_x^2 \tag{3.89}$$

from which the pressure can be easily obtained as

$$p = nm v_x^2 \tag{3.88}$$

However, there are other molecules with different velocity component values in the x direction and these must be considered in the same way. The total pressure is therefore due to

n_1 molecules, velocity v_{x1}
n_2 molecules, velocity v_{x2}
n_3 molecules, velocity v_{x3}
$\vdots$
n_i molecules, velocity v_{xi}

the sum of n_1 to n_i making up the total number of molecules.
    Therefore, the total pressure is

$$p = n_1 m v_{x1}^2 + n_2 m v_{x2}^2 + n_3 m v_{x3}^2 + \cdots = m \sum n v_x^2 \tag{3.90}$$

But the total number of molecules is

$$\sum n = n_1 + n_2 + n_3 + \cdots = N(\text{say}) \tag{3.91}$$

Multiplying (3.90) by the pressure term by $N/\sum n$ (which is unity) then gives

$$p = \frac{mN\sum(n\boldsymbol{v}_x^2)}{\sum \boldsymbol{n}} = mN\bar{\boldsymbol{v}}_x^2 \tag{3.92}$$

where $\bar{\boldsymbol{v}}_x^2$ is the root mean square velocity.

For large numbers of particles, the pressure at a point is equal in all directions and using a similar equation to x for pressure for the y and z axes then gives

$$\bar{\boldsymbol{v}}_x^2 = \bar{\boldsymbol{v}}_y^2 = \bar{\boldsymbol{v}}_z^2 \tag{3.93}$$

But because

$$\bar{\boldsymbol{v}}^2 = \bar{\boldsymbol{v}}_x^2 + \bar{\boldsymbol{v}}_y^2 + \bar{\boldsymbol{v}}_z^2 \tag{3.94}$$

it can be seen that

$$\bar{\boldsymbol{v}}_x^2 = \bar{\boldsymbol{v}}^2/3 \tag{3.95}$$

Also the product of m and N is equal to (mass per molecule) times (molecules per unit volume) which is the density, ρ or $1/v$ where v is the specific volume.

Therefore

$$p = \frac{\bar{\boldsymbol{v}}^2}{3v} \tag{3.96}$$

or

$$pv = \frac{\bar{\boldsymbol{v}}^2}{3} \tag{3.97}$$

The isolated molecule postulate of the above theory fits the ideal gas assumptions and the equation, $pv = RT$ applies. Comparing these two equations for pv then shows that the temperature of a gas is related to its mean molecular kinetic energy, i.e.

$$\bar{\boldsymbol{v}}^2 = 3RT \tag{3.98}$$

If it is now assumed that the internal energy per unit mass, u, of isolated molecules is unaffected by intermolecular forces and is due to their kinetic energy only, the basic equation from classical mechanics states that

$$u = \frac{\bar{\boldsymbol{v}}^2}{2} \tag{3.99}$$

Substituting (3.99) into (3.98) then gives

$$u = \frac{3}{2}RT \tag{3.100}$$

It should be noted that this equation gives the anticipated result that internal energy is a function of temperature only for an ideal gas.

Note that this equation gives the total molecular kinetic energy (internal energy) per unit mass. The specific gas constant $R(kJ\,kg^{-1}\,K^{-1})$ in this equation can be replaced by the universal gas constant $R_0(kJ\,kmol^{-1}\,K^{-1})$ or k, the Boltzmann constant $(kJ\,K^{-1})$ for the molecule which is the universal gas constant divided by Avogadro's number to give other forms of the specific energy.

Now the specific heat at constant volume is, by definition

$$c_v = \left(\frac{\partial u}{\partial T}\right)_v = \frac{3}{2}R \qquad (3.101)$$

while, specific heat at constant pressure for an ideal gas is

$$c_p = c_v + R = \frac{5}{2}R \qquad (3.102)$$

from which the isentropic index γ can now be determined as

$$\gamma = \frac{c_p}{c_v} = \frac{5}{3} = 1.667 \qquad (3.103)$$

This can be compared with actual values for monatomic gases, some examples being

Helium $\qquad\qquad\qquad \gamma = 1.659$ ⎫
$\qquad\qquad\qquad\qquad\qquad\qquad\qquad\qquad$ ⎬ at 300 K
Argon $\qquad\qquad\qquad \gamma = 1.668$ ⎭

Obviously the theory in these cases gives very good results.

3.5.4 Partition of energy

The kinetic theory so far has been developed for particles which only have translational momentum. Compact molecules with no separation of the atoms which form them will have negligible rotational inertia and thus monatomic particles are likely to fit the theory well thereby anticipating the agreement with these measured values. For larger molecules with separated atoms, the theory needs expansion and this is achieved by the use of an assumption called **equipartition of energy**. Here, what appears to be a sweeping assumption is made and it is postulated that each motion in which energy can be stored (called a **degree of freedom**) carries the same amount of energy. These include both translational and rotational motions. It is only intrinsically justifiable for the three translational modes where the large numbers of molecules provides an extremely high probability of each direction being the same. The reason for equality between the translational and rotational modes is more obscure. Nevertheless, it provides reasonable answers in a number of cases suggesting that it is, at least for these, correct. Its validity can now be demonstrated by the modern methods of statistical thermodynamics but these were not developed until well after the original concept was put forward. The possible degrees of freedom are shown on Figure 3.5 and listed on Table 3.1.

For a monatomic gas, $u = \frac{3}{2} RT$ and therefore the energy per degree of freedom is $\frac{1}{2} RT$. If this applies to each degree of freedom for all gases, it can be seen that a diatomic gas has an internal energy of $\frac{5}{2} RT$. Its specific heats can then be obtained as before and are

$$c_v = \frac{5}{2}RT \tag{3.104}$$

$$c_p = c_v + R = \frac{7}{2}RT \tag{3.105}$$

from which the isentropic index γ is 1.4.

When this is compared with values for some typical diatomic gases which are

Oxygen $\gamma = 1.394$
Nitrogen $\gamma = 1.400$
Hydrogen $\gamma = 1.405$ at 300 K
Carbon
Monoxide $\gamma = 1.400$

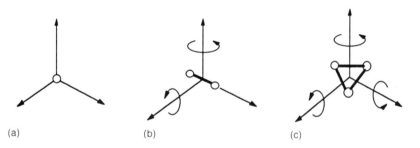

(a)   (b)   (c)

Figure 3.5 Possible molecular degress of freedom for translation and rotation. (a) monatomic: three translations, (b) diatomic: three translations, two rotations, (c) triatomic or greater: three translations, three rotations.

Table 3.1 Degrees of freedom

Molecule type	Energy storage modes	Degrees of freedom
monatomic gas	translation in three directions	3
diatomic gas	translation in three directions, rotation about three axes, but one is negligible	5
triatomic gas (or greater)	translation in three directions, rotations about three axes	6

it can be seen that again the agreement is good. This indicates that, at low temperatures for widely spaced molecules, the equipartition of energy assumption is likely to have some validity.

Repeating the above for a triatomic gas with six degrees of freedom, the equations become

$$u = 3RT, \quad c_v = 3R, \quad c_p = 4R \qquad (3.106)$$

from which γ equals 1.333. Typical triatomic gases have the following measured values:

Carbon Dioxide $\qquad\qquad \gamma = 1.29$ ⎫
Hydrogen Sulphide $\qquad\quad \gamma = 1.21$ ⎬ at 300 K
Sulphur Dioxide $\qquad\qquad \gamma = 1.25$ ⎭

Here, the agreement, although still reasonable, has deteriorated substantially. To account for the greater inaccuracy for large numbers of molecules, additional modes of energy storage to those of translation and rotation need to be included. That of particular interest at present is the energy stored by vibration between the atoms making up the molecule. This may occur in several ways, the particles vibrating on an axis towards or tangentially to each other. This vibrational energy increases with the number of atoms in the molecule and the temperature.

To determine property values of high-temperature non-monatomic gases requires the introduction of a means of evaluating the vibrational modes. This is based on the theory of statistical thermodynamics, details of which can be found in specialist texts such as Sonntag and Van Wylen (1991). It introduces a **partition function**, Z, from which the molecule's energy can be partitioned into a number of modes. These can be evaluated from the distribution functions which apply to the model of the substance under consideration. The overall partition function is made up of individual partition functions, those usually considered being translational Z_{TR}, rotational Z_{ROT}, vibrational Z_{VIB}, and what are sometimes called the more exotic forms of electronic, nuclear spin, chemical, etc. labelled here as Z_{EX} most of which are not important in the ranges to be considered here. Then

$$Z = Z_{TR} Z_{ROT} Z_{VIB} Z_{EX} = Z_{TR} Z_{INT} \qquad (3.107)$$

All but the translational form are considered to be internal modes of energy storage because they do not involve relocating the molecule in space.

It can be demonstrated from statistical thermodynamics that the translational and rotational modes for an ideal gas are equipartitioned at normal temperatures, consistent with the classical kinetic theory. The internal energy equation in terms of the partition function is

$$u = RT^2 \left(\frac{\partial \ln Z}{\partial T} \right)_v \qquad (3.108)$$

Once the internal energy is evaluated, the normal ideal gas equation for enthalpy, $h = u + RT$, applies for the translational mode only with additional components from all other (internal) forms adding $h = u$. This is because the pv term in the enthalpy equation (or RT for ideal gases) is related to the molecule's position in space, which does not alter the internal modes.

The translational partition function can be shown to be a function of the total volume of the system V and its absolute temperature $T^{3/2}$ while the rotational one is a function of the moment of inertia of the molecule, I, and its temperature, T. For fixed volume and moment of inertia, substitution into (3.108) gives $u = (3/2) RT$ and $u = RT$ for the complete addition of energy from these two modes which is identical to the kinetic theory approach. The translational mode is always complete but this is not necessarily so for the internal modes. A quantity T_R, called the **characteristic rotational temperature** can be used to normalize the temperature for different gases to determine whether the full rotational contribution has occurred and it has values of 2.08 K for oxygen, 2.86 K for nitrogen and 87.2 K for hydrogen (Table 3.2), the last being the highest of the diatomic gases. Normalized internal energy and specific heats are plotted against T/T_R on Figure 3.6 and it can be seen that the rotational mode is close to complete (i.e. $u = RT$) when this is greater than about five. For most diatomic gases, complete rotational energy is therefore imparted to the molecule for all temperatures above what are very low values (e.g. 10.4 K for oxygen). The specific heat contributions can be seen to be complete at the even lower value of T/T_R of around two with a corresponding very low temperature.

The vibrational partition function is more important in that it expresses a frequently encountered change from the values of the kinetic theory. The expression for it has been developed for what is termed an **harmonic oscillator** and is then

$$Z_{VIB} = \left[1 - \exp\left(\frac{-hv}{kT} \right) \right]^{-1} = [1 - \exp(-x)]^{-1} \qquad (3.109)$$

Table 3.2 Characteristic rotational temperatures, T_R and characteristic vibrational temperatures, T_V for some common gases

Gas	T_R (K)	T_V (K)
H_2	87.2	6340
N_2	2.86	3390
CO	2.76	3120
NO	2.44	2740
O_2	2.08	2270

Here hv/k is called the **characteristic vibrational temperature**, T_v, and it can be used in the same way as the characteristic rotational temperature. The equivalent plot is also shown on Figure 3.6. Here, the internal energy contributions approach RT (i.e. the complete contribution) at a little above eight to nine times the characteristic vibrational temperature and the specific heats approach R at around three. While these values are a little higher than in the rotational case, the main difference lies in the fact that the characteristic vibrational temperature is much higher being 2270, 3390 and 6340 K for the same three gases considered before (calculated using $h =$ Planck's constant $= 6.625 \times 10^{-34}$ Js, $k =$ Boltzmann's constant $= 1.380 \times 10^{-23}$ Jk^{-1}, $v = c\omega$ where $\omega =$ vibrational wave number of the molecule, m^{-1} and $c =$ velocity of light $= 2.998 \times 10^8$ ms^{-1}). The vibrational contribution to internal energy is minimal at 300 K ($T/T_R = 0.15$ to 0.05 for the above gases) but becomes significant as the temperature rises. The full vibrational contribution is not complete until extremely high temperatures

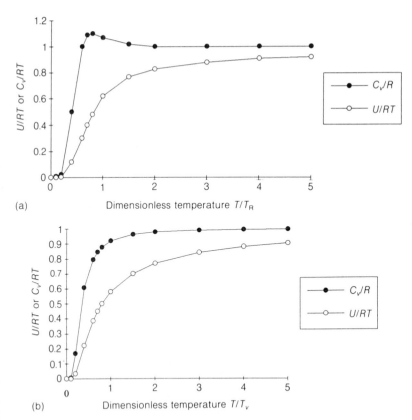

Figure 3.6 Rotational and vibrational contributions to the internal energy and specific heat at constant volume. (a) rotational effects, (b) vibrational effects.

are attained at which time, it should be noted, the gases have also reached a high level of dissociation. A progressive contribution occurs during this range.

In summary, for the kinetic theory, equipartition is adequate at temperatures above 300 K for most diatomic gases to be considered here but an additional vibrational contribution must be added using the above equation as the temperature increases. With larger molecules, the vibrational effect is noticeable earlier with temperature increase and this explains the differences noted previously against the measured data.

The contribution of the electronic states to internal energy, enthalpy and specific heat is insignificant for substances in the temperature ranges considered here although it has a small contribution to entropy. It will not be further considered although the reader, if interested, is referred to Sonntag and Van Wylen for details. Because the datum for energy contributions of molecules larger than monatomic has been selected as the fully dissociated state, a chemical contribution must be added to the internal energy and enthalpy for the undissociated molecule. This is a constant, ho_f which is the enthalpy of formation. Because specific heat is the differential of internal energy or enthalpy, the chemical contribution to specific heat is therefore equal to zero. Note that the ho_f for a substance is often written as $\Delta H_{0,298}$ which is tabulated for the substance. These values will be discussed in the next chapter.

It should be noted that, as the temperature of a gas rises, the molecules first dissociate and later ionize. This, put in simplistic terms, is because the vibration starts to shake the molecule apart. That is, a gas originally made up of particles which are substantially the same throughout, now consists of some of the original molecules and some dissociated components, the latter increasing in proportion as the temperature rises. At even higher temperatures, the electrons are stripped from the molecule and the process is known as **ionization**. Additional electrons and positively charged particles start to appear and the gas progressively becomes a more and more complex mixture. While the methods of calculation for evaluating the amount of dissociation and ionization will be considered later, it should be noted that the gas must now be treated as a mixture of several ideal gases each with properties which vary with temperature in order to find its average properties.

3.6 THERMODYNAMICS OF GAS MIXTURES

Some readers of this text may be familiar with gas mixtures as the subject is sometimes dealt with in first courses in thermodynamics. However, because mixtures are of great importance in combustion and engine calculations, this section is devoted to a brief review of the topic.

Tabulated properties of many individual gases are readily available in the moderate pressure and temperature ranges. Even at high temperatures (e.g. for hydrogen up to perhaps 30 000 K) some tabulated properties exist and can be found, for example, in NASA reports, the very high temperature values being necessary for such space research as planetry re-entry. Most of these tabulations were determined by statistical thermodynamic calculations. For mixtures, tabulations are not generally feasible due to the infinite variation in the possible constituents and air is the only one whose properties are usually available in this form. It is therefore necessary to examine how mixture properties can be obtained from the data, either tabulated or calculated, for the individual component gases. In general, ideal gas mixtures will predominate in the combustion work of this text and the greatest emphasis will be in that direction although a brief consideration of non-ideal gas mixtures is included at the end of the section.

3.6.1 Ideal gas mixtures

Basic mass, kilomole and molecular mass relationships

These are all obvious relationships and all follow simple summation techniques. As such, they only need a brief description here.

The total mass of a mixture being the sum of the masses of the individual components is given by

$$m = m_1 + m_2 + m_3 + \cdots + m_k = \sum_{i=1}^{k} m_i \tag{3.110}$$

If the mass fraction of any component i, called m_{fi}, is defined as:

$$m_{fi} = \frac{m_i}{m} \tag{3.111}$$

then

$$\sum_{i=1}^{k} m_{fi} = 1 \tag{3.112}$$

Using a similar analysis, based on the number of kilomoles rather than the mass of each gas in the system, gives similar equations. That is

$$n = n_1 + n_2 + n_3 + \cdots + n_k = \sum_{i=1}^{k} n_i \tag{3.113}$$

and a mole fraction of any component, called y_i, is defined as

$$y_i = \frac{n_i}{n} \tag{3.114}$$

Therefore

$$\sum_{i=1}^{k} y_i = 1 \qquad (3.115)$$

Now the molecular mass, $\bar{M}$, is by definition the mass per kilomole* and this applies either to the mixture or to any component. It is therefore given by

$$\bar{M}_i = \frac{m_i}{n_i} \qquad (3.116)$$

Hence

$$m = n_1 \bar{M}_1 + n_2 \bar{M}_2 + \cdots + n_k \bar{M}_k = n\bar{M} \qquad (3.117)$$

where $\bar{M}$ is the average molecular mass of the mixture. Therefore, by rearrangement and substitution from (3.114)

$$\bar{M} = y_1 \bar{M}_1 + y_2 \bar{M}_2 + \cdots + y_k \bar{M}_k = \sum_{i=1}^{k} y_i \bar{M}_i \qquad (3.118)$$

p, v, T Relationships for ideal gases

In a mixture, it is not possible to measure the individual contributions of each component to the total pressure and volume because the particles are fully distributed. Thus, some basic laws are required and two are used. These are Dalton's law of partial pressures and Amagat's law of partial volumes. While both give identical results for ideal gases, they do not do so for real gases. Hence both are discussed here.

Dalton's law
Dalton's law states that each component gas exerts a partial pressure equivalent to that which it would exert if it alone occupied the whole volume at the mixture temperature with all other gases absent. The total pressure is the sum of these partial pressures:

$$p = p_1 + p_2 + p_3 + \cdots + p_k = \sum_{i=1}^{k} p_i \qquad (3.119)$$

This applies to all gases and vapours, ideal or otherwise. For ideal gases, the equation of state allows an easy substitution for p in terms of the other properties, T and V. That is

$$p_i = n_i \frac{R_0}{V} T \qquad (3.120)$$

---

*Throughout this section a bar on a variable (for example, $\bar{M}$) represents that quantity per k mole.

where T and V are the mixture temperature and volume which are the same for all components. Hence

$$p = (n_1 + n_2 + n_3 + \cdots)\frac{R_0 T}{V} = n\frac{R_0 T}{V} = \frac{mR_0 T}{\bar{M}V} = \frac{RT}{v} \tag{3.121}$$

showing that the ideal gas equation applies to the mixture where v is its specific volume of the mixture and R_0 its specific gas constant. Now

$$p_i = \frac{n_i R_0(T/V)}{nR_0(T/V)} = \frac{n_i}{n} = y_i \tag{3.122}$$

That is, the ratio of partial to total pressure is identical to the mole fraction. For any two components, i and j, it is then obvious that

$$\frac{p_i}{p_j} = \frac{y_i}{y_j} \tag{3.123}$$

Note that for non-ideal gases, their more complex state equations do not allow the temperature and volume to be cancelled and the equality of (3.123) does not apply.

Amagat's law
This is sometimes known as the Amagat–Leduc law. It relies on a similar principle to Dalton's law but works by summing individual component volumes rather than pressures. More precisely, it states that, if the volume of each component gas was measured individually at the mixture pressure and temperature, the total volume would be the sum of these volume components. This gives

$$V = V_1 + V_2 + V_3 + \cdots V_k = \sum_{i=1}^{k} V_i \tag{3.124}$$

Again, this basic formulation applies to all gases and vapours. Substituting the ideal gas equation as before, gives

$$V = (n_1 + n_2 + n_3 +)\frac{R_0 T}{p} = n\frac{R_0 T}{p} = m\frac{RT}{p} \tag{3.125}$$

That is, the ideal gas equation, as with Dalton's law still applies to the mixture when a specific gas constant of R equal to the universal gas constant R divided by the mixture molecular mass is used. For ideal gases only, the Dalton and Amagat laws are identical and, starting with either one, the other can be shown to be valid. The ratio of the component volume to the total volume is, from Amagat's law with ideal gases

$$\frac{V_i}{V} = \frac{n_i}{n} = y_i \tag{3.126}$$

which is identical to the ratio of partial to total pressure for the same situation from Dalton's law. It should be noted that the molar specific volume, v_i, of each component is identical to the molar specific volume of the mixture. That is

$$\bar{v}_i = \frac{V_i}{n_i} = \frac{y_i}{y_i} \frac{V}{n} = \bar{v} \tag{3.127}$$

This is not so for specific volumes on a mass basis as each component then has a different value of R.

Other properties of gas mixtures

The Gibbs–Dalton law
For each component, two intensive properties are now known, these being either T and p_i or T and v_i. Other properties, particularly u, h and s, are required and can be found using the Gibbs–Dalton law which states that any property of a mixture acts as if it occupied the whole volume, V, at the temperature, T, of the mixture. The total value of the property required is then the sum of those of the components of the mixture at these conditions. This law again has general application. Taking, for example, internal energy as the property under consideration, the total internal energy is

$$U = \sum_{i=1}^{k} U_i \tag{3.128}$$

where u_i is evaluated at the temperature and volume of the mixture. The specific internal energy, u_i, may be obtained from either

$$mu = \sum_{i=1}^{k} m_i u_i \tag{3.129}$$

or

$$n\bar{u} = \sum_{i=1}^{k} n_i \bar{u}_i \tag{3.130}$$

Therefore, the internal energy values (intensive) are, on a mass basis

$$u = \frac{\displaystyle\sum_{i=1}^{k} m_i \bar{u}_i}{\displaystyle\sum_{i=1}^{k} m_i} = \sum_{i=1}^{k} m_{fi} \bar{u}_i \tag{3.131}$$

and on a molar basis

$$\bar{u} = \frac{\sum\limits_{i=1}^{k} n_i \bar{u}_i}{\sum\limits_{i=1}^{k} n_i} = \sum\limits_{i=1}^{k} y_i \bar{u}_i \qquad (3.132)$$

Similarly

$$h = \sum\limits_{i=1}^{k} m_{fi} h_i \quad \text{or} \quad \bar{h} = \sum\limits_{i=1}^{k} y_i \bar{h}_i$$

$$s = \sum\limits_{i=1}^{k} m_{fi} s_i \quad \text{or} \quad \bar{s} = \sum\limits_{i=1}^{k} y_i \bar{s}_i \qquad (3.133)$$

Specific heats may also be summed in the same way giving

$$c_v = \sum\limits_{i=1}^{k} m_{fi} c_{vi} \quad \text{or} \quad \bar{c}_v = \sum\limits_{i=1}^{k} y_i \bar{c}_{vi}$$

$$c_p = \sum\limits_{i=1}^{k} m_{fi} c_{pi} \quad \text{or} \quad \bar{c}_p = \sum\limits_{i=1}^{k} y_i \bar{c}_{pi} \qquad (3.134)$$

However, the isentropic index cannot be obtained by a summation of its individual values in this way because it is a ratio of the two specific heats. Its average value for the mixture must be found by first evaluating the individual specific heats and dividing c_p by c_v.

Note that, in all the above equations, the use of tabulated values for internal energy and enthalpy require no pressure correction as long as they are ideal gas values because these vary only with temperature. However, the entropy change of an ideal gas varies with pressure (or specific volume) as well as temperature and data tabulated on a temperature base only will require correction for any pressure change from the value, usually of one atmosphere specified for the tabulation. A further point of note is that, in a closed non-reacting system, the mole fraction or mass fraction of the component gases does not change. When reactions occur in a closed system as happens during combustion, the gases and their mole fractions change requiring new partial pressures to be calculated. The total number of kmoles in the system may alter although the total mass is conserved.

A common problem is the mixing of an ideal gas and a slightly super-heated vapour– in particular, the air, water vapour mixture in the atmos-phere. Combustion products normally also include water vapour and so fall into this category. While it is not proposed to treat this area, known as psychrometry, here and the reader is referred to specialist texts, it needs to be noted that although water vapour is strictly not an ideal gas, approxi-mating it as such does not normally result in a significant error due to its low proportion in the mixture, usually 5% or less. That is, the above formulations are still revelant for these calculations.

3.6.2 Real gas mixtures

For mixtures which contain a large proportion of vapour or vapours or mixtures of gases which are all at high pressures, the ideal gas equation of state is a poor approximation and real gas equations or data are necessary. Mixtures of real gases cannot be handled as easily as ideal gas mixtures although the common rules, noted above, still apply. That is, using Dalton's law, the total pressure is still the sum of the partial pressures and, using the Amagat–Leduc law, the total volume is the sum of the component volumes. Each component is then evaluated by the appropriate methods (e.g. complex state equation) at the mixture temperature and total volume in the former case and at the mixture temperature and pressure in the latter. Note that the two methods do not now give identical results as was the case with ideal gases. Since Amagat's law considers the total pressure for all species, it accounts more correctly for the intermolecular forces between species and between molecules of the one species. In general it gives superior results except at very low pressures. Dalton's law, however, is usually easier to apply although neither is simple. For engineering approximations, an additional equation is often used because of its easy application for real gases. This is called Kay's rule. It gives a critical temperature and critical pressure for the gas mixture as

$$T_c = y_1 T_{c1} + y_2 T_{c2} + y_3 T_{c3} + \cdots \tag{3.135}$$

$$p_c = y_1 p_{c1} + y_2 p_{c2} + y_3 p_{c3} + \cdots \tag{3.136}$$

Here the subscript c represents critical temperatures and pressures for the mixture and components. The above equations may then be used to obtain reduced mixture properties for substitution into appropriate real gas equations.

3.7 REFERENCES

Hottell, H., Williams, G. and Satterfield, C. (1936) *Thermodynamic Charts for Combustion Processes, Vols. 1 and 2*, Wiley, NY.
Sonntag, R. E. and Van Wylen, G. J. (1991) *Introduction to Thermodynamics*, Wiley, New York.

PROBLEMS

3.1 A fuel cell operating on a closed system at atmospheric pressure and a constant temperature produces a maximum possible quantity of internal work equal to the change in the Helmholtz function. Show that, when the expansion against the external atmospheric conditions is taken into account, that this becomes equal to the change in the Gibbs function.

3.2 An experimental fuel cell (either as in Problem 3.1 or under steady flow conditions) operates at a constant temperature of 350 K in a constant temperature atmospheric environment of 300 K. The fuels are hydrogen and oxygen and liquid water is produced. The measured work output is 173 750 kJ kmol^{-1} (of the water) and it is estimated that the efficiency compared with the best possible value is currently at 75%. The total enthalpy value for the oxygen and hydrogen prior to the reaction is known to be 4750 and that of the water after the reaction $-284\,000$ (both in units of kJ kmol^{-1} of the water produced). Estimate (a) the best possible work output and hence the change in the Gibbs function values for the reaction (per kmol of water), (b) the change in energy (enthalpy) from inlet to outlet, (c) the theoretical best possible efficiency and hence the efficiency of the current cell, (d) under the best possible operation, the heat transfer from the cell (hint: relate this to the constant temperature entropy change), (e) the entropy change at the cell for condition (d), (f) for condition (d), the entropy change of the atmosphere from the heat transferred to it. Hence determine the total entropy change of the system and surroundings and comment on whether the process is reversible or irreversible.

3.3 A piston-cylinder arrangement contains 0.001 kg of air initially at 300 °C, 1.6 MPa. During a reversible, isothermal process, 2.0 kJ of heat is added to the air. The temperature and pressure of the environment is 300 K, 101 kPa. Assume that air has an isentropic index of 1.4 and $R = 0.287\,\text{kJ}\,\text{kg}^{-1}\,\text{K}^{-1}$.

Determine the change of entropy for the process, the work done, the available energy in the heat transferred to the air and the change in the internal and net availability of the air in the closed system during the process.

3.4 Air (an ideal gas, $R = 0.287\,\text{kJ}\,\text{kg}^{-1}\,\text{K}^{-1}$, $c_p = 1.0045\,\text{kJ}\,\text{kg}^{-1}\,\text{K}^{-1}$) flows in the intake manifold of an internal combustion engine at 95 kPa, 40 °C with a velocity of 40 ms^{-1}. It then passes over a throttle valve undergoing an adiabatic pressure drop of 60 kPa. Determine (a) the stream availability before the throttle, (b) the change in the stream availability caused by the throttling process, (c) the stream availability after the throttle.

Assume that kinetic and potential energy changes are negligible across the throttle. The local ambient conditions are 20 °C, 101 kPa.

3.5 A particular air standard cycle analysis of an IC engine follows the dual cycle and assumes that half the heat is transferred to the cycle at constant volume while the remaining half is transferred at constant pressure. The other processes are conventional, isentropic compression and expansion and constant volume heat rejection.

The following data applies:

temperature at the end of compression	722.5 K;
temperature at the end of expansion	1507 K;
temperature at the end of heat rejection	300 K;
ambient temperature	300 K;
total head added	12 kJ;
mass of working substance	0.006 kg;
cycle thermal efficiency	56.7%;
working substance (air)	$R = 0.287, c_p = 1.0045\,\text{kJ}\,\text{kg}^{-1}\text{K}^{-1}$

Determine (a) the total available energy supplied in all the heat added to the cycle, (b) the total available energy removed in the heat rejected from the cycle, (c) the change in the availability of the working substance due to the net work extracted during the cycle, (d) the net change in the availability of the working substance between start and finish of a cycle. Comment on your answer, (e) Compare an efficiency based on total heat added to the cycle and one based on the energy available in the heat addition process.

3.6 Show, starting from the general thermodynamic relations, that for an ideal gas ($pv = RT$) the following relationships hold for any process

$$du = c_v dT$$

$$dh = c_p dT$$

Also develop the equation for the difference in specific heats and the well-known ideal gas equations for entropy change ds from these relationships.

3.7 For an ideal gas, the specific heat is assumed to vary only with temperature according to the relation

$$c_v = A + BT$$

where A and B are constants. Develop, for an isentropic process, equations for dp and dv in terms of p, v, T, R, A, B and dv and hence prove that the temperature volume relationship is given by

$$\ln[Tv^{R/A}] = \text{const} - BT/A$$

3.8 For a gas which obeys the van der Waals equation of state, develop expressions for the variation from the ideal gas du, dh and $c_p - c_v$ relationships of Problem 3.6 in terms of R, p, v, a and b.
When

$$c_v = A + BT$$

show that, for an isentropic process

$$\ln[T(v-b)^{R/A}] = \text{const} - BT/A$$

and, if $B=0$ (i.e. c_v is constant) that

$$(p + a/v^2)(v-b)^\gamma = \text{const}$$

3.9   Air in a closed system at 25 °C and 1 atmosphere pressure is heated from a constant temperature reservoir ($T = 1500$ K) to a final temperature of 1150 K. The variation in specific heat is significant for air over this range and is given by

$$\frac{c_p}{R} = 3.653 - 1.334 \times 10^{-3}\,T + 3.291 \times 10^{-6}\,T^2$$
$$- 1.910 \times 10^{-9}\,T^3 + 0.275 \times 10^{-12}\,T^4$$

where T is the absolute temperature.

   Evaluate the ideal gas change in enthalpy term $h_2 - h_1$ and compare it with that predicted by an ideal gas with the constant value of $c_p = 1.0045$ kJ kg^{-1} K^{-1} which applies at the low end on the range. From the above, calculate the constant value of c_p over the temperature interval which would give the correct answer and compare it with c_p determined from the mean of that at the two temperature points. Using the full formula for c_p above, determine the change in entropy for the process and compare it with those calculated for the constant c_p values.

3.10 (a) By considering the critical isotherm passing through the critical point, show that the constants a and b in the Berthelot equation of state

$$p = \frac{RT}{v-b} - \frac{a}{Tv^2}$$

are given by

$$a = \frac{9}{8}RT_c^2 v_c \quad \text{and} \quad b = \frac{1}{3}v_c$$

3.11 A gas has an equation of state given by

$$\frac{pv}{RT} = 1 + \frac{b}{vT}$$

where $R = 188$ J kg^{-1} K^{-1} and $b = 0.51$ m^3 K kg^{-1} (a) Develop an equation for the change in internal energy between two state points 1 and 2 respectively. (b) Determine the internal energy and enthalpy change between state 1:440 K and 4.5×10^{-3}m^{3}kg^{-1} and state 2: 340 K and 6.0×10^{-3}kg^{-1}. (c) For an adiabatic, steady flow process, determine the work done in magnitude and direction. Ignore kinetic and potential energy changes.

$$c_p = 0.24 + 11.5 T^{-0.5} \text{ kJ kg}^{-1} \text{ K}^{-1},$$

$$c_v = 0.19 + 8.91 \, T^{-0.5} \text{ kJ kg}^{-1} \text{ K}^{-1}.$$

Hint: use $h = u + pv$ for enthalpy calculations.

3.12 The behaviour of a particular gas is described by the van der Waals equation of state. During a particular process which is described by the process equation

$$p(v - b)^{1.2} = \text{const.}$$

the gas, initially at 3 MPa, $0.007 \text{ m}^3 \text{ kg}^{-1}$ is compressed in a closed system to one half its initial volume. (a) Determine the pressure and temperature at the end of the process, (b) Calculate the work supplied for the compression per kg of gas, (c) Calculate the heat transfer per kg in magnitude and direction.

Molecular mass: $M = 58$, $c_v = -4.4 + 0.37 T$ kJ mol^{-1} K^{-1}, van der Waals constants: $R = 8.3143$ J kmol^{-1} K^{-1}, $a = 0.74$ MPa m^{6}kmol^{-2}, $b = 0.085$ m^3 kmol^{-1}.

3.13 Show that the vibrational contribution to internal energy and specific heat of an ideal gas is given by the following relationships. Start your derivation from the vibrational partition function, Z_{VIB}

$$\frac{u}{R} = \frac{xT}{(e^x - 1)} \qquad \frac{c_v}{R} = \frac{x^2 e^x}{(e^x - 1)^2}$$

where $x = hv/kT$.

3.14 For oxygen, nitrogen and hydrogen gases, determine the internal energy, enthalpy, specific heats (constant volume and pressure) and isentropic index at temperatures of 3000 K and 6000 K respectively. Assume that the molecules remain stable (i.e. do not break up into atoms) at these temperatures and that contributions occur from translation, rotation and vibration of the molecule. Comment on the contributions of each mode in relation to its maximum.

3.15 (a) A mixture of gases consists of 60% oxygen and 40% helium by mass. Determine the isentropic index for the mixture given that for oxygen, c_p and c_v are 0.918, 0.658 kJ kg^{-1} K^{-1}, for helium, c_p, c_v are 5.19, 3.12 kJ kg^{-1} K at the conditions considered. (b) Determine the molar ratio (of each to the total mixture) in which helium and carbon dioxide should be mixed in order to give a gas with the same molecular mass as diatomic oxygen.

Molecular masses are: helium, 4.003; carbon dioxide, 44.01; oxygen, 32.00.

3.16 A mixture of 60% carbon dioxide and 40% nitrogen by volume is at a pressure of 20 MPa, temperature 325 K. Determine, using Kay's rule the critical constants that apply to the mixture. Assuming that the gas state can be described by the van der Waals equation, calculate its temperature if its specific volume is increased at 20% by constant pressure. Compare the result with that for carbon dioxide or nitrogen alone.

Molecular mass, critical temperature, pressure and volume are:

for nitrogen, 28.013, 126.2 K, 3.39 MPa, 0.0899 m^3 kmol^{-1};
for carbon dioxide, 44.01, 304.2 K, 7.39 MPa, 0.0943 m^3 kmol^{-1}.

3.17 A mixture of air and methane gas (CH_4) at 100 kPa has an air/fuel mass ratio of 15:1. Air may be considered to consist of 3.76 mol of nitrogen to 1 mol of oxygen. Determine, assuming ideal gas behaviour, the partial pressure of each component. Molecular masses for methane, oxygen and nitrogen are, respectively, 16.04, 32.00 and 28.01.

4

Compressible flows for engines

4.1 THE WORKING SUBSTANCE

In any engine, the working substance, almost always a fluid of some type, is the essential substance that allows the interchange of energy to take place from the supply source to the work output and heat rejection processes. The useful output is obtained by developing pressure in this fluid which then can be used to apply a force that moves appropriate components of the engine thereby providing the work. Two things are important. Firstly, the quantity of energy that can be transferred per unit volume of the working substance must be high, otherwise the dimensions of the engine will become excessively large, bringing problems associated with cost, versatility and high losses. These losses increase because large surface areas increase both the heat transfer and friction. The second important factor is that, because engines operate on a cyclical basis, the rate at which the working substance can be introduced to and removed from the work-producing components is critical in obtaining a high power output. A working subtance which is easily moved is therefore essential and it is for this reason that fluids are preferable.

In an internal combustion engine, the working substance also exists simultaneously as both the reactants and products of the combustion process. The essential feature is that it must then be basically air (although the use of pure oxygen is possible as in the case of rocket engines) and must be able to be mixed completely and very rapidly with the fuel either just prior to or within the combustion chamber. This is called the **fuel preparation** (i.e. the preparation for combustion) process. The preparation needs to be complete in the sense that all the fuel must be able to come into contact with oxygen for the combustion and, in order to achieve this, it should provide either a uniform, homogeneous or a non-uniform but correctly stratified charge. For a given fuel/air ratio, the energy which can be handled by the engine depends on the through-put of the mixture. Therefore, either this mixture or the air alone, depending on the engine type, must be introduced into the engine so that the highest possible densities are achieved at the start of the thermodynamic cycle. For good cycle thermodynamics, pressure developed during compression and combustion must be retained for the expansion (i.e. work-producing) process which should be as extensive as possible given the engine's physical constraints, after which the gases need to be expelled from the system as completely as possible so that they do not interfere with the following incoming charge. Maximizing the amount of gas exchange in an engine is therefore critical for its good operation. The more rapid the exchange process, the faster the engine can operate with good efficiency and the higher is the power output. Thus, the gas exchange and fuel preparation processes need to be studied in some detail. This is not usually the subject of the engine analysis chapters of basic thermodynamic texts as they are concerned fundamentally with the thermodynamic cycle and therefore assume some appropriate state conditions as the starting point for it. It must be remembered, however, that the pressure and temperature conditions for the state point at the commencement of the thermodynamic cycle is dependent on the gas exchange process. In turn, the complete cycle conditions which follow and the cycle output then depend on this point.

As with most aspects of engines, a complete analysis requires a text in its own right and an appropriate one for gas turbines is Hill and Peterson (1992) while for reciprocating engines, Annand and Roe (1974) is recommended. Their total scope cannot be encompassed here. However, the aim of this chapter is to give the reader the basic background knowledge necessary to comprehend the problems which are unique to internal combustion engine gas flows both steady and unsteady and to the fuel preparation processes.

4.2 BASIC DETAILS OF THE FLOW

4.2.1 Types of flows in engines

It is assumed that the reader of this text is familiar with the fundamentals of incompressible one-dimensional flow. Real flows are always much more complex with completely accurate models of multi dimensional unsteady, compressible, turbulent, multiphase, stratified flows being generally intractable although some modern CFD (computational fluid dynamic) codes incorporate many of these aspects. Fortunately, for most engine analysis, this degree of complexity is not required although, to construct reasonable mathematical models, some advance over the most simple type of steady, compressible, single-phase flow is necessary. Appropriate flows will need to be considered for several different regimes, these being basically for gas turbines, and for the inlet and exhaust manifold flow and for the in-cylinder flows in reciprocating engines.

In engines of all types, the vast majority of fluids are mixtures of gases which are compressible and hence density changes must be considered. In addition, they may contain liquid fuel in droplet form. Let us consider what type of analysis we need to apply for either a highly accurate or for what

Table 4.1 Typical engine flows for analysis

Engine and Application	Dimension	Compres-sibility	Time Dependence	Viscous Effects	Phases
Gas Turbine					
Inlet, Diffuser, Compressor	1, 2 or 3D	C	S	I or T	1P
Combustion Chamber	3D	C	S	T	1P or SM
Turbine, Nozzle	1, 2 or 3D	C	S	I or T	1P
Reciprocating Engines [SI and CI]					
Inlet, Manifold and Ports	1, 2 or 3D	C	QS or U	I or T	1P or PS*
Exhaust Ports and Manifold	1, 2 or 3D	C	U	I or T	1P
In-Cylinder	3D	C	U	T	1P or PS or SM**
Turbo-chargers	1D	C	QS or U	I or T	1P
Combustion	3D	C	U	L or T	SM**

Code: 1D, 2D, 3D = one, two or three dimensions; C = compressible; S, QS, U = steady, quasi-steady, unsteady; I, L, T = inviscid, viscous laminar, turbulent; 1P, 2P = single, two phase; PS, SM = two phase separated, single phase with spray mixing; * = throttle body and port injected petrol engines; ** = Diesel engines and DISC petrol engines.

could be termed a minimal solution. In gas turbines, each component is essentially a separate piece of machinery requiring its individual model for the flow through it. The components are the inlet diffuser, the compressor, the combustion chamber, the turbine and the nozzle sections. In each component, the flow is basically steady and single phase except for the very small region where the fuel as droplets is sprayed into the combustion chamber. Here, it is multidimensional and turbulent. The exhaust gases through the turbine and nozzle are, of course, of a different composition to those in the inlet diffuser and compressor due to the intervening combustion process. Also, in the turbomachinery associated with the gas turbine, modern analysis requires multidimensional viscous flow assumptions. While one-dimensional analysis may suffice in other regions, the availability of CFD codes to cope with a multidimensional turbulent flow makes these a reasonably efficient option.

Even though the basic compression, combustion and expansion processes takes place within the cylinder of a reciprocating engine, we also require the flow to be considered in the separate sections of the inlet tracts, the in-cylinder region and the exhaust ducts. In reciprocating engine inlet manifolds, the flow is basically unsteady, being a pulsatile flow driven by the regular motion of the pistons and a one-dimensional unsteady analysis is often used although for steady state engine operation it may be approximated by an even simpler quasi-steady flow to provide a quick solution.

In some manifold regimes, for example, the flow around throttle (butterfly) and inlet valves is multidimensional and turbulent with noticeable wake regions being formed. Modern CFD codes are generally then required. Under load change (i.e. acceleration or deceleration) conditions, a transient flow superimposed over the assumed pulsatile or steady flow base must be considered. The fuel introduction and transport in the inlet manifold and/or ports of spark ignition engines is complex, consisting of air, fuel vapour and droplets which are airborne and the accumulation of fuel on surfaces may occur forming moving fuel films. Interchange occurs between these fuel forms and the two-phase separated, multicomponent flows are extremely complicated. The in-cylinder flows in all reciprocating engines are multidimensional, turbulent and unsteady and may still be two-phase, particularly in diesels and in the developing direct in-cylinder stratified charge (DISC) petrol engine types. Exhaust flows are single phase but, because of the high pressures at exhaust valve opening, they have a more pronounced pulsatile nature than intake flows. Thus, a one-dimensional, pulsatile analysis is required.

The list on Table 4.1 classifies flows in various parts of engines in relation to that normally required to produce useful results. Where two classifications are given, the first refers to the simplest approach commonly used while the second refers to that required for a sophisticated analysis.

It can be seen that there are many possible combinations with the one common feature being compressible flow. It is not the intention here to treat all possible flow types. However, a background in compressible flow is necessary for all engine work and this will be dealt with briefly in this chapter.

4.2.2 The basic flow equations

All fluid mechanics start with the equations of continuity, momentum and energy. These, in three-dimensional time-dependent form, can be found in many fluid mechanics texts, for example Streeter (1971) where their development is given in detail. They are as follows.

Continuity

The equation of continuity is

$$\frac{\partial}{\partial t} \int_{cv} \rho dV + \int_{cs} \rho(\boldsymbol{v} \cdot d\boldsymbol{A}) = 0 \qquad (4.1)$$

The first term represents the accumulation of mass with time within the control volume and is the difference between the mass inflow and the outflow rates through the boundary (the control surface) given by the second term. In the steady flow case, the second term only exists and it states that the net mass flow rate through the boundary must be zero.

Momentum

Conservation of momentum gives

$$\frac{\partial}{\partial t} \int_{cv} \rho \boldsymbol{v} \, dV + \int_{cs} \rho \boldsymbol{v} (\boldsymbol{v} \cdot d\boldsymbol{A}) = \sum F \tag{4.2}$$

The first term is the time rate of increase of momentum within the control volume while the second is the effect of the momentum flux across the boundary. Together, they give the net time rate of change of momentum associated with the control volume. The term on the right-hand side is the total force acting on the fluid in the control volume and includes pressure terms acting from either solid surfaces or fluid at the boundary, body (gravitational) terms and, in the case of viscous flow, the shear effects of the fluid at the boundary which can be called **dissipation terms**. When these last are included, the momentum equations are called the Navier–Stokes equations.

Energy

Conservation of energy gives

$$\frac{\partial}{\partial t} \int_{cv} \rho (u + e_{ext}) \, dV + \int_{cs} (h + e_{ext}) \rho (\boldsymbol{v} \cdot d\boldsymbol{A}) = \frac{\delta Q}{\delta t} - \frac{\delta W}{\delta t} \tag{4.3}$$

Here e_{ext} is used to represent the sum of the kinetic and potential energies.

$$e_{ext} = \frac{v^2}{2} + gz$$

Again, the first term is the accumulation of energy in the control volume with time while the second term is the rate at which energy flows in the fluid through the boundary. The terms on the right-hand side are the time rate of energy transfer through the boundary independent of the fluid flow.

These need to be used with appropriate state and process equations to obtain a solution. It can be seen that, as a system, they are likely to be hard to solve analytically for any other than the most simple situations without considerable simplification although numerical solutions are now frequently used. They apply basically to viscous laminar flows which have only limited application and require the incorporation of turbulence models to be able to deal with most real flows which are turbulent in nature. A further discussion of the nature of turbulence and the modelling of turbulent flows will be undertaken later in this chapter although it needs to be noted at this stage that it is an area still undergoing considerable development. Thus, it cannot be dealt with here in any but a cursory manner. A considerably less complex flow than that described above is usually the subject of a first course in fluid mechanics and the reader will be familiar with the simplest form to which these equations can be reduced. This is, for one-dimensional steady incompressible flow without dissipation terms and the equations are:

- Conservation of mass (continuity equation)

$$A\rho v = \text{const} \qquad (4.4)$$

- Conservation of momentum (Euler's equation)

$$dp/\rho + v dv + g dz = \text{const} \qquad (4.5)$$

which can be integrated along a streamline to give a form of the conservation of energy equation which has no energy transfers. This is the well-known Bernouilli's equation.

$$p/\rho + v^2/2 + gz = \text{const} \qquad (4.6)$$

- Conservation of energy

$$\Delta(h + v^2/2 + gz) = \delta q - \delta w \qquad (4.7)$$

While, as pointed out above, the inclusion of dissipation caused by fluid friction is handled by the more complex Navier–Stokes equations, it can be incorporated into (4.6) by introducing a pressure loss of the form $-kv^2/2$ which is added to the left-hand side. Here, the term k is a loss coefficient for general purposes and may have a more specialized form such as that associated with wall friction losses in pipes. This is the well-known friction factor, f, k then being given by fL/d where L/d is a characteristic ratio, the length to diameter of the pipe. In (4.7), dissipation is already included in the enthalpy, h, and the heat transfers, q. Equation (4.7) will be recognized as the thermodynamic steady flow energy equation.

4.3 COMPRESSIBLE FLOWS

4.3.1 Disturbances in a compressible substance

The basic difference between a compressible and incompressible substance is that a wave motion can exist in the former which carries information at a finite rate to other parts of the flow field. Depending on the local conditions, the whole field may not always be accessible to this wave. In an incompressible substance, the wave velocity is infinite and adjustment takes place instantaneously. Any disturbance in a compressible substance results in the waves moving away from the disturbance, carrying information about it to more distant particles. For a stationary line or point disturbance in a free field, these, as shown on Figure 4.1, are a set of cylindrical or spherical waves respectively. A typical example is the wave pattern caused by dropping a stone into a pool of water where the free surface waves are analogous to these pressure waves in the compressible fluid. For a very small disturbance, they are waves of extremely small amplitude and hence are sound or acoustic waves. The velocity at which any point on them is travelling is thus the local acoustic velocity, designated here as a.

Now consider the one-dimensional problem of a disturbance equally distributed over the cross-section of a tube such as, for example that which would be created by a small movement of the piston shown in Figure 4.2 which has a velocity change dv_p. Here again, an acoustic wave is generated and, in the absence of viscous effects, will move down the tube as depicted. An alternative example, also illustrated, is the wave created by a small pressure difference dp which might be initiated by the bursting of a diaphragm. In either case, the wave is acoustic only in the limit. For larger velocities and pressure differences, the wave is more complex and will move much faster if it is a compression wave or will be lengthened if it is a rarefaction. This will be dealt with at a later stage.

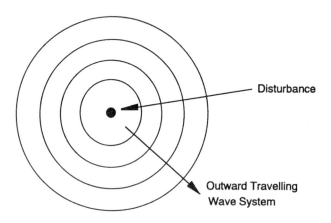

Figure 4.1 Wave system from a disturbance.

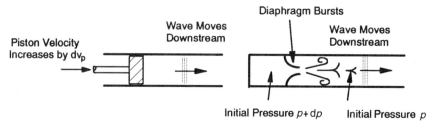

Figure 4.2 Acoustic waves created in a one-dimensional system.

The important point at present is to relate the value of the acoustic velocity, a, to the properties of the substance. Consider the situation as shown in Figure 4.3 where a control volume is placed around the wave and is considered to move with it, with the acoustic wave velocity a. Ahead of the wave, the fluid has properties given by a, p, ρ etc. (with $v = 0$) while after it these have changed by a small amount to $a + da, p + dp, dv, \rho + d\rho$. If we now transform the system to a stationary one by superimposing a velocity equal to a but in the opposite direction to the wave, a simple analysis follows.

Applying mass continuity to the control volume gives

$$A\rho a = A(\rho + \delta\rho)(a - \delta v) \qquad (4.8)$$

from which, neglecting second order terms, gives

$$\rho\delta v = a\delta\rho \qquad (4.9)$$

The momentum principle gives

$$pA - (p + \delta p)A = \dot{m}(a - \delta v) - \dot{m}a = A\rho a[(a - \delta v) - a] \qquad (4.10)$$

Hence

$$\delta v = \frac{\delta p}{\rho a} \qquad (4.11)$$

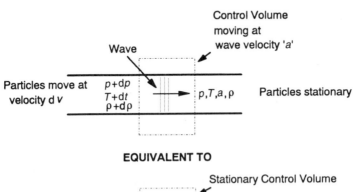

EQUIVALENT TO

Figure 4.3 Transformation of flow for an acoustic wave.

Eliminating δv from (4.9) and (4.11) gives

$$a^2 = \frac{\delta p}{\delta \rho} \qquad (4.12)$$

It should be noted that the wave is very small in amplitude. It is isentropic because no heat transfer to or from the control volume has been assumed and the constant wave velocity infers that frictional dissipation is absent. Thus, the equation is

$$a^2 = \left(\frac{\partial p}{\partial \rho}\right)_s \qquad (4.13)$$

This is a most important equation because it indicates the fastest way a signal can travel through any medium. All substances are in fact compressible to some extent, the difference being that those classified as incompressible have a very small density change for a large applied pressure and so their acoustic velocity approaches infinity.

Because the wave is isentropic

$$\frac{p}{\rho^\gamma} = \text{const} \qquad (4.14)$$

By differentiating and substituting into (4.13)

$$a = \sqrt{\frac{\gamma p}{\rho}} \quad \text{in the general case} \qquad (4.15)$$

$$= \sqrt{\gamma R T} \quad \text{for an ideal gas} \qquad (4.16)$$

For gases, the relatively slow sound speed (e.g. $347\,\mathrm{m\,s^{-1}}$ in air at $300\,\mathrm{K}$) allows major compressibility effects to be felt in the flow at low velocities. This can be categorized in terms of a dimensionless group, the Mach number, M, which is defined as

$$M = \frac{v}{a} \qquad (4.17)$$

where v is the velocity of the fluid or body under consideration. Consider a point disturbance moving with a velocity v as shown on Figure 4.4. Here the sound waves emanating from the disturbance are closer together ahead of the body than behind it. As long as $v < a$, the disturbance can be detected at any point some time before it arrives, at $v = a$ the disturbance and the initial sound wave arrive simultaneously while at $v > a$, the disturbance arrives first forming a zone of silence and a zone of action behind in the wedge or cone bounded by Mach lines. It is worth noting that the angle that these lines form with the direction of motion is called the **Mach angle** and has a value

$$\alpha = \sin^{-1}(1/M) \qquad (4.18)$$

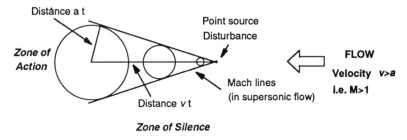

Figure 4.4 Effect of a disturbance in supersonic flow.

4.3.2 Equations for the solution of compressible flows

For incompressible frictionless flows, the choice of equations to obtain an adequate solution usually does not present any difficulties. Typically, the known upstream conditions might be the velocity, density and one other property, usually the pressure, while the requirement is to find the downstream conditions. Because of the incompressibility, one property, the density, is known downstream also and only two equations are necessary to find the remaining property and the velocity. These are the continuity and the Bernouilli equations, the latter being developed from either momentum or energy considerations which are identical as long as there are no heat and work transfers to be included. The disadvantage of this easy choice is that the student does not become attuned to examining the physics of the problem in order to choose the correct equation, momentum or energy, to properly model the system. Compressible flows are different in that two downstream properties and the velocity must be evaluated. Here, for one-dimensional flow without work transfer, the choice of three equations must be made rationally from the following basic equations.

1. Mass continuity;
2. Momentum

   (a) without external friction,
   (b) with external friction;

3. Energy

   (a) without heat transfer (adiabatic)
   (b) with heat transfer;

4. State path laws, particularly the isentropic relationship;
5. The state equation.

The flows to be examined here are isentropic flows, flows in which shock waves occur, flows with friction and flows with heat transfer. These can be combined where necessary into more complex flows. The equations to be chosen for each case are listed in Table 4.2 without, at this stage, an explanation. This will follow as the development proceeds.

The state equation(s) is a further relationship used to find an additional property when necessary either upstream or downstream.

Combination of equations (1) and 2(a) give what is known as a Rayleigh line while (1) and 3(a) gives a Fanno line both being for constant area, constant mass flow rate cases. These represent lines rather than points on a solution diagram because a full solution is not possible until an additional equation is specified and hence one property must vary for selected values of another. It can be seen that Rayleigh lines apply to isentropic flow, shock waves, flow with heat transfer, Fanno lines apply to isentropic flow, shock waves, flow with friction. The intersection of Rayleigh and Fanno lines provides a solution to the trivial constant area isentropic flow case and, of much more significance, to that of normal shock waves.

4.3.3 Steady compressible one-dimensional flows

Isentropic flows

One-dimensional isentropic flows are, of course, idealized as friction always exists to some extent in practice. However, an isentropic solution gives reasonably accurate values in many cases and allows a good understanding of the basic flow with a simple analysis of the effect of area changes. Using it in gas turbines, an initial assessment can be carried out of the flow in any of the connecting ductwork in the engine and in the intake diffuser and the exhaust nozzle in particular. In reciprocating engines, a preliminary understanding of the flow in the intake and exhaust manifolds is also possible. It must be remembered however that, while the one-dimensional assumption may be a reasonable approximation of reality in a wide range of cases, that of isentropic flow may be less accurate because substantial heat transfer may sometimes occur. This applies particularly to exhaust flows.

Table 4.2

Isentropic flow	Shock waves	Flow with friction	Flow with heat transfer
i	i	i	i
ii(a) or iii(a)	ii(a)	ii(b)	ii(a)
iv	iii(a)	iii(a)	iii(b)

Consider now the variation of area in a one-dimensional, isentropic flow. The differential forms of the continuity and Euler equations are

Continuity

$$\frac{dA}{A} + \frac{d\rho}{\rho} + \frac{dv}{v} = 0 \tag{4.19}$$

Euler

$$\frac{dp}{\rho} + v\,dv = 0 \tag{4.20}$$

At constant entropy, the sound speed, as given by (4.13) can be substituted into (4.20) giving

$$\frac{a^2\,d\rho}{\rho} = -v\,dv \tag{4.21}$$

Further substitution into (4.19) now gives

$$\frac{dA}{A} - \frac{v\,dv}{a^2} + \frac{dv}{v} = 0 \tag{4.22}$$

from which

$$\frac{dA}{dv} = \frac{A}{v}\left(\frac{v^2}{a^2} - 1\right) = \frac{A}{v}(M^2 - 1) \tag{4.23}$$

As A must be positive, different cases emerge for $M < 1$, $M = 1$, $M > 1$ as shown on Figure 4.5. The cases are:

• for subsonic flow, $M < 1$, $dA/dv < 0$ and so as A increases, v decreases (this is a subsonic diffuser) and as A decreases, v increases, this is a subsonic nozzle;
• for sonic flow, $M = 1$, $dA/dv = 0$ and so A is constant as v either increases or decreases;
• for supersonic flow, $M > 1$, $dA/dv > 0$ and so as A increases, v increases (this is a supersonic nozzle) and as A decreases, v decreases (this is a supersonic diffuser).

To create a nozzle which changes a subsonic flow to a supersonic one, the shape must first be convergent and of sufficient area ratio in order to bring the flow Mach number to one. This position is then the throat as specified in the case of sonic flow above and the subsequent flow will then continue to increase in velocity if the nozzle diverges. Conversely, an initially supersonic flow can only be reduced to subsonic (in the absence of a shockwave, to be discussed later) by using first a convergent supersonic diffuser followed after the throat by a divergent subsonic one. That is, both nozzle and diffuser for converting flows from one regime to the other are of a convergent–divergent pattern. A nozzle at which the throat velocity is such that the local Mach number is one at that point is said to be choked because it cannot accommodate a further increase in mass flow rate by reducing the exit pressure.

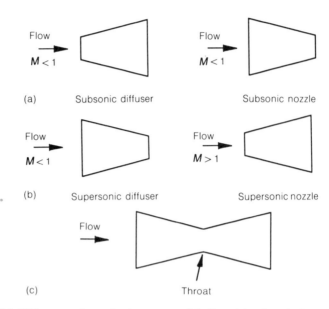

Figure 4.5 Diffusers and nozzles in compressible flow. (a) subsonic flow, $dA/dV < 0$, (b) supersonic flow, $dA/dV > 0$, (c) to change subsonic to supersonic or supersonic to subsonic flow, a convergent–divergent nozzle is required.

It is of interest to try to understand at a very simple physical level the reason for the above geometries. In a subsonic flow, signals or disturbances can travel upstream at the local sound speed. Hence, the changed velocity and pressure downstream modify the upstream flow progressively. That is, the velocity will always increase towards a reduced area or decrease towards an enlarged one. Once the flow has reached sonic velocity at any one point in the flow, upstream movement of these waves is no longer possible. Small elements of the gas are controlled only by the pressure immediately in front of them. An increasing area can then only result in an increased velocity, a diminishing area in a reducing one. Adjustment to the overall pressure field must take place separately.

Consider two reservoirs joined by a convergent–divergent passage as shown on Figure 4.6. If the nozzle exit (or back) pressure is above the UCBP (upper correct back pressure) value, the throat Mach number is below one and the nozzle acts as a venturi with decreasing velocity in the divergent section. At the UCBP, the flow just reaches $M = 1$ at the throat and then decelerates. If the exit pressure is below the LCBP (lower correct back pressure) value, the flow accelerates from $M = 1$ at the throat and is supersonic throughout the divergent section. Except when the exit pressure is exactly at the LCBP value, further expansion must occur outside the nozzle causing the flow to diverge further in a complex pattern. A nozzle performing in this way is said to be **underexpanded** because additional

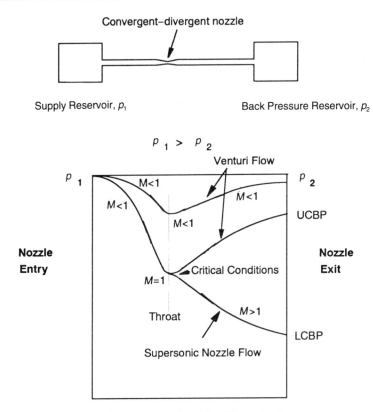

Figure 4.6 Convergent–divergent nozzle with different back pressures.

diverging length would be required to complete expansion within the nozzle itself. If the exit pressure falls between the UCBP and the LCBP values, the nozzle is said to be **overexpanded**. This is because the velocity has reached too high a value within the diverging section and the pressure has consequently fallen below the exit pressure. A shock wave is required somewhere in this section of the nozzle to change the flow from supersonic to subsonic so that the pressure rises in the subsequent flow to the correct exit pressure. This is illustrated on Figure 4.7. The shock wave in practice may not be a simple normal shock in all cases but generally is assumed to be so in simple calculations. When the exit pressure is just above the LCBP, the shock pattern emerges as a series of interacting oblique shocks from the exit. Below the LCBP, an expansion fan forms at the nozzle exit, essentially increasing the expansion area ratio. Fitting a normal shock into a nozzle system will be considered later.

Now consider an isentropic flow throughout the convergent-divergent nozzle. From Euler's equation (4.5) and the isentropic relation (4.14), the

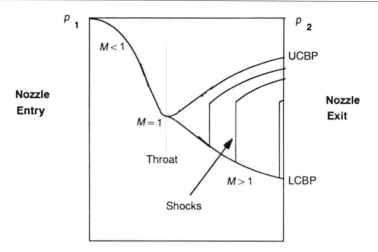

Figure 4.7 Convergent–divergent nozzle with back pressures between the upper correct back pressure (UCBP) and lower correct back pressure (LCBP).

following can be formed.

$$\frac{\mathrm{d}p}{\mathrm{const}\, p^{1/\gamma}} + v\mathrm{d}v = 0 \tag{4.24}$$

Integrating and resubstituting for the constant from (4.14) gives

$$\frac{\gamma}{(\gamma-1)}\frac{p}{\rho} + \frac{v^2}{2} = \mathrm{constant} \tag{4.25}$$

This is the basic relationship for compressible, isentropic flow.

Now consider the steady flow energy equation for the same flow with gravitational terms neglected, i.e.

$$q + h_1 + \frac{v_1^2}{2} = w + h_2 + \frac{v_2^2}{2} \tag{4.26}$$

Assuming no external work transfer and an adiabatic flow, both q and w are equal to zero. Substituting $h = c_p T$ for an ideal gas gives

$$c_p T_1 + \frac{v_1^2}{2} = c_p T_2 + \frac{v_2^2}{2} \tag{4.27}$$

or

$$c_p T + \frac{v^2}{2} = \mathrm{constant} \tag{4.28}$$

But

$$c_P = \frac{\gamma R}{(\gamma - 1)} \tag{4.29}$$

Hence

$$\frac{\gamma RT}{(\gamma - 1)} + \frac{v^2}{2} = \text{constant} \tag{4.30}$$

or

$$\frac{\gamma p}{(\gamma - 1)\rho} + \frac{v^2}{2} = \text{constant} \tag{4.31}$$

which is the same equation as obtained from Euler's momentum considerations. It should be noted that (4.30) and (4.31) were both derived assuming that the flow was adiabatic but the first derivation required the additional assumption of reversibility (i.e. isentropic flow). The anomaly can be resolved by realizing that first term in (4.30) and (4.31) is essentially a function of temperature only. While non-isentropic flow under identical conditions will result in a greater fall in pressure than isentropic flow, its energy, being adiabatic, is still retained in the flow. That is, it is conserved by conversion of some of the energy associated with the pressure to a higher internal energy at the increased temperature which it exhibits and to an increased kinetic energy due to its lower density.

Now substituting from (4.15) for the acoustic velocity into (4.25) gives

$$\frac{a^2}{(\gamma - 1)} + \frac{v^2}{2} = \text{constant} \tag{4.32}$$

which is a convenient form for many calculations.

For example, taking stagnation conditions (sometimes called **reservoir** conditions because they apply to an isentropic flow from a container where the gas is originally stationary) as designated by a subscript 0, the energy equation becomes

$$\frac{a^2}{\gamma - 1} + \frac{v^2}{2} = \frac{a_0^2}{(\gamma - 1)} \tag{4.33}$$

This can be rearranged to give

$$\frac{a_0^2}{a^2} = 1 + \frac{(\gamma - 1)}{2} M^2 \tag{4.34}$$

For an ideal gas, a^2 is proportional to T. Hence

$$\frac{T_0}{T} = 1 + \frac{(\gamma - 1)}{2} M^2 \tag{4.35}$$

This equation applies to any adiabatic flow, reversible or irreversible. If however, the flow is isentropic, the remaining properties can be also determined. That is

$$\frac{p_0}{p} = \left(1 + \frac{\gamma - 1}{2} M^2\right)^{\gamma/(\gamma-1)} \tag{4.36}$$

and

$$\frac{\rho_0}{\rho} = \left(1 + \frac{\gamma - 1}{2} M^2\right)^{1/(\gamma-1)} \tag{4.37}$$

If flow in a duct occurs from zero velocity in a reservoir to supersonic conditions, it must pass through the critical or sonic condition where the Mach number, $M = 1$. The critical conditions are often designated as T^*, p^*, ρ^*, a^*, etc. Substituting $M = 1$ in the above equations, gives the following

		for $\gamma = 1.4$	for $\gamma = 1.667$
$\dfrac{T^*}{T_0} = \dfrac{2}{\gamma + 1}$		$= 0.833$	$= 0.750$
$\dfrac{p^*}{p_0} = \left(\dfrac{2}{\gamma + 1}\right)^{\gamma/(\gamma-1)}$		$= 0.528$	$= 0.487$
$\dfrac{\rho^*}{\rho_0} = \left(\dfrac{2}{\gamma + 1}\right)^{1/(\gamma-1)}$		$= 0.634$	$= 0.650$
$\dfrac{v^*}{a_0} = \dfrac{a^*}{a_0} = \left(\dfrac{T^*}{T_0}\right)^{1/2}$		$= 0.913$	$= 0.866$

Now applying the continuity equation for any point in the duct and the critical point, a relationship between these areas can be obtained in terms of the Mach number. That is

$$\frac{A}{A^*} = \frac{\rho^* v^*}{\rho v} = \frac{1}{M}\left\{\left[\frac{2}{\gamma + 1}\right]\left[1 + \frac{(\gamma - 1)}{2} M^2\right]\right\}^{(\gamma+1)/(2(\gamma-1))} \tag{4.38}$$

This is plotted on Figure 4.8. It can be seen that for any value of M, the value of A/A^* is always greater than one, demonstrating that the shape of a duct connecting a subsonic to a supersonic flow or vice versa must be convergent–divergent as indicated earlier.

Stationary shock waves in one-dimensional flow

Stationary shock waves may exist in the flow in gas turbines. These are of importance in both the intake system and in the thrust nozzles of vehicles designed for supersonic and hypersonic flight. In the intake diffuser, the

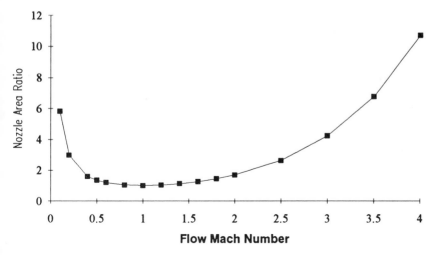

Figure 4.8 Area ratio A/A^* of a convergent–divergent nozzle as a function of the flow Mach number M. A^* is the throat area.

flow must be slowed from the supersonic velocities entering the engine to subsonic ones suitable for combustion. Standing shock waves in the diffuser section often occur and, while these maintain the stagnation (or total) temperature as can be seen from equation (4.28), they result in a loss in stagnation pressure. Intake systems are designed to minimize this loss either by the construction of a 'shock free' diffuser or by use of one where a series of weak shock waves replace a single strong one. Also, shock waves may form in the thrust nozzles of gas turbines due to incorrect expansion for the prevailing pressure ratio across the nozzle. This will be discussed later in this section. Finally, while a lesser problem, the rapid opening of exhaust valves in reciprocating engines causes a step pressure wave to form and this has been recently shown to often be a weak shock wave. It is the cause of the exhaust noise associated with these engines. Thus, a good understanding of shock waves is required for an analysis of engine flows.

A compression wave moving in a one-dimensional flow in the absence of viscous interaction will rapidly become steeper and eventually will 'break' forming a shock wave. This essentially represents a sudden discontinuity between the compressible, often isentropic flows, on either side with a step increase in both pressure and temperature occurring across it. A shock wave is extremely thin, typically being about one mean free path or 2.5×10^{-4} mm. Because a very rapid rearrangement of the particles in the flow occurs in such a small distance, there is extensive shearing with consequent viscous dissipation and the shock process itself is not isentropic although, due to its small surface area, the heat transfer from it is negligible and it may be closely approximated as adiabatic. Moving waves can be rendered stationary by the imposition of a flow velocity equal in magnitude but opposite in direction to their own.

In steady supersonic flows, stationary shock waves occur facing upstream and, in the normal shock wave case, convert the locally supersonic flow to a locally subsonic flow at higher pressure. The term, locally, is important as these stationary waves are based on the acoustic velocities on either side of the shock which are different due to the rise in temperature across it. Without the convergent–divergent nozzle mentioned in the previous section, this is the only way a supersonic flow can slow. Thus, it is a frequent occurrence. These stationary normal shock waves may be considered as a special case of moving shock waves. That is, if a supersonic flow reaches a condition that does not allow it to be perpetuated, the particles will bank up and the trailing edge of this bank is essentially the wave which now progresses back upstream. This also occurs in subsonic flow but here the acoustic signals generated can progress back upstream and slow the incoming flow progressively. In supersonic flow, this is not possible and the shock wave so formed will move upstream against the flow until it reaches a position where it is moving into a gas flow with equal and opposite velocity. Thus, it comes to rest and becomes a normal stationary shock. The shock movement before this stabilization may be viewed as a starting process in the unsteady period before steady flow is obtained.

The relationships for a shock wave, moving or stationary, are essentially the same and will be derived in the next section on moving waves. It should be noted that oblique shock waves can also form and are important in two or three-dimensional flow. These can be simply analysed by taking the flow component normal to the shock and treating it as instrumental in providing the pressure, temperature and density rise. The velocity component normal to the shock on the downstream side becomes locally subsonic, again according to the normal shock relationships. However, the velocity parallel to the shock is unchanged across it and the total velocity may therefore be supersonic in this downstream region.

In supersonic nozzles, such as those in the thrust nozzles of a supersonic aircraft, the flow can only remain completely isentropic throughout if the pressure at the outlet (termed the **backpressure**) is correct in relation to the inlet pressure and the nozzle geometry. If the backpressure is too high, the flow is said to be 'overexpanded' as explained previously. A normal shock can be fitted, as shown in Figure 4.7, so that a supersonic flow which is increasing in velocity but declining in pressure reaches it. The shock wave not only raises the pressure immediately but produces a subsonic flow which will further increase in pressure towards the outlet as it slows in this diverging portion of the nozzle. The higher the backpressure, the more this shock will move upstream and weaken but the more extensive subsonic flow now downstream of it will provide the higher outlet pressure. The shock cannot move upstream past the throat as a sonic flow results in a shock of zero pressure increase and this corresponds to the nozzle becoming subsonic. If the shock occurs right at the nozzle outlet, the backpressure will be noticeably higher than LCBP. For pressures between these values, a series of oblique shocks occur outside the nozzle exit.

Another important case where shocks must be considered is when pitot tubes are used to measure velocities in steady supersonic flows. Here, the obstruction of the pitot tube causes a shock to form just upstream of it. This shock is normal just in front of the pitot tube opening but becomes oblique at the sides as shown on Figure 4.9. The normal shock formulas must be used to determine the downstream subsonic flow which is then further brought to rest in the usual way between the shock and the entrance to the tube. This detached shock will also be found in front of other blunt bodies in supersonic flows.

Equations for shock waves

A shock wave, being extremely thin, is essentially a constant area event. Therefore, the continuity equation can be expressed simply as

$$\rho_1 v_1 = \rho_2 v_2 \qquad (4.39)$$

The energy equation as given by (4.31) applies and can be rewritten as

$$v_1^2 - v_2^2 = \frac{2\gamma}{\gamma - 1}\left(\frac{p_2}{\rho_2} - \frac{p_1}{\rho_2}\right) \qquad (4.40)$$

Note that the only restriction is that the process is adiabatic. It does not have to be isentropic.

The momentum equation per unit area can be written as

$$p_2 - p_1 = \rho_1 v_1 (v_1 - v_2) = \rho_2 v_2 (v_1 - v_2) \qquad (4.41)$$

where the ρv combination represents the mass flow rate per unit area.

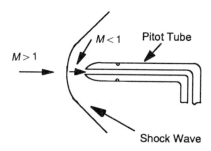

Figure 4.9 Pitot tube in a supersonic flow.

Adding the two versions of (4.41) gives the momentum equation in the following form.

$$\left(\frac{1}{\rho_1}+\frac{1}{\rho_2}\right)(p_2-p_1)=v_1^2-v_2^2 \tag{4.42}$$

Substituting (4.40) into this gives

$$\left(\frac{1}{\rho_1}+\frac{1}{\rho_2}\right)(p_2-p_1)=\frac{2\gamma}{\gamma-1}\left(\frac{p_2}{\rho_2}-\frac{p_1}{\rho_1}\right) \tag{4.43}$$

On multiplying by ρ_2/ρ_1 and rearranging gives

$$\frac{p_2}{p_1}=\frac{\dfrac{\gamma+1}{\gamma-1}\dfrac{\rho_2}{\rho_1}-1}{\dfrac{\gamma+1}{\gamma-1}-\dfrac{\rho_2}{\rho_1}} \tag{4.44}$$

or

$$\frac{\rho_2}{\rho_1}=\frac{\dfrac{\gamma+1}{\gamma-1}\dfrac{p_2}{p_1}+1}{\dfrac{\gamma+1}{\gamma-1}+\dfrac{p_2}{p_1}} \tag{4.45}$$

Finally, substituting into the ideal gas equation for p_2/p_1 and ρ_1/ρ_2 allows the temperature ratio to be determined. This is

$$\frac{T_2}{T_1}=\frac{\dfrac{\gamma+1}{\gamma-1}+\dfrac{p_2}{p_1}}{\dfrac{\gamma+1}{\gamma-1}+\dfrac{p_1}{p_2}} \tag{4.46}$$

These three equations (4.44), (4.45) and (4.46) are called the Rankine–Hugoniot equations. They relate properties on one side of a shock wave in an ideal gas to those on the other side in terms of two other property ratios.

Property ratios across a shock in terms of Mach number
The momentum (4.41) may be reexpressed in terms of the Mach number as follows.

$$\frac{p_2-p_1}{\rho_1}=v_1^2\left(1-\frac{v_2}{v_1}\right)=v_1^2\left(1-\frac{\rho_1}{\rho_2}\right) \tag{4.47}$$

Multiplying the denominator and numerator on the right hand side by a_1^2 and $\gamma p_1/\rho_1$ respectively gives

$$\frac{p_2 - p_1}{\rho_1} = M_1^2 \frac{\gamma p_1}{\rho_1}\left(1 - \frac{\rho_1}{\rho_2}\right) \tag{4.48}$$

from which on rearrangement

$$\frac{\rho_1}{\rho_2} = 1 - \frac{1}{\gamma M_1^2}\left(\frac{p_2}{p_1} - 1\right) \tag{4.49}$$

Substituting ρ_1/ρ_2 in the second Rankine–Hugoniot relationship (4.45), rearranging and solving the quadratic equation for p_2/p_1 so obtained gives

$$\frac{p_2}{p_1} = 1 + \frac{2\gamma}{\gamma + 1}(M_1^2 - 1) \tag{4.50}$$

The other root of the equation is $p_2/p_1 = 1$ which is, of course, a trivial solution.

Equation (4.50) now gives the pressure ratio across the shock as a unique function of the Mach number of the flow approaching it on the upstream side. Alternatively, if the shock is not stationary but moving into a fluid at rest, the same relationship will apply except that the velocity will be that of the shock and the Mach number is then called the shock Mach number.

Similar equations for the temperature and density ratios across the shock can now also be obtained as functions of Mach number only. Taking (4.49) and substituting for p_2/p_1 from (4.50) gives

$$\frac{\rho_2}{\rho_1} = \frac{(\gamma + 1)M_1^2}{(\gamma - 1)M_1^2 + 2} \tag{4.51}$$

Now using the ideal gas equation and substituting for p_2/p_1 and ρ_1/ρ_2 gives

$$\frac{T_2}{T_1} = 1 + \frac{2(\gamma - 1)}{(\gamma + 1)^2}\left[\gamma M_1^2 - \frac{1}{M_1^2} - (\gamma - 1)\right] \tag{4.52}$$

Alternatively, using (4.32) and (4.39) and (4.41) with the ideal gas equation gives

$$\frac{p_1}{p_2} = \frac{1 + \gamma M_2^2}{1 + \gamma M_1^2} \tag{4.53}$$

or

$$\frac{T_1}{T_2} = \frac{1 + \left(\dfrac{\gamma - 1}{2}\right)M_2^2}{1 + \left(\dfrac{\gamma - 1}{2}\right)M_1^2} \tag{4.54}$$

Rearranging and solving gives either $M_2 = M_1$, again a trivial solution or

$$M_2 = \left[\frac{1 + \left(\frac{\gamma - 1}{2} \right) M_1^2}{\gamma M_1^2 - \frac{\gamma - 1}{2}} \right]^{1/2} \tag{4.55}$$

Characteristics of shock waves
A comparison between the isentropic and shock temperature ratios for a range of pressure ratios is of interest. The graphs of Figure 4.10 are obtained using the Rankine–Hugoniot relationship ((4.44) to (4.46)) with an isentropic index of 1.4 for pressure ratios ranging from zero to five as an example. It can be seen that above a pressure ratio of one, the shock relationship gives a higher temperature ratio while below this value the isentropic relationship is the highest.

Now for identical pressure ratios, the entropy change is related directly to these temperature ratios from the equation

$$s_2 - s_1 = c_p \ln \frac{T_2}{T_1} - R \ln \frac{p_2}{p_1} \tag{4.56}$$

Hence, when the pressure ratio exceeds unity, the entropy change from upstream to downstream is greater for the shock wave than for an isentropic pressure wave of equivalent magnitude. That is, the entropy rise is positive for the adiabatic shock which is consistent with the second law of Thermodynamics. If the pressure ratio is less than unity, the entropy change for the 'shock' is less than constant entropy and so is negative. This violates

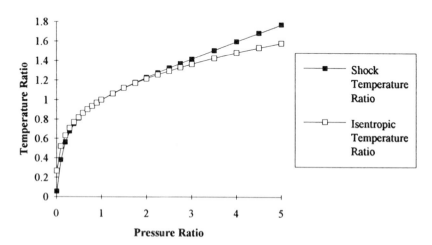

Figure 4.10 Comparison of the temperature ratio due to an isentropic and shock wave pressure change.

the second law and is not possible. That is, shock waves can only occur if they cause a pressure increase relative to the shock, from upstream to downstream. In other words, for a shock p_2 must be greater than p_1.

Equation (4.50) for the pressure ratio across the shock can be rewritten as

$$M_1 = \left[1 + \frac{(\gamma+1)}{2\gamma}\left(\frac{p_2}{p_1} - 1\right)\right]^{1/2} \tag{4.57}$$

If $p_2/p_1 > 1$, then M_1 is also greater than 1. That is, the flow approaching the shock must be supersonic. For a moving shockwave, this means that the shock velocity must exceed the local acoustic velocity in the region ahead of it. Shock waves can only be supersonic or exist in supersonic flows.

Equation (4.55) gives M_2 terms of M_1. Now the value of γ is always greater than 1 for a gas and this relationship shows that, for $M_1 > 1$, M_2 must be always less than 1. That is, relative to the shock, the flow behind it is subsonic. This applies only to a normal shock, not to an oblique shock where the additional velocity component parallel to the shock is carried through unchanged and must be added vectorially to the normal component. It should also be noted that, for a **moving shock** wave, supersonic flow in the region behind it can occur because the relative velocity (away from the shock) must be added to that of the shock to give the absolute flow velocity. In this case, the absolute flow velocity becomes

$$\frac{v_2}{a_1} = \frac{2}{\gamma+1}\left(M_s - \frac{1}{M_s}\right) \tag{4.58}$$

where M_s, the shock velocity, is equal and opposite to M_1 and v_2 is the absolute flow velocity behind the moving wave. Substituting for a_2/a_1, which can be obtained from (4.52) for the temperature ratio allows the value of v_2/a_2 to be determined as a function of γ and M_s. With $\gamma = 1.4$, it becomes greater than 1 (i.e. the flow becomes locally supersonic) for $M_s > 2.068$ and, with $\gamma = 1.667$ for $M_s > 2.539$.

Another important difference between stationary and moving shock waves is the stagnation condition. For a stationary shock wave, the stagnation temperature is conserved across the shock as can be readily seen by examination of the adiabatic energy equation with zero work. However, the stagnation pressure falls due to the irreversibility. For a moving shock, the stagnation temperature and pressure behind the wave are both higher than those ahead of it. This is because the moving wave is carrying additional energy with it from its source, this being either a moving body, a high pressure reservoir or an energy discharge.

Compressible flows with friction

The flow of a gas is strongly affected by both friction and heat transfer and this applies to those flows that occur in engines. Friction losses manifest themselves as pressure losses and occur in intake and exhaust systems as well as in communicating passages between the various components of a gas turbine. They are of most significance where the velocities are high and the passages narrow. In the former case, the friction modifies the momentum equation by the inclusion of a term containing shear stress while heat transfer has a direct input into the energy equation. Frictional adiabatic flow is best analysed using the Fanno line concept which is a temperature relationship obtained from the energy and continuity equations plotted against entropy.

The basic energy equation is as previously

$$h + \frac{v^2}{2} = h_0 \tag{4.59}$$

It should be noted that, because the enthalpy of an ideal gas is a function of temperature only, this equation implies that the 'total' temperature is conserved in the absence of heat transfer even though friction exists. Replacing h by $c_p dT$ as appropriate, rearranging and differentiating gives

$$\frac{dv}{v} = -\frac{dT}{2(T_0 - T)} \tag{4.60}$$

But from continuity

$$\rho v = \dot{m}_a \tag{4.61}$$

where $\dot{m}_a$ is the mass flow rate per unit of cross-sectional area normal to the flow. Differentiating gives

$$\frac{d\rho}{\rho} = -\frac{dv}{v} \quad \text{for } \dot{m}_a \text{ constant}$$

Now the entropy is given by

$$ds = c_v \frac{dT}{T} - R \frac{d\rho}{\rho}$$

$$= c_v \left[\frac{dT}{T} - (\gamma - 1) \frac{d\rho}{\rho} \right]$$

$$= c_v \left[\frac{dT}{T} - \frac{(\gamma - 1)}{2} \frac{dT}{(T_0 - T)} \right] \tag{4.62}$$

From which, on integration, gives

$$s - s_0 = c_v \left\{ \ln T + \left(\frac{\gamma - 1}{2} \right) \ln(T_0 - T) \right\} \tag{4.63}$$

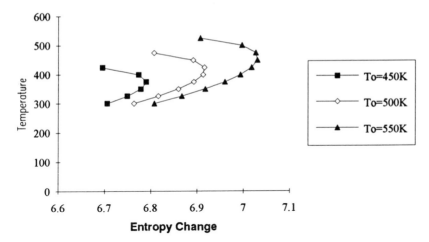

Figure 4.11 Fanno lines for compressible flow with friction: temperature–entropy plots for constant mass flow rate. Upper branches: subsonic flow; lower branches: supersonic flow. Entropy scale is a relative value $(s/c_v - s_0/c_v)$; $\gamma = 1.4$; temperature scale is in K; $M = 1$ at maximum entropy position.

A temperature–entropy plot from this equation is shown on Figure 4.11, a series of similar curves being shown for different values of the stagnation temperature T_0. These curves are termed **Fanno lines**. They have a maximum value of entropy which can be found by equating ds/dT to zero. From (4.62), this is

$$\frac{1}{T} = \frac{\gamma - 1}{2(T_0 - T)} \tag{4.64}$$

But from the energy equation (4.27)

$$c_P(T_0 - T) = \frac{v^2}{2} \tag{4.65}$$

Hence, substituting in (4.64) gives

$$T = \frac{v^2}{(\gamma - 1)c_P} = \frac{v^2}{\gamma R} \tag{4.66}$$

That is

$$v^2 = \gamma RT = a^2$$

which shows that the maximum value of entropy is at the sonic condition, $M^* = 1$.

For $T > T^*$, the energy equation shows that $v < v^*$ while the acoustic velocity, temperature relationship gives $a > a^*$. In other words, the upper part of the loop is subsonic flow. For the case where $T < T^*$, the reverse is the case and the lower part of the loop represents supersonic flow. Now in the adiabatic case, friction can only increase the entropy. This implies the following.

- for flows that are originally subsonic, the flow Mach number increases towards one, this being the maximum possible.
- for flows that are originally supersonic, the flow Mach number decreases towards one, this being the minimum possible.
- the values of one are only reached if the duct is of sufficient length for the friction and flow conditions.

Consider a subsonic flow. Here, the flow chokes for a duct length corresponding to the point where $M = 1$. This can be called L^*. For a duct which is longer than this, the flow rate must reduce. It can be readily shown by including the mass flow rate as a variable, that for T constant

$$ds = -c_v \frac{(\gamma - 1)}{2} \frac{dm_a}{m_a} \tag{4.67}$$

This shows that a reduction in the mass flow rate will allow an increase in entropy.

In supersonic flow, an increase in duct length above that which will produce a Mach number reduction to one must also result in some accommodation to the flow. Here the conditions relating to the length cannot be fed back upstream by acoustic signals. A similar phenomenon to that in a supersonic nozzle then occurs with a shock wave forming in the duct which changes the supersonic to a subsonic flow. That is, on the Fanno line curves, a jump takes place from the lower to the upper branch. The flow downstream of the shock is now subsonic and will then increase its Mach number as it continues downstream. An increase in the back pressure or a further increase in a pipe length will move this shock upstream and, if sufficient, through the original nozzle required to produce the supersonic flow in the first place creating an overall subsonic flow.

Note that in the adiabatic case, friction reduces the total pressure but not the total temperature. For a given pipe of fixed diameter and friction factor, the drop in total pressure will occur more rapidly in the supersonic case. Thus, the change via the shock wave to subsonic flow allows a greater length of pipe with the same value of mass flow rate per unit area before the sonic point is reached.

Friction can be introduced into the momentum equation by considering the shear stress at the wall. For a short length at duct, dL, this gives

$$Adp - \tau P = A\rho v dv \qquad (4.68)$$

where P is the perimeter of the duct.

Substituting for the perimeter in terms of the hydraulic diameter D_H

$$P = 4A/D_H \qquad (4.69)$$

and the shear stress in terms of a friction factor (Darcy friction factor)

$$f = \tau / \tfrac{1}{2}\rho v^2 \qquad (4.70)$$

gives

$$dp + \left(\frac{4f}{D_H}dL\right)\frac{\rho_2 v^2}{2} + \rho v dv = 0 \qquad (4.71)$$

Also, substituting from the adiabatic energy, continuity and ideal gas equations, the following equations can be obtained:

$$\frac{4f\,dL}{D_H} = \left\{ \frac{1 - M^2}{\left[1 + \left(\frac{\gamma - 1}{2}\right)M^2\right]\gamma M^2} \right\}\frac{2dM}{M} \qquad (4.72)$$

$$\frac{dp}{p} = -\left[\frac{1 + (\gamma - 1)M^2}{1 + \left(\frac{\gamma - 1}{2}\right)M^2}\right]\frac{dM}{M} \qquad (4.73)$$

$$\frac{dT}{T} = -\left[\frac{(\gamma - 1)M^2}{1 + \left(\frac{\gamma - 1}{2}\right)M^2}\right]\frac{dM}{M} \qquad (4.74)$$

It is usual to integrate these equations between the original (zero length) position of Mach number M and that of maximum length L^* where the

Mach number is one. The equations thus obtained are

$$\frac{4fL^*}{D_H} = \frac{1 - M^2}{\gamma M^2} + \left(\frac{\gamma + 1}{2\gamma}\right)\ln\left\{\frac{(\gamma + 1)M^2}{2\left[1 + \left(\frac{\gamma - 1}{2}\right)M^2\right]}\right\} \tag{4.75}$$

$$\frac{p}{p^*} = \frac{1}{M}\left\{\frac{\gamma + 1}{2\left[1 + \left(\frac{\gamma - 1}{2}\right)M^2\right]}\right\}^{1/2} \tag{4.76}$$

$$\frac{T}{T^*} = \left\{\frac{\gamma + 1}{2\left[1 + \left(\frac{\gamma - 1}{2}\right)M^2\right]}\right\} \tag{4.77}$$

$$\frac{\rho}{\rho^*} = \frac{v^*}{v} = \frac{1}{M}\left\{\frac{2\left[1 + \left(\frac{\gamma - 1}{2}\right)M^2\right]}{\gamma + 1}\right\}^{1/2} \tag{4.78}$$

$$\frac{p_0}{p_0^*} = \frac{1}{M}\left\{\frac{2\left[1 + \left(\frac{\gamma - 1}{2}\right)M^2\right]}{\gamma + 1}\right\}^{\gamma + 1/2(\gamma - 1)} \tag{4.79}$$

Compressible flows with heat transfer

While friction in compressible flows is of only moderate significance in the engine examples mentioned in section 4.1, heat transfer, particularly in the exhaust, is important because of the high gas temperatures. It also often plays some role in the intake system of SI reciprocating engines where some manifold heating is used to aid fuel evaporation. The fuel evaporation itself requires latent heat and so also has an additional effect, acting like a heat transfer from the flow.

Flow with heat transfer follows a similar pattern of analysis to that with friction except the basic momentum and continuity equations are used to construct Rayleigh lines on a T–s diagram instead of Fanno lines. Because of the similarity, it is possible to deal with the theory briefly.

The momentum equation is

$$(p_1 - p_2) = \dot{m}_a(v_2 - v_1) \tag{4.80}$$

from which it can be shown that

$$p(1 + \gamma M^2) = \text{constant} \tag{4.81}$$

Hence

$$\frac{p}{p^*} = \frac{1 + \gamma}{1 + \gamma M^2} \tag{4.82}$$

The heat transfer can be related to the total temperatures from the energy equation

$$q = c_\mathrm{P}(T_{02} - T_{01}) \tag{4.83}$$

Also, from the mass flow rate equation together with the acoustic velocity relationship

$$\rho = \frac{\dot{m}_\mathrm{a}}{Av} = \frac{\dot{m}_\mathrm{a}}{AMa} = \frac{\dot{m}_\mathrm{a}}{AM\sqrt{(\gamma RT)}} \tag{4.84}$$

By replacing ρ using the ideal gas equation, it can be seen that, for constant mass flow rate

$$T = \text{constant}(pM)^2 \tag{4.85}$$

relating conditions at any point to those where $M = 1$, again designated by*

$$\frac{T}{T^*} = \left(\frac{p}{p^*}\right)^2 M^2 = \left(\frac{1 + \gamma}{1 + \gamma M^2}\right)^2 M^2 \tag{4.86}$$

Also,

$$\frac{T_0}{T_0^*} = \left(\frac{1 + \gamma}{1 + \gamma M^2}\right)^2 M^2 \left[\frac{1 + \dfrac{\gamma - 1}{2}M^2}{\dfrac{\gamma + 1}{2}}\right] \tag{4.87}$$

and

$$\frac{p_0}{p_0^*} = \left(\frac{1 + \gamma}{1 + \gamma M^2}\right) \left[\frac{1 + \dfrac{\gamma - 1}{2}M^2}{\dfrac{\gamma + 1}{2}}\right]^{\gamma/(\gamma - 1)} \tag{4.88}$$

Now

$$\frac{s - s^*}{c_\mathrm{P}} = c_\mathrm{P} \ln\frac{T}{T^*} - R \ln\frac{p}{p^*} \tag{4.89}$$

from which by substitution from (4.82) and (4.86) is

$$\frac{s-s^*}{c_P} = \ln\frac{T}{T^*} - \frac{(\gamma-1)}{\gamma}\ln\left\{\frac{(\gamma+1)}{2} \pm \sqrt{\left[\left(\frac{\gamma+1}{2}\right)^2 - \frac{\gamma T}{T^*}\right]}\right\} \qquad (4.90)$$

When plotted on the T–s diagram, this also gives a looped curve with a maximum value of entropy. The maximum value occurs at T/T^* equal to 1.0 and it can easily be shown from (4.86) that $M = 1$ at this point. Other points of interest on the curve are shown on Figure 4.12. Again, the upper loop is subsonic and the lower is supersonic, both increasing to sonic at the maximum entropy value.

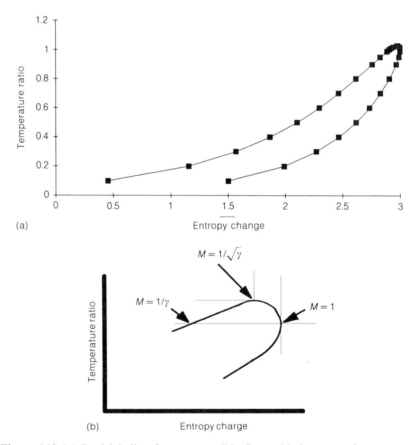

Figure 4.12 (a) Rayleigh line for compressible flow with heat transfer: tempera-ture–entropy plots for constant mass flow rate. Upper branch: subsonic flow; lower branch: supersonic flow. Entropy scale is a relative value $(s/c_P - s^*/c_P)$; temperature scale is T/T^* where $*$ represents conditions at $M = 1$; $\gamma = 1.4$. (b) details of the Mach number at the maximum temperature and entropy points on the Rayleigh curve.

Heating increases the entropy while cooling (without frictional effects) decreases it. Thus the following points should be noted.

If a subsonic flow is heated, the Mach number and temperature both increase until a Mach number of $1/\sqrt{\gamma}$ is reached. This is the maximum T/T^* condition. Further heating continues to increase the Mach number until the sonic condition but the temperature now decreases. The flow is choked at $M = 1$, which is the maximum entropy point a little to the right of the maximum T/T^* point on Figure 4.12. Any further heating can now only reduce the mass flow rate.

For a supersonic flow, heating causes both the temperature and Mach number to rise towards the choked condition. Now for a given value of T_{01}, there is a certain quantity of heat, q_{max}, required to raise the final temperature to T_0^*. This is the same, supersonic or subsonic. A normal shock does not change the total temperature, T_0, and so, unlike the flow with friction, no benefit can be gained in supersonic flow from shock development in the system if the heat transfer exceeds q_{max}. Any shock which forms is unstable. Only a shock within the nozzle region of the flow where the flow initially becomes supersonic can have an effect and this is essentially a separate function. If then the flow is supersonic, heat transfer in excess of q_{max} causes the flow to change to subsonic right from the nozzle.

4.4 WAVE MOTION IN COMPRESSIBLE FLUIDS

4.4.1 Formation of moving waves

General considerations

Moving waves are created by disturbances to the initial equilibrium and carry information about the new conditions from their source to the remote parts of the substance allowing an overall new equilibrium to be established after a short time lapse. Typical sources of waves in fluids are a moving body (e.g. a piston moving in a cylinder, a moving vehicle etc.), the bursting of a container at a different pressure than its surroundings or an energy input such as a chemical explosion or an electrical discharge. They can be compression waves at a higher pressure than the medium into which they are travelling or rarefaction waves at a lower pressure. With compression waves, the particles of fluid behind the wave move in the wave direction while, with a rarefaction wave, they move in the opposite direction. An energy input generates a higher pressure and therefore the former wave type. That is, an accelerating piston will create compression waves as will the bursting of a high-pressure container or an energy discharge. A decelerating piston or the bursting of a partially evacuated container will produce rarefactions. However, it is common for both wave types to appear with the one phenomenon. An accelerating piston will eventually decelerate and the fluid during this second phase will tend to move away from the piston face thereby creating a low-pressure region which must be filled by a reduction in flow velocity thereby creating a rarefaction wave. A high-pressure vessel on bursting will have compression waves moving outwards but rarefaction waves moving inwards to the centre caused by the outward flow. The reverse will be true of the burst of a low-pressure vessel which will have inwardly imploding compression waves and outwardly progressing rarefactions. Thus a total wave motion needs to be considered with any phenomenon.

Waves generated by any particular event distribute their information widely and create a new equilibrium but they do not continue indefinitely. They can be attenuated by growth in their frontal area and by viscous dissipation, amplified by reducing area and modified by reflections, such as at a closed end in a pipe or a corner. Reflections are important and may be examined initially as follows. Consider a plane, normal compression wave with a uniform pressure and particle velocity for some distance behind it moving down a pipe and impinging on a closed end. The particles behind the wave are moving in the wave direction and, when they strike the rigid boundary they must come to rest, building up an even higher pressure region of increasing size at the end. This creates a backward moving wave superimposed over the original flow. A partially closed end will create a reflected compression wave similarly but of smaller magnitude because some of the flow can escape. In a two-dimensional flow, a concave corner (i.e. a corner turning towards the flow) will have a similar effect building up a high-pressure region in the corner. For a shallow corner with a shock wave, this will have the effect of driving a segment of the wave faster. If the same compression wave in the pipe now impinges on an enlargement or open end, the particles are free to expand and will move more rapidly. Thus a reduced pressure region will be created near the pipe end which will move back upstream as a rarefaction. In the two-dimensional case, a convex corner (a corner turning away from the flow) will also produce a rarefaction.

Consider now the parallel cases to the above with a rarefaction wave moving down the pipe. The particles flow against the wave end, when it reaches the closed (or partially closed) end, these have no new (or less new) gas to draw from. Hence the pressure is further reduced and a rarefaction wave will move back upstream reducing the pressure further. If the end is open (or of increasing size) however, the supply of particles is increased and the pressure will also increase. Thus a compression wave will now move upstream.

Wave reflection can therefore be categorized fairly simply, at least for the one-dimensional cases. Multidimensional flows are a little more complex and will not be described here but, using the same general line of reasoning, can also be understood. Simple rules for one-dimensional wave motion are given in Table 4.3.

Table 4.3

Incident wave	Boundary	Reflected wave
Compression	Closed	Compression
Compression	Open	Rarefaction
Rarefaction	Closed	Rarefaction
Rarefaction	Open	Compression

In summary, waves are reflected as waves of the same type at closed or reducing ends and of the opposite type at open or increasing ends.

With waves, the wave pressure ratio is uniquely related to both its own velocity and the velocity of the particles behind it. Waves may, of course, move into a fluid which is itself moving which can be termed the reference fluid. In this case, the wave pressure ratio and velocities remain as before except that the latter must now be expressed relative to the moving reference fluid. If the reference fluid moves in the wave direction, the absolute wave velocity increases whereas if it moves in the opposite direction, the wave velocity decreases. In the latter case, when the fluid velocity becomes equal and opposite to that of the wave (when determined independently), the wave has an absolute velocity of zero and must become stationary. This is the steady flow case.

Description of some aspects of moving waves in engines

Moving waves are an important part of engine flows. In reciprocating engines, the inlet and exhaust systems are driven by the piston motion and hence a complex wave motion develops in the manifolds. This is most important in the exhaust where the pressure in the cylinder relative to the exhaust system pressure is high and the wave system generated can be quite strong. A typical pressure ratio might be about 3:1 to 4:1. It should be noted that this ratio refers to the driving pressures and the wave itself will have a lower pressure ratio. On the inlet side, the cylinder pressure is often marginally above that in the manifold and the waves are weak, if anything causing an initial backflow. As the piston moves down, the pressure ratio reverses because the manifold is now at the higher value. Maximum driving pressure ratios are less than about 1.5:1 and a Helmholtz resonator technique is sometimes used for design purposes. In addition, the throttle control used on spark ignition engines is a valve that can be rapidly opened or shut and is therefore a source of an additional wave motion during transients which is superimposed over the pulsating piston driven flow. Again, driving pressure ratios are not high being less than 2:1. The unsteady combustion process can be a source of waves itself, in particular during abnormal combustion. Here, under some circumstances, an explosion of a proportion of the charge near the end of combustion occurs and this is sufficient to create a rapidly oscillating compression or shock wave motion throughout the cylinder volume. Some recent experiments have noted the focusing of these waves which can exacerbate combustion damage to the piston and cylinder. In gas turbines, transient effects also cause wave motion. However, as the flow is often close to steady, this is perhaps not of major importance and is of much less significance than the standing shock waves in the diffuser section of the intake in supersonic and hypersonic flight discussed previously.

All the above subjects cannot be dealt with in detail here. As examples, the pulsating flow in reciprocating engine inlets and exhausts and the superimposed transient throttle opening will be considered. The gas turbine diffuser standing shock wave system has already been considered in section 4.2.3.

The amplitude of the waves in the exhaust system of a reciprocating engine are significant because the absolute pressure in the cylinder, an exhaust valve opening is three to four times that of the exhaust system itself. The waves so generated are equivalent to a shock wave of about $M = 1.2$ to 1.3 although generally less and, while a valve or port opening is somewhat slower than a bursting disc thereby inhibiting shock formation, it is interesting to note that modern laser holographic interferometry has been used to demonstrate that shock waves do, indeed, form. Hence, exhaust systems have been extensively studied for many years, both because of the noise inherent with shock waves of that magnitude and because of the potential reinforcement of the scavenging process. That is, by judicious design of the exhaust system it is possible to properly utilize any reflected waves to provide low pressure at the appropriate time. This is of particular importance with two-stroke engines. Consider a compression wave formed by the blowdown pulse at exhaust opening. It first moves rapidly downstream interacting with previous reflections but remaining substantially as a strong compression or perhaps a weak shock wave. On reaching the pipe end or an enlargement which may be part of the muffler system, it will be reflected as a rarefaction. If this arrives back at the exhaust port towards the end of the outflow, it will lower the pressure downstream of the valve and enhance the final outflow from the cylinder. Thus the quantity of residual gases remaining in the cylinder will be lessened. It should be noted that the process is not a simple one as the reflected wave will be moving back across the tail of the initial compression which is itself of varying amplitude. A typical diagrammatic layout of the waves is shown on Figure 4.13. It should be noted that the appropriate arrival time of this low-pressure wave is a function of both the length of the exhaust pipe, which dictates the time between the initial pulse leaving the cylinder and its rearrival at the engine, and the engine speed which determines the total duration of the exhaust process. It should also be noted that the wave velocities in the exhaust pipe are substantially independent of the engine speed. Thus, the times only correctly match at one engine speed. To extend this speed range, a variable pipe length can be used although this is cumbersome. An alternative is to use a

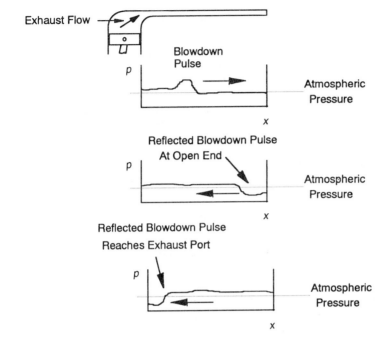

Figure 4.13 Diagrammatic representation of the blowdown wave motion in an engine exhaust system.

gradually increasing area exhaust pipe which provides a weaker but extended low pressure reflection. With two-stroke engines, the closure of the port is critical in that an early closure prevents adequate scavenging while a late closure allows the intake mixture to flow directly to the exhaust. By using a tapered system as described above with decreasing area pipe on the end, the transmitted compression wave will eventually reflect as a compression. Correct sizing of the system can allow this to return to the port right at the end of exhaust at lower speeds thereby reducing the fresh mixture outflow. This is, again, depicted on Figure 4.14.

The method of analysis of this wave interaction is quite complex and will not be described here. The one-dimensional problem requires conservation of mass, momentum and entropy equations with distance and time as independent variables. This set of partial differential equations may be solved by finite-difference techniques or, because they are what are mathematically termed **hyperbolic**, may be converted to ordinary differential equations along '*characteristics*' or lines on the x–t plane. This latter is convenient and is often used under its general title: **method of characteristics**.

With the inlet manifold, a similar wave system occurs with the exception that the initial wave, moving back upstream, is now a rarefaction with a smaller amplitude than the compression previously described. Its initial

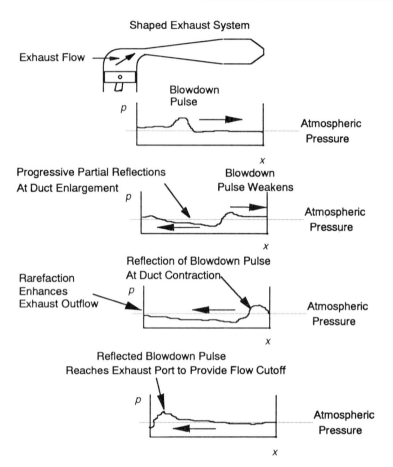

Figure 4.14 Effect of exhaust system shape on wave interaction with the exhaust flow.

pressure ratio is somewhat less than 1.5:1. It will reflect at the manifold entrance as a compression wave, which will, if it arrives at the inlet valve while the valve is still open, provide a small increase in new charge to the cylinder. Of course, as with the exhaust, there are many possible interactions with bends, branches and area changes in a real system. Also heat transfer and, for the inlet manifold, two-phase flow effects need to be considered for a completely accurate analysis. Nevertheless, modern computing techniques are approaching the stage where they can provide an inlet manifold design which maximizes the volumetric efficiency of an engine.

The inlet transient processes are also worth considering. Here the basic cases may be regarded as follows.

1. Throttle valve opens with engine speed constant. Here a compression wave moves down the manifold to the engine.

2. Throttle valve closes with engine speed constant. A rarefaction wave moves downstream to the engine.
3. Engine speed increases with the throttle valve constant. A rarefaction wave moves upstream from the engine.
4. Engine speed decreases with throttle valve constant. A compression wave moves upstream from the engine.

These are superimposed over the pulsating flow and their magnitude will depend on the rapidity of the transient. They are subject to similar reflections and interactions to those discussed above and, because both throttle change and engine speed may change simultaneously, can be complex interactions of the various types. With systems where the fuel is introduced into the manifold, they interact with the condensation and evaporation processes, and cause transient fuel/air ratio excursions which can cause engine stumble, reduced economy and poor emission levels.

4.4.2 Isentropic waves

In section 4.2.1, equations (4.12) and (4.11), it was shown that, for an acoustic wave, the following apply.

$$a^2 = \frac{\gamma p}{\rho} \tag{4.91}$$

and

$$dv = \frac{dp}{\rho a} \tag{4.92}$$

from which, by eliminating ρ

$$dv = \frac{adp}{\gamma p} \tag{4.93}$$

Now, from the equation for isentropic compression of an ideal gas

$$\frac{p^{(\gamma - 1)/2\gamma}}{a} = \text{constant} \tag{4.94}$$

Differentiating gives

$$\frac{(\gamma - 1)}{2\gamma} \frac{dp}{p} = \frac{da}{a} \tag{4.95}$$

which, on substitution into (4.93) becomes,

$$dv = \frac{2da}{\gamma - 1} \tag{4.96}$$

where da is positive for a compression and negative for a rarefaction wave.

That is, for a compression wave, the velocity change, v, for the particles is in the wave direction while for a rarefaction it is in the opposite direction.

For an isentropic wave of finite amplitude, the change in particle velocity and temperature may be considerable. Let us consider a 'non-steep' wave as being made up of a series of acoustic wavelets immediately following each other. Each point on the wave may be regarded as an acoustic wavelet moving into the gas immediately ahead of it which has conditions which themselves have been modified from the basic prewave values which, for the time being, we will assume to be the stationary gas at the initial conditions. That is, each wavelet is only propagated as an acoustic wave relative to that immediately ahead. The absolute propagation velocity of the wave point under consideration is obtained by adding its propagation velocity, its local value of a to the gas velocity v ahead of it. That is

$$c = a + v \qquad (4.97)$$

where c, a and v are the wavepoint, acoustic and particle velocities respectively.

If the initial conditions prior to the passage of the wave can be designated as p_0, a_0, T_0, etc., the local values of a and v just ahead of the wavepoint can be substituted. Now, the particle velocity v can be found from (4.93). That is

$$dv = \frac{a \, dp}{p} = \frac{a_0}{\gamma} = \left(\frac{p}{p_0}\right)^{(\gamma-1)/2\gamma} \frac{dp}{p} \qquad (4.98)$$

Integrating between the limits of p_0 and p remembering that p_0 and a_0 are constant, gives

$$\frac{v - v_0}{a_0} = \frac{2}{\gamma - 1}\left[\left(\frac{p}{p_0}\right)^{(\gamma-1)/2\gamma} - 1\right] \qquad (4.99)$$

Assuming that the initial velocity, $v_0 = 0$ and substituting in (4.97) to find the wave velocity gives

$$c = a_0 \left(\frac{p}{p_0}\right)^{(\gamma-1)/2\gamma} + v \qquad (4.100)$$

or

$$\frac{c}{a_0} = \frac{2}{(\gamma - 1)}\left[\frac{(\gamma + 1)}{2}\left(\frac{p}{p_0}\right)^{(\gamma-1)/2\gamma} - 1\right] \qquad (4.101)$$

Note that, for $p > p_0$, $c/a_0 > 1$ and $v/a_0 > 0$, and, for $p < p_0$, $c/a_0 < 1$ and $v/a_0 < 0$.

From the above equations, it can be seen that waves will change shape as they progress. For a compression wave, the term p/p_0 is always greater than one and becomes larger from the leading to the trailing edge of the wave front. That is, at any point in the front, the local wave velocity must be higher than that of points preceding it. Hence these later points must move faster and therefore eventually overtake points ahead of them. That is, an isentropic compression wave will become steeper as it progresses.

The opposite applies to a rarefaction wave. Here, p/p_0 is less than one (but greater than zero) and its value becomes smaller from the leading to the trailing edge of the wave front. The local wave point velocities will therefore gradually decrease through the wave front causing the wave to lengthen as it progresses. Both the compression and rarefaction wave development are depicted in Figure 4.15.

It should be noted that the particle velocity becomes negative as soon as p/p_0 falls below one whereas the wave velocity only becomes negative for much lower pressure ratios (these are 0.279 for $\gamma = 1.4$ and 0.237 for $\gamma = 1.667$). That is, for compression waves, the particles and wave both move in the positive x direction. For rarefaction waves over a wide range of values of p/p_0 less than one, the particles move negatively while the wave moves in the positive direction. However, once the pressure ratios become very low for rarefaction waves, the particles and wave both move negatively. The last case is because the local particle velocity now exceeds the local acoustic velocity thereby carrying the wave backwards.

4.4.3 Wave diagrams

It is usual to plot the wavepoints (or wavelets) on an x–t diagram as shown on Figure 4.16. A wavepoint may be regarded as having a particular

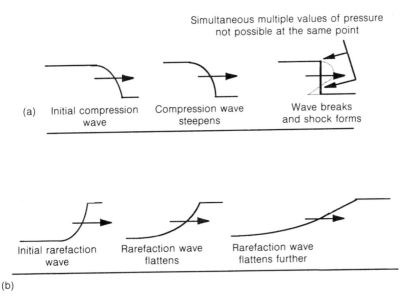

Figure 4.15 Change in shape of compression and rarefaction waves as they progress. (a) shape of a compression wave as it progresses, (b) shape of rarefaction wave as it progresses.

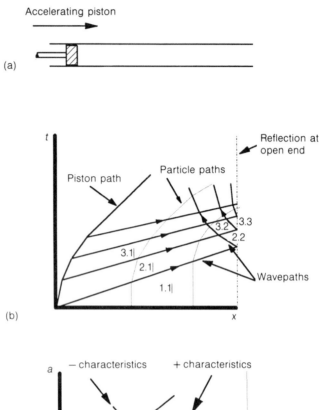

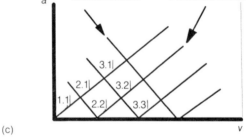

Figure 4.16 Wave diagram for the example of an accelerating piston (a) configuration diagram, (b) position diagram, (c) state diagram (source: Bannister, F. K., 1958, *Pressure Waves in Gases and Pipes*, University of Birmingham, Birmingham, UK).

pressure ratio and will therefore have a constant value of both wave and particle velocity. These can both be represented by lines of constant slope on the *x–t* diagram. For a compression wave, they converge due to the later wavelets overtaking the earlier ones as discussed previously while for a rarefaction wave, they diverge. Compression and rarefaction waves make two 'families' of lines, termed **characteristics**. Note that, along the wave lines

$$\frac{dx}{dt} = a \pm v \tag{4.102}$$

They may also be shown on a diagram using axes of v and a. Integrating (4.96) gives

$$v = \pm \frac{2}{\gamma - 1} a = \text{constant} \tag{4.103}$$

The positive sign designates a set of right-running/compression wavelets while the negative sign indicates the left-running rarefaction wavelets. These are termed **positive** and **negative characteristics** respectively. Sets of each can be drawn for different initial wavepoints using different constants. Note that, while the slope of each set is constant on the v–a (state) diagram, they may be curved on the x–t (position) diagram if they cross other characteristics carrying non-uniform information.

The above derivation giving the fundamentals of the concept of 'characteristics' is essentially descriptive applied to this particular physical phenomenon. The method of characteristics is much more general and is useful for the solution of many situations described by hyperbolic partial differential equations. Indeed, a more rigorous approach starting from the fundamental mass and momentum conservation equations coupled with the isentropic condition can be used to derive the above characteristic relationships. This will not be attempted here.

4.4.4 Shock waves

Formation of moving shock waves

It is perhaps, easier to understand the formation of shock waves in the unsteady (moving) case than it is in steady flow although application of the former to problems may be more difficult. While some brief discussion of moving shocks was undertaken in section 4.2.3 and the reader will be aware that the only mathematical difference between the moving and stationary situation is the imposition of a velocity to the latter equal and opposite to the approaching flow to give it the shock motion, it is useful to look at the formation of moving shock waves.

Consider the case of an accelerating piston as shown on Figure 4.17. As the piston undergoes an initial acceleration, the particles in the gas just ahead of it will have, at the piston face, a velocity equal to that of the piston in the same direction which will be greater than that of those further away. Due to this particle velocity, it can be seen that the particles will close up on those ahead of them, generating a positive pressure which gives rise to compression wavelets travelling away from the piston. Each wavelet travels at the local speed of sound which will normally be substantially higher than that of the piston. Assuming that the piston motion is such that the pressure is maintained in the region between the wave and the piston, the gas temperature in that region will rise due to the compression. If now the piston is further accelerated, the new wave generated will be travelling through a gas with a raised acoustic velocity and must therefore overtake the first wave. A series of disturbances of this nature can therefore build up from the continued acceleration of the piston causing a very steep pressure gradient at the wave front. If the later particles attempt to overtake the earlier ones, a situation will exist where there is more than one particle velocity at the same point in the flow which is not possible. This is shown on Figure 4.15. That is, the wave will be modified so that all the particles throughout the wave jump suddenly to the one set of pressure, temperature and velocity conditions. This is much like the wave build up at a beach when a wave breaks. The equivalent to the broken water wave in a gas flow is a shock wave. In a compressible flow, the wave is extremely thin and may, for analytical purposes, be regarded as a single discontinuity or jump in the conditions in the flow. The shock wave velocity will exceed the acoustic velocity of the gas ahead of it which is the minimum wavelet velocity in the original compression wave and is thus supersonic with respect to it. There can be no subsonic moving shock waves.

While it is obvious from the above that moving bodies can cause this sudden pressure buildup, it is also possible as discussed in section 4.2.3 for shock waves to occur in stationary flows. Consider now a moving shock wave with a velocity imposed on the gas ahead of it in a direction towards the shock. The velocity of the shock relative to the gas will be the same as before the addition of this velocity and therefore the overall effect is, in an absolute frame of reference, to slow the shock. A stationary shock is therefore the special case when the superimposed velocity is equal and opposite to the shock velocity thereby causing it to be held stationary relative to the surroundings. Note that this flow velocity may also be in the shock direction giving a greater shock velocity in absolute coordinates than in those relative to the flow ahead of it. In most flows, even ones which will eventually become steady, unsteady starting conditions occur with moving waves, often shock waves, which adjust their position as the flow stabilizes until they eventually become stationary.

Shock waves have, in reality, a small but finite thickness. The thickness will vary with the gas and the conditions but is of the order of about one mean free path of the gas molecules which is say about 2.5×10^{-4} mm as a rough approximation. During the passage of the gas through this distance, the gas molecules must adjust to the higher pressure and temperature conditions on the other side. There is, consequently, a considerable interaction between them and the fluid friction is large. However, because the surface area bounding the shock is so small due to its very thin nature, there is little opportunity for heat to escape from within a control volume place immediately around it. Thus it is reasonable to assume that a shock wave is adiabatic but it is definitely not isentropic. That is, a shock wave is a highly irreversible event. To analyse it, then, we can only use the continuity, momentum and energy equations, not the isentropic law as in many other compressible flows.

The appropriate equations for the pressure temperature and density change across a shock wave and the shock wave velocity have been given in section 4.2.3. These are most convenient as functions of the shock Mach number, M. Tracking a shock wave during its motion is relatively easy as the above parameters are uniquely defined. For a one-dimensional shock wave moving through an area change, the shock strength increases as the area decreases and vice versa. The change in shock strength with area can be found (Whitham, 1958; Milton, 1975) by

$$\frac{\mathrm{d}A}{A} = \{2M/[k(M)(M^2 - 1)] + \eta/M\}\mathrm{d}M \tag{4.104}$$

where

$$K(M) = 2\{1 + [2(1 - \mu^2)]/[(\gamma + 1)\mu]\}^{-1}\{2\mu + 1 + 1/M^2\}^{-1} \tag{4.105}$$

and

$$\mu^2 + [(\gamma - 1)M^2 + 2]/[2\gamma M^2 - (\gamma - 1)]$$

For expanding and smoothly contracting area changes, $\eta = 0$ while for sharply contracting ones, it is given by

$$\eta = 1/(2\gamma)\{[\gamma(\gamma - 1)/2]^{0.5} + 1\}\{1 - (M_0/M)^2\} + 0.5\ln(A_0/A) \tag{4.106}$$

In order to relate finite values of the area change to that of Mach number, (4.104) must be integrated and it can be seen that this is somewhat difficult. A reasonable approximation of the integral of the first term (Duong and Milton, 1985) of (4.104) is

$$A(M - 1.00)^{\alpha}(M + 1.00)^{\beta}(M - \delta)^{\varepsilon} = \text{constant} \tag{4.107}$$

where symbols α, β, δ, ε have values of 2.004, 2.719, 0.493 and 0.354 for $\gamma = 1.4$ and 2.000, 2.234, 0.867 and 0.203 for $\gamma = 1.667$.

4.5 THREE-DIMENSIONAL TURBULENT COMPRESSIBLE FLOWS

Most flows are, in fact, three-dimensional and turbulent. While the one-dimensional approach discussed so far is adequate in many cases, it is sometimes important to examine the flows using more complete models. The solutions are then much more complex and, as they generally require a numerical approach, are time consuming. They do not give as clear a physical understanding of the flow as the simple one-dimensional models which have analytical solutions. Therefore, the specific information that is required needs to be carefully considered before going to a fully multi-dimensional compressible turbulent flow. Even then, some simplifications may often be justifiable and these include a two- rather than a three-dimensional approach and the use of incompressible rather than compressible flow equations. Typical examples where the one-dimensional approach is inadequate are the flow within the blades of turbomachines and in the combustion chambers of gas turbines and the flow around butterfly valves, through inlet and exhaust valves and the in-cylinder flows in reciprocating engines.

In order to describe the above flows mathematically, the equations given in section 4.1.2 (equations (4.1), (4.2) and (4.3)) are required. These require some form of turbulence model. Turbulence, which consists of rapidly fluctuating velocities about the mean value of flow velocity is quite variable in scale. The vortices produced may be categorized according to several scales. These range from the largest, called the **integral** scale, which are of the order of size of the flow field because they are controlled by its boundaries to the smallest, called the **Kolmogorov** scale, which are of a molecular size and are where the energy dissipation occurs. Various models of turbulence are now available. These include direct modelling of the Reynold's stresses which are the fluctuating momentum terms and the k–ε model which is widely used. The latter examines the turbulent kinetic energy (k) in relation to its production due to the fluctuating velocities and its dissipation (ε) and develops a turbulent viscosity which can then be applied to the basic flow equations. Turbulence modelling is complex and the interested reader is referred to specialist texts on the subject for further details.

While researchers may sometimes wish to develop their own flow codes, many proprietary CFD packages are now available. All have their various strengths and weaknesses and the specific application needs examining before a final choice of a code is made. Usually, different turbulence models are available and the user may select the most appropriate. If the code does not completely model the situation under study, it is often possible to use their output as input data for the development of further computational codes.

4.6 TWO-PHASE SEPARATED FLOWS

Engine flows are not always only air but are sometimes air and fuel mixtures. This does not present a problem when the fuel is fully vaporized because all substances are then in the gas phase. However, there are circumstances in engine operation when the flow consists of two phases, these being liquid and gas. An obvious example is in the region near the liquid fuel jet in any engine, be it a gas turbine or reciprocating type. Further, in the latter type, an extended two-phase fuel/air flow region may exist under some operating regimes in the induction system of either carburetted, throttle body or port injected SI engines and the in-cylinder injection and mixing processes of direct in-cylinder injection SI and all CI engines. In manifolds and induction tracts, the fuel droplets may separate from the main gas stream to form wall films which then move with a much slower flow velocity than the gas stream and this can cause air/fuel ratio excursions, particularly during acceleration and deceleration transients. This film formation is due either to gravitational effects, deposition from the turbulent core flow or direct impact on surfaces such as throttle plates, inlet valve stems or bends in the manifold because of the linear momentum of the fuel drops. Two-phase, separated flows therefore need some consideration although a full treatment is again not possible here.

Two-phase flows are generally categorized as either bubbly flows or mist flows, the former being essentially a gas or vapour carried within a basic liquid stream while the latter is the reverse, with the liquid being transported by the gas. Bubbly flows are not of great concern in engine modelling although they may exist in the fuel flow when it becomes more than usually heated. Problems in the fuel supply line such as vapour lock, percolation or foaming are all well known and are examples of bubbly flow. The mist or airborne droplet flows are of more importance once the fuel has been introduced into the airstream. Maximum fuel/air ratios in modern engines of any type rarely exceed the stoichiometric values of about 1:15 by mass and may be substantially less in gas turbines or CI types. Also, the fuel is multicomponent and the lighter fractions are extremely volatile, the middle fractions moderately so and the heavy fractions relatively involatile. Hence, shortly after introduction, a reasonable proportion of the fuel has vaporized and fully joined the airstream. It can, however, recondense if it passes through a lower temperature region which may be possible in, say the expansion around a restriction such as a throttle plate. Nevertheless, liquid to gas stream ratios are likely to be noticeably less than stoichiometric for much of the flow and could be around 1:30 by mass (about 1:18 000 by volume) except close to the fuel injection point. Under these circumstances, some simplification can be made by assuming that the flows are only coupled in one direction, air stream to liquid stream. That is, the effect of the liquid on the gas stream may be neglected with little loss in accuracy. The complexity of a fully coupled three-dimensional, turbulent, compressible flow can therefore be avoided.

One-directional coupling requires a set of droplet equations of motion additional to the normal set of equations for the gas stream. These are based on the droplet accelerations imparted by the drag force on them. Typical droplet drag coefficient, C_D, and forces, F_D, are:

$$C_D = K + \alpha(\text{Re})^{-\beta} \qquad (4.108)$$

where K, α, β have values of about 0.48, 28 and 0.85 respectively (Ingebo, 1954).

The drag force is then

$$F_D = C_D \tfrac{1}{2} A \rho_a v^2 \tag{4.109}$$

A fuel evaporation model needs to be incorporated into the model as droplets will decrease in size as they progress obtaining different drag forces. Droplet acceleration can then be determined, followed by velocity and position. By assuming a distribution function for the droplet flow from the injector and discretizing it into droplets or droplet groups, the liquid fuel parcels can be tracked through the flow field. In the dense regions near the injector, this is more difficult and allowance has to be made for droplet collision. Using these techniques, it is possible to determine where the fuel is distributed away from the injector and where it deposits on surfaces to form films.

Greater accuracy can be obtained using a fully coupled model but the calculations are then greatly extended. Here, the exchange of mass due to condensation and evaporation, momentum due not only to the airstream force on the droplets but also that of the droplets on the airstream and energy associated with the heat and mass transfer between streams needs to be included.

Wall films form a separated stream which requires similar techniques. Again mass, momentum and energy exchange occurs between this and the airstream. Evaporation and condensation can be readily handled. Additional problems which have received little attention are the value of the interfacial shear stress between the gas stream and the film which provides the driving force for the film motion and re-entrainment of the film as droplets into the gas stream. Typical experimental values of wall film velocities in manifolds are about 100 to 200 mm s^{-1} in gas streams of 10 to 30 m s^{-1}.

4.7 REFERENCES

Annand, W. J. D. and Roe, G. E. (1974) *Gas Flows in the Internal Combustion Engine*, Haessner, Newfoundland, NJ.

Duong, D. Q. and Milton, B. E. (1985) The Mach reflection of shock waves in converging cylindrical channels, *Expt. Fluids*, **3**, 161–8.

Hill, P. G. and Peterson, C. R. (1992) *Mechanics and Thermodynamics of Propulsion*, 2 ed., Addison-Wesley, Reading, MA.

Ingebo, R. D. (1954) NACA Tech. Note 3265, Washington D.C.

Milton, B. E. (1975) Mach reflection using ray shock theory, *AIAAJ*, **13**, 11, 1531–33.

Streeter, V. L. (1971) *Fluid Mechanics*, McGraw-Hill, New York.

Whitham, G. B. (1958), On the propogation of shock waves through regions of non-uniform area or flow, *J. Fluid Mech.*, **2**, 145–171.

PROBLEMS

4.1 Air at a temperature of 50 °C and a pressure of 0.2 MPa flows in a duct at a velocity of 100 m s^{-1}. Further downstream, the duct reduces in cross-sectional area and the pressure falls to 0.19 MPa. Assuming that the process from the larger to the smaller area is isentropic, what is the flow velocity and the area (as a percentage of the first area) in the second section? For air, assume that $\gamma = 1.4$ and $R = 0.287$ kJ kg^{-1} K^{-1}.

4.2 Air flows in a 0.1 m^2 cross-sectional area duct and is initially at a pressure of 0.5 MPa, temperature 100 °C and velocity 80 m s^{-1}. Downstream, the pressure and temperature are measured at 0.3 MPa and 55 °C respectively. The flow is adiabatic. What is the new velocity and the new duct cross-sectional area? What is the change in entropy between the two measuring points?

A theoretical comparison assuming isentropic flow (instead of having an entropy change as in the case above) is carried out using the same pressure drop. What would the calculated temperature now have been? Explain why the actually measured temperature downstream differs from the isentropic value. For this isentropic flow, what would the velocity at the second position then have been? Is this consistent with the previously calculated area? If not, explain what would have happened in the isentropic case with the actual area as calculated at the second position.

4.3 The Mach cone generated by a supersonic projectile is measured as having an included angle of 60°. The local air conditions 50 kPa, −23 °C. Calculate the local acoustic velocity and estimate the Mach number and the velocity of the projectile.

4.4 An air supply flows through convergent–divergent nozzle to the atmospheric pressure which may be assumed to be 0.1 MPa. Assuming that isentropic flow occurs throughout, estimate the velocity at the throat, the velocity and Mach number at the exit of the nozzle, the ratio of exit to throat areas and the area at both the throat and exit per unit (kg s^{-1}) of mass flow rate.

If the system is replaced by one containing helium (molecular mass, 4.003; isentropic index, 1.667), flowing from a helium filled reservoir at 1 MPa, 323 K to 0.1 MPa, recalculate the above.

4.5 A rocket uses a converging exhaust nozzle of 12 500 mm^2 cross-sectional area. The pressure and temperature in the rocket combustion chamber are 0.8 MPa and 1500 K and these may be assumed to be reservoir (stagnation) conditions. Assuming isentropic flow, determine the exhaust velocity and the thrust of the rocket if it is at (a) take-off, ambient pressure equals to 0.1 MPa, (b) outside the earth's atmosphere, ambient pressure equals zero.

The rocket exhaust gas may be assumed to have an isentropic index of 1.3 and a molecular mass of 20.

4.6 A fuel tank of volumetric capacity, V of $1\,m^3$ containing compressed methane gas at 17 MPa experiences damage to a valve which causes a hole of $10\,mm^2$ area to appear. The hole acts as a converging nozzle to the atmosphere outside which is at 0.1 MPa, 300 K. Show that the differential relationship between time and pressure for the outflow is

$$dt = \frac{-V}{\gamma A^*}\left(\frac{\gamma+1}{2}\right)^{(\gamma+1)/(2(\gamma-1))} \frac{p_0^{(\gamma-1)/2\gamma}}{a_0}\frac{dp}{p^{1+(\gamma-1)/2\gamma}}$$

Here A^* is the throat area of the hole and subscript o represents the initial conditions at $t = 0$. Calculate the time taken for the pressure in the tank to fall to: (a) one-half of its original value, (b) the level where the flow velocity through the rupture just falls below sonic. For methane, molecular mass $= 16.04$, isentropic index $= 1.32$.

4.7 An air stream at 70 kPa, 260 K and with a velocity of $460\,m\,s^{-1}$ experiences a sudden pressure rise through a shock wave normal to the flow. What is the Mach number of the flow approaching the shock? Calculate the expected air velocity, static and stagnation pressure and temperature conditions just downstream of the shock. Determine also the downstream Mach number. What is the loss in stagnation temperature and pressure through the shock wave? Is this result expected?

4.8 A jet engine for supersonic flight uses a convergent–divergent thrust nozzle. The ratio of exit to throat areas is 4:1 and the stagnation conditions of the gas upstream of the nozzle are 3.0 MPa, 1200 K. Determine the velocity at the nozzle exit plane and the local ambient pressure if (a) a normal shock exists at the exit plane, (b) a normal shock occurs at the section with twice the throat area. The gas may be assumed to have a molecular mass of 29 and an isentropic index of 1.3.

4.9 A jet engine has a supersonic diffuser just upstream of the compressor. An aircraft travelling at a Mach number of 2.5 into air at 15 kPa, 220 K has a normal shock standing at the diffuser entrance. The diffuser then expands in area to a value three times that of its inlet. Calculate the static and stagnation pressure at the diffuser exit together with the Mach number. Will it make any difference if the shock is located differently, at say the diffuser exit. Repeat the calculations for these conditions and compare the results. In both cases, determine the loss in stagnation pressure.

4.10 A pitot-static tube on an aircraft records a static pressure of 30 kPa (gauge) and a total (pitot) pressure of 40 kPa. Calculate the air stream velocity approaching the tube (i.e. the aircraft velocity) and Mach number if the local conditions are 20 kPa, 220 K. If the aircraft velocity is increased so that the pitot pressure becomes 150 kPa, what now are these values?

4.11 An air flow in a constant area, 15 mm diameter, insulated duct initially has a Mach number of 0.2. The Fanning friction factor has a value of 0.02. What duct lengths are required for the flow to reach Mach numbers of 0.6 and 1.0? If, in the latter case, an additional 0.75 m length of the same type pipe is added to the end, calculate the percentage reduction in mass flow rate that will occur.

4.12 An experimental jet engine has a convergent–divergent exhaust nozzle with a throat area of 8000 mm² and an exit area of 16 000 mm². During a stationary test, the exhaust gas then directly enters into a parallel pipe of the same diameter as the exit and which is 5 m long before finally discharging to a reservoir in which the pressure can be controlled. The flow in the nozzle may be assumed to be isentropic while the duct has a Fanning friction factor of 0.0025. The stagnation conditions upstream of the nozzle are 0.5 MPa, 1000 K. Determine the range of back pressures in the reservoir over which a normal shock will appear in the duct. The gas may be assumed to have a molecular mass of 29 and an isentropic index of 1.3.

4.13 Air enters a simple, straight pipe type combustion chamber at a velocity of $100\,\mathrm{m\,s^{-1}}$ and a static temperature of 400 K. Heat is then added but, for the purposes of calculation, it may be assumed that the gas properties remain those of air. Calculate the amount of heat that can be added without reducing the mass flow rate and hence find the air to fuel ratio if it is assumed that the heating value of the fuel transferrable to the air is $40\,\mathrm{MJ\,kg^{-1}}$.

If the fuel to air ratio is increased by 10% and it is assumed that the combustion efficiency remains the same, estimate how much the mass flow rate will reduce for the same stagnation conditions.

4.14 For moving compression waves of pressure ratio 2.5, 5, 10 compare the temperature ratio for the wave, wave velocity and particle velocity behind the wave (expressed as a ratio to the acoustic velocity ahead of the wave) if the wave is assumed to be: (a) isentropic; (b) a shock wave. The particle velocity ahead of the wave may be assumed to be zero. The gas may be assumed to be air with $R = 287\,\mathrm{J\,kg^{-1}\,K^{-1}}, \gamma = 1.4$.

4.15 A small compression wave (considered to be isentropic) of pressure ratio 1.5:1 occurs in the exhaust system in an engine. The isentropic index may be considered to be 1.3 and the molecular mass 29. Determine the wave velocity and the particle velocity behind it. The wave can be considered to be moving into a gas at 110 kPa, 400 K.

If the gas ahead of the wave is now assumed to have an increasing temperature gradient of $T = 400 + 13\,s$ where s is the distance in metres along the exhaust pipe, determine the time taken for the wave to reach the end of the pipe 5 m away.

4.16 The wave of the first part of Problem 4.15 (gas temperature 400 K) is now returning in the exhaust system towards the engine at a time when the exhaust valve is shut. If it is assumed that the reflection at the engine occurs as at a plane closed end normal to the flow, calculate the maximum pressure ratio reached at this closed end and the velocity with which this wave now returns towards the open end of the exhaust assuming that (a) the wave is short in length and after reformation at the reflection, it travels into a stationary gas, (b) that the wave is long with uniform particle velocity and gas conditions for a considerable length behind it and the returning wave travels into this flow.

Neglect any other possible motion of the exhaust gas.

4.17 A moving, normal shock wave is generated in a parallel-sided duct where it travels with a velocity of $600\,\mathrm{m\,s}^{-1}$ into air at $80\,\mathrm{kPa}$, $260\,\mathrm{K}$. Calculate the shock Mach number (relative to the stationary gas ahead of it) and determine the flow velocity and the static and stagnation pressure and temperature conditions behind it. Is the stagnatioin temperature conserved in this case? Does the stagnation pressure fall across the shock? If not, why not? Compare with the stationary shock in a steady flow as in Problem 4.7.

4.18 The shock of Problem 4.17 reaches the end of the duct where it is reflected by a plane wall normal to the flow direction. Noting that the reflected shock moves back over the incoming particles behind the original shock rendering them stationary, calculate the Mach number of the shock (relative to the incoming particles), its absolute velocity and the pressure and temperature conditions existing between the reflected shock and the end wall.

4.19 A shock tube is a device widely used to create short duration, high enthalpy flows for aerodynamic, combustion and other forms of testing. In its simplest form, it is a parallel tube with a diaphragm placed part of the way along it. The section upstream of this, the 'driver' section is at a high pressure, that downstream, the 'driven' section being at low pressure. Bursting the diaphragm allows a shock wave to move downstream with uniform velocity, pressure and temperature conditions behind it.

A simple shock tube containing air in both driver and driven sections is used with a diaphragm pressure ratio of $100:1$. By matching the pressure and particle velocity of the downstream moving shock to those of the upstream moving rarefaction wave formed at the burst, calculate the shock Mach number obtained. Also determine the flow (particle) velocity, pressure and temperature available for the test (i.e. the conditions behind the shock). The air in the driven section is at $10\,\mathrm{kPa}$, $300\,\mathrm{K}$ while the driver section is at $300\,\mathrm{K}$. The normal data for air may be assumed to be relevant.

If the driver section is now heated to $500\,\mathrm{K}$ affecting the acoustic velocity but not the isentropic index while the driver gas and diaphragm pressure ratio remains as before, calculate the new shock Mach number and test conditions for the driven section air.

If the air in the driver is replaced by helium, (molecular mass 4.003, isentropic index 1.667), while the driver gas and diaphragm pressure ratio remains as before, calculate the new shock Mach number and test conditions for the driven section air.

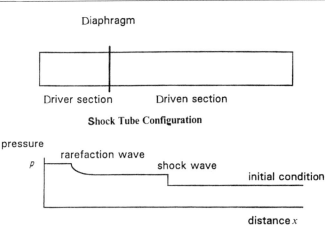

Diaphragm

Driver section   Driven section

Shock Tube Configuration

pressure

rarefaction wave

p   shock wave

                              initial condition

                              distance x

Pressure–Distance Diagram After Burst

4.20 A piston situated in a uniform, frictionless pipe initially containing air with acoustic velocity of $340\,\mathrm{m\,s^{-1}}$ and a pressure of $100\,\mathrm{kPa}$ is accelerated uniformly from rest at $t=0$ to attain a velocity of $102\,\mathrm{m\,s^{-1}}$ at $t=0.03\,s$. Subsequently the piston moves at constant velocity.

Using only the piston displacements and velocities given below construct a piston, wave piston diagram for the waves from the four discrete piston points given up to and including the second wave reflection from the open end. Estimate (a) the time at which the motion of the piston is first observable at the closed end, (b) the particle velocity with which each wave section initially travels after leaving the piston and the acoustic velocity and pressure associated with it.

For the waves 1 and 2 sections only, determine the following for the condition immediately after reflection, (c) the particle velocity at the closed end, (d) the wave speed just after reflection from the closed end, (e) the pressure on the closed end immediately after the wave is reflected.

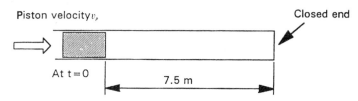

Piston velocity v,   Closed end

At t = 0   7.5 m

PISTON DATA

Wave	1	2	3	4	
$t(s)$	0	0.01	0.02	0.03	$t > 0.03$
$x(m)$	0	0.17	0.68	1.53	> 1.53
$v_p(\mathrm{m\,s^{-1}})$	0	34	68	102	102

5 | Turbomachinery for engines

Turbomachinery is a widely used, integral part of the operation of a great many internal combustion engines. In particular, all gas turbines use turbomachinery for both the compressors required to produce the high combustion pressures and for the turbines needed to drive them. In a large proportion of these engines, particularly stationary gas turbines and aircraft engines with bypass type jet propulsion, the main power output is via an additional turbine unit which drives a power shaft, fan or propeller. With reciprocating engines, turbocharging is now also widely used, both on spark ignition and compression ignition types, but particularly on the latter because there are fewer associated limitations in that application as will be discussed later. Thus, a knowledge of the fundamentals of turbomachinery is essential to the engineer or technologist who is associated with the design, development or operation of modern internal combustion engines.

5.1 BACKGROUND ON TURBOMACHINES

Turbomachines belong to a class of devices which exchange mechanical work with the energy of the fluid stream. The interchange of fluid and mechanical work and power can be achieved with two basic types of devices, these being positive displacement and rotodynamic machines. The former, positive displacement machines, have a direct relationship between the movement of their surfaces and the volume of fluid which passes through them. That is, their volume flow rate is directly proportional to an increase in speed although the mass flow rate may vary nonlinearly due to changes in density. These are usually of the piston–cylinder type although there are some geometrical variations of this theme. Turbomachines, on the other hand, are rotating devices where there is an interchange of shaft and fluid power without such a direct movement/volume relationship and a variation in speed can cause considerable, nonlinear variations in the volume flow rate. These machines move the fluid and obtain or absorb their power by use of moving vanes which affect the momentum of the fluid stream. As a class of devices, they include fans, propellers centrifugal and axial flow pumps, inward flow or axial flow turbines and may be either hydraulic (incompressible fluid) or gas (compressible fluid) types. While the hydraulic types are extremely common, this chapter will concentrate on the gas types because they are used directly in the generation of power in internal combustion engines.

5.1.1 Velocity triangles for turbomachine analysis

To obtain an understanding of turbomachines at any level, it is necessary to start with a consideration of the absolute velocities of the flow both entering and leaving the blades. The analysis is aimed at obtaining an assessment of the mass flow rate through the machine and the change in the velocity and hence momentum of the fluid imparted by the blades. Each configuration will require a slightly different approach although the general principles are the same. For example, a centrifugal machine where the flow is through the machine in a radial direction will obtain or use power through the change in the tangential component of the absolute velocity of the fluid flowing along the blade between inlet and exit. In an axial flow machine, the important changes are those of the flow across the blade. The mass flow rate in a centrifugal machine is provided by the radial component of the absolute velocity while in an axial flow machine it is the axial component. These are illustrated on Figure 5.1. The essential aim of the analysis is the determination of the absolute flow velocity vectors at inlet and exit as all other required velocities can be derived from them by breaking them into their component vectors. It should be noted that, while these are the two limiting types, a range of intermediate (mixed flow) machines exists in between.

Turbomachines can therefore be understood at a simple level by the use of velocity vector diagrams at the inlet and exit to the blade. Generally two velocity triangle diagrams can be drawn at each position. The first of these diagrams sums the local blade velocity and the fluid velocity, relative to the

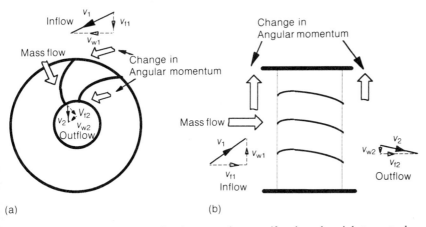

(a) (b)

Figure 5.1 Flow and torque development in centrifugal and axial type turbomachines (a) centrifugal flow machine, (b) axial flow machine. Absolute velocity $= v$; radial or axial component $= v_f$; tangential component $= v_w$.

blade, vectorially to give the absolute fluid velocity. That absolute velocity may also be broken, using a second vector triangle, into two components, radial and tangential in the case of the radial flow machines and axial and tangential in the case of the axial flow types. This is illustrated on Figure 5.2. The two triangles are often superimposed giving a diagram of complex appearance but, as long as the origin of their construction is understood, this single diagram is simple to use. In both the above cases, the tangential component provides the rotational momentum (i.e. moment of momentum) while the other component generates the flow through the machine. The change in the moment of momentum of the absolute fluid velocity between inlet and outlet of the blade uses or produces the torque and hence the shaft power.

Construction of these basic diagrams is very simple and examples are shown on Figures 5.3 and 5.4 respectively for the radial and axial flow cases. One assumption made is that the fluid velocity relative to the blade impinges and exits in the direction determined by the blade angle. If the fluid enters at a different angle to the blade, a separated region will occur just behind the leading edge resulting in flow losses as depicted in Figure 5.5. This can and does occur, particularly in off-design conditions. Therefore an important aspect of turbomachine design, particularly in devices where the power transfer is high, is the minimization of these losses throughout the rotational speed, flow rate range. This is achieved by appropriately adjusting the flow direction by, for example, the use of inlet guide vanes. Simple turbomachinery analysis does not take such separated flow into account and the required loss-free inlet condition is usually termed **shockless**. This has nothing to do with the formation of the shock waves discussed in

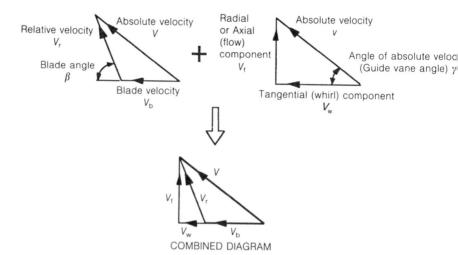

Figure 5.2 Construction of velocity triangles for turbomachinery flow.

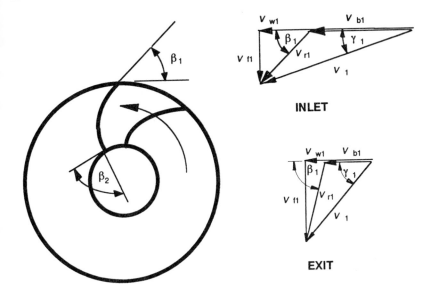

CENTRIFUGAL FLOW MACHINE

Figure 5.3 Velocity triangles for an inward flow radial turbine. Absolute velocity $= v$; blade velocity $= v_b$; radial or axial component $= v_f$; relative velocity $= v_r$; tangential component $= v_w$; blade angle $= \beta$; absolute flow direction angle $= \gamma$.

Chapter 4 although it is possible for true shocks to form in turbomachinery flows. The term is perhaps unfortunate but it is in common usage and therefore will be adhered to here.

Another assumption that is often made is that the flow from inlet to exit is frictionless which implies that the flow velocity can be taken as constant (i.e. the velocity profile is a 'plug' flow) across any particular area under consideration. It is also often, although perhaps less frequently, assumed that the blade thickness is very small and the total space through which the fluid flows is a function of only the radius and the blade width (for radial machines) or the inner and outer radii (for axial machines). However, these areas can easily be modified by simple geometric calculations for the effects of blade blockage to give a more accurate analysis.

Given the above assumptions, the triangles drawn on Figures 5.3 and 5.4 represent a vectorial addition of the velocities:

$$v = v_b + v_r \qquad (5.1)$$

where v is the absolute fluid velocity, v_b the blade velocity and v_r the fluid velocity relative to the blade.

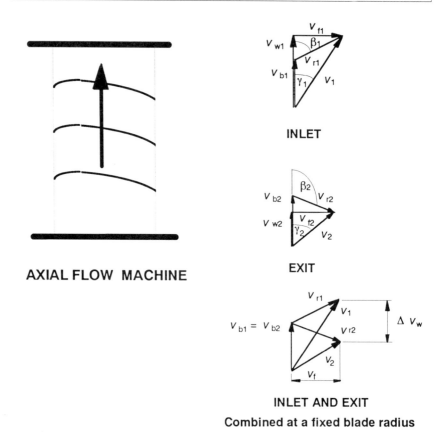

AXIAL FLOW MACHINE

Figure 5.4 Velocity triangles for an axial flow turbine. Absolute velocity $= v$; blade velocity $= v_b$; radial or axial component $= v_f$; relative velocity $= v_r$; tangential component $= v_w$; blade angle $= \beta$; absolute flow direction angle $= \gamma$.

When the absolute velocity is further broken into the two components normal to each other, (these being the tangential and radial components for radial machines or the tangential and axial components for axial machines as discussed above), the absolute velocity can again be expressed by a vectorial addition:

$$\boldsymbol{v} = \boldsymbol{v}_w + \boldsymbol{v}_f \tag{5.2}$$

where $\boldsymbol{v}$ is the absolute fluid velocity, $\boldsymbol{v}_w$ the tangential velocity (whirl) component and $\boldsymbol{v}_f$ the radial or axial velocity (flow) component.

The whirl component rotates the fluid about the axis and therefore transfers the moment of momentum to shaft torque but it does not cause the flow to progress through the machine. This is a result of the flow component normal to it. The equations are normally expressed graphically and, because information required for their construction may not be sufficient

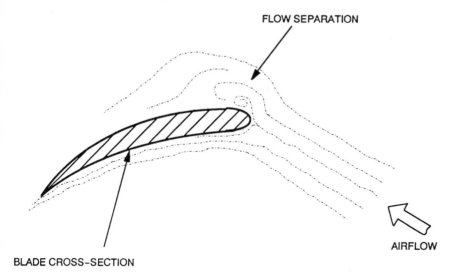

FLOW SEPARATION

AIRFLOW

BLADE CROSS-SECTION

Figure 5.5 Incorrect flow incidence on a blade.

for either to be completed individually, the superposition of one on the other provides a simple solution. It should be noted that three quantities, either one length and two angles, two lengths and one angle or three lengths are necessary for the triangle set to be completed.

The mass flow rate through the machine is related to v_f in the usual way.

$$\dot{m} = A\rho v_f \tag{5.3}$$

where A is the cross-sectional area normal to the velocity component v_f.

The blade velocity can be found by using the angular velocity of the rotor and the radius to the appropriate point on it

$$v_b = r\omega \tag{5.4}$$

In addition, a blade angle may be specified. Thus, three pieces of information are available although other sets may be specified in different cases.

The torque τ created by the whirling fluid can now be determined

$$\tau = (\dot{m}rv_w)_{out} - (\dot{m}rv_w)_{in} = \dot{m}[(rv_w)_{out} - (rv_w)_{in}] = \dot{m}[(r_2 v_{w2}) - (r_1 v_{w1})] \tag{5.5}$$

That is, the torque can be related to the change in the whirl velocity, the mass flow rate and the geometrical factors (radii) associated with the machine. The first two can be determined from the velocity triangles as described above. Note that for a pump or compressor, it is negative while for a turbine it is positive. Other quantities, power $\dot{W}$ and theoretical head (called the Euler head, H_e) across the machine, can now be easily derived:

$$\dot{W} = \tau\omega = \dot{m}\omega(r_2 v_{w2} - r_1 v_{w1}) = \dot{m}(v_{b2}v_{w2} - v_{b1}v_{w1}) \tag{5.6}$$

$$H_e = \frac{\dot{W}}{\dot{m}g} = \frac{1}{g}(v_{b2}v_{w2} - v_{b1}v_{w1}) \tag{5.7}$$

The actual head across the machine may differ somewhat from the Euler head due to fluid flow losses, being slightly more in the case of a compressor and slightly less in the case of a turbine. The ratio (of the lower to the higher value in each case) is known as the hydraulic efficiency and represents the ability of the machine to utilize the moment of momentum change of the fluid. It is not the same as mechanical efficiency which is due to friction losses in such things as the bearings and which must be accounted for in a similar manner but by a separate term.

5.1.2 Turbomachinery types

Centrifugal (radial) flow machines

These are commonly used in the small to medium size range and are popular because of their simplicity and robust construction. Compressors (or pumps) have the flow entering at the centre and flowing radially outwards while turbines are usually radial inflow devices. The blades may be backward swept, straight or forward swept as appropriate for the design purpose. These are illustrated on Figure 5.6 with a compressor as an example. It can be seen that, with the backward swept blades, the absolute velocity becomes relatively smaller. That is, the energy is converted from

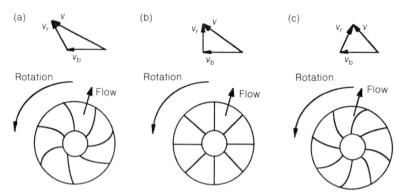

Figure 5.6 Forward, straight and backward swept blades on a radial flow compressor (a) forward swept blades, (b) straight blades, (c) backward swept blades. The velocity triangles show the difference in absolute velocity for the same blade velocity and flow rate.

kinetic to pressure energy which results in fewer losses in the downstream diffuser section. In most cases, a high pressure is preferred to a high velocity and backward swept blades are preferable. Straight blades are simpler to construct and are mechanically strong and so are frequently used while forward facing blades are uncommon and are only used where a high exit velocity at low pressure, as for example in a fan, is required.

Axial flow machines

These are more commonly used for high pressure ratios because they are very suited to multiple staging. Also, in aircraft engines, they keep the cross-sectional area low which is important in drag reduction. The stages usually consist of a stator row followed by a rotor, the former directing the flow at the appropriate inlet angle to the latter. The stator rows may be adjusted to cope with different flow rates. The advantage of the axial machine is that the flow can pass directly from one stage to another without having to pass through long passages with associated losses. The blade rows are often referred to as impulse or reaction blading which are the two extremes of the range. With impulse stages, the velocity relative to the blade remains the same as it passes over it, with the work being obtained or added by the change in flow direction. That is, from the energy equation, a pressure change occurs across the stage. With reaction staging, the pressure remains the same and the velocity changes. This is achieved with a change in the cross-sectional area of the flow passage through the blades by shaping them.

5.1.3 Dimensionless groups

Groups for compressible and incompressible flow machines

Similarity techniques are often applied to turbomachines. In order to do this, the machines must be geometrically similar and have similar velocity triangles for the flow at inlet to and exit from the blades. The basic variables are shown in Table 5.1 although it should be noted that others may be considered for specific purposes such as cavitation.

The dimensionless groups (it will be remembered from basic fluid mechanics that these are called Πs) are obtained from these. In this case there are five of these groups, labelled Π_1 to Π_5 respectively. They are

$$\frac{\vartheta}{ND^3}, \quad \frac{g}{N^2D}, \quad \frac{H}{D}, \quad \frac{\dot{W}}{\rho D^5 N^3}, \quad \frac{ND^2}{v}$$

The second group is often replaced by $\Pi_6 = gH/N^2D^2$ which is obviously a combination of Π_2 and Π_3. The Π_4 group is an efficiency term and the denominator can be also recognized, noting that v is proportional to ND, as $\rho D^2 v^3$. This last may be further modified to $\dot{m}v^2$ which is itself proportional

Table 5.1

Variable	Symbol	Unit
Volume flow rate	ϑ	$\mathrm{m^3\,s^{-1}}$
Rotational speed	N	$\mathrm{rad\,s^{-1}}$
Size (diameter)	D	m
Head	H	m
Power	$\dot{W}$	W
Fluid density	ρ	$\mathrm{kg\,m^{-3}}$
Fluid viscosity (absolute)	μ	$\mathrm{kg\,(ms)^{-1}}$
Fluid viscosity (dynamic)	ν	$\mathrm{m^2\,s^{-1}}$
Gravitational constant	g	$\mathrm{m\,s^{-2}}$

to $\dot{m}gh$. The fifth group, $\prod_5$, is the Reynolds number which is used to account for viscous effects. Plots of head and efficiency are often given against flow rate using these groups. These plots can therefore describe the characteristics of similar machines of different sizes and speeds.

As well as the rearrangement of $\prod_2$, it is often convenient to express speed in terms of either a volume flow rate (for a pump or compressor) or the power (for a turbine) and head (for both) all having one unit value. That is, the size of the machine is not considered. When this approach is used, the group designating the machine speed is referred to as the **specific speed**. To produce an appropriate group of this type, the $\prod_6$ and either the $\prod_1$ or $\prod_4$ groups are combined to eliminate D. This gives*

$$\prod_7 = \text{specific speed for a pump, } N_{sp}$$

$$= \frac{N\,\vartheta^{0.5}}{(gH)^{0.75}}$$

$$\prod_8 = \text{specific speed for a turbine, } N_{st}$$

$$= \frac{N(\dot{W})^{0.5}}{\rho^{0.5}(gH)^{1.25}}$$

Finally, in practice, the gravitational constant is often neglected simply for convenience because it is unlikely to vary in any significant way. This gives groups which, although having dimensions, can be used similarly for comparative purposes. These are

$$\prod_9 = \frac{N\,\vartheta^{0.5}}{(H)^{0.75}}$$

and

$$\prod_{10} = \frac{N(\dot{W})^{0.5}}{\rho^{0.5}(H)^{1.25}}$$

---

*Either group could be used for a pump (compressor) or a turbine; however, the most important parameter for pumps is volume flow, for turbines it is power.

Typical plots of turbomachinery characteristics are shown on Figure 5.7. Classification of turbomachinery by use of specific speed is shown on Figure 5.8.

Compressible flow machines

Compressible flow turbomachines are of primary interest in engines as these are the types associated directly with the production of power. Here, an analysis gives an identical set of groups to those discussed above for incompressible machines with the addition of other groups which will be now developed from the extra variables related to the incompressible flow. These are shown in Table 5.2.

The R and T terms are usually placed together as RT. In addition, the volume flow rate ϑ used with the incompressible machine is best replaced by the mass flow rate $\dot{m}$. An efficiency term is usually included as η which is already dimensionless. In conjunction with rotational speed N and size (diameter) D, the additional groups are then

$$\prod_a = \frac{\dot{m}\sqrt{(RT)}}{pD^2}, \quad \prod_b = \frac{\sqrt{(RT)}}{ND}, \quad \prod_c = r_P, \quad \prod_d = r_T, \quad \prod_e = \eta$$

As the gas is normally either air or a mixture predominantly consisting of air with vapours, the gas constant is often neglected giving groups with dimensions:

$$\prod_a = \frac{\dot{m}\sqrt{T}}{pD^2}, \quad \prod_b = \frac{ND}{\sqrt{T}}, \quad \prod_c = r_P, \quad \prod_d = r_T, \quad \prod_e = \eta$$

Note that $\prod_b$ is proportional to the Mach number of the blades, the most critical position being their tips, whereas $\prod_a$ is related to the flow. For a given machine, the diameter is fixed and so it also is often deleted from the above groups when they are used for conventional plots of the machine

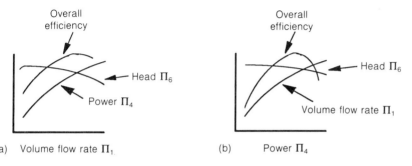

Figure 5.7 Some typical turbomachinery characteristics (a) typical pump characteristics, (b) typical turbine characteristics.

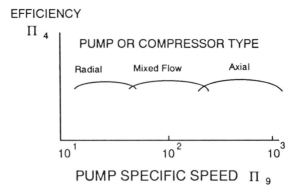

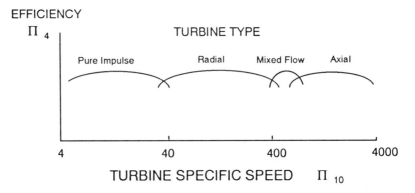

Figure 5.8 Compressor and turbine types: typical classification by specific speed.

Table 5.2

Variable	Symbol	Unit
Inlet pressure	p	pa, kPa
Inlet temperature	T	K
Pressure ratio for the machine	r_p	
Temperature ratio for the machine	r_T	
Gas constant	R	$J\,kg^{-1}\,k^{-1}$

characteristics. That is, the groups most commonly used have dimensions and are

$$\prod_a = \frac{\dot{m}\sqrt{T}}{p}, \quad \prod_b = \frac{N}{\sqrt{T}}, \quad \prod_c = r_P, \quad \prod_d = r_T, \quad \prod_e = \eta$$

Typical plots (or maps) for a compressor and turbine are shown on

Figures 5.9 and 5.10. It is usual to plot either the pressure or temperature ratio against $\prod_a$ as the axes with lines of constant $\prod_b$ shown as contours.

On the compressor map of Figure 5.9, a surge line is shown. For operation to the left of this line, an instability in the flow manifests itself. That is, the surge line represents a demarcation between steady flow (to the right) and an unsteady condition (to the left). The unsteadiness, called surge, exists only where the Mach number is high enough for the flow to have compressible characteristics which is when the Mach number is above about $M = 0.2$ to 0.3. The cause is the flow separating and reversing from the impeller surfaces causing an oscillation or surge which reduces performance. In a

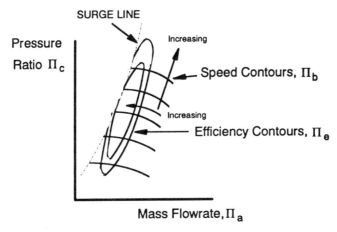

Figure 5.9 Typical map of compressor characteristics.

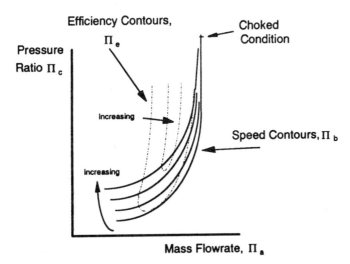

Figure 5.10 Typical map of turbine characteristics.

multistage machine, surge will propagate backwards from the last, high pressure, stage which stalls first thereby exposing the previous stages to the high pressure. This induces stall progressively in the blades forward through the system. As the downstream pressure then falls, the flow becomes re-established in the normal direction. This phenomenon can then repeat as surging. It is well known, however, that surging is not confined to multistage machines and exists similarly on single stage compressors, the stalling occurring progressively back across the single stage.

On the right side of the compressor map, another limiting phenomenon occurs. Here a high mass flow rate increases the velocity and the flow moves towards the choked condition. At choking, any increase in rotational speed cannot increase the mass flow rate further and the speed contours therefore tend towards a common, vertical line. As choking is approached, the efficiency falls off creating efficiency contours which lie with their axes close to the surge line. These contours appear on the map as islands because, at or very near to surge, a decrease in mass flow rate causes a fall in efficiency and to the right, away from the surge line, the high Mach numbers which occur close to choking also reduce efficiency. Thus the contours represent efficiency which increases towards the centre of the map. A high turbomachinery efficiency is most important for gas turbine (and turbocharged reciprocating engine operation) as discussed in Chapter 2.

The turbine map looks somewhat different (Figure 5.10) although there are many similarities. There is no surge line for a turbine because the flow has a favourable pressure gradient rather than the adverse one for compressors, but choking is an even more important phenomenon. This may occur in the turbine nozzle which is either the connecting passage from the volute casing to the rotor in the case of inward flow radial turbines with no guidevanes or in the stator nozzle blades. The latter are used on some radial flow turbines, particularly the larger variety and on axial flow types. Indeed, in the axial flow turbine, the convergence of the $N/\sqrt{T}$ contours towards a single vertical choked line is the dominant feature. The contours are close together throughout which restricts the ability to use the same diagram for efficiency islands. The radial turbine characteristics are not so tightly packed and more flexibility in their diagrammatic presentation is possible.

5.1.4 Determination of real flow effects

The cross-sectional shape of a blade of a turbomachine is an aerofoil profile. For flow over an aerofoil, whether it is stationary, moving in simple translation or rotating, real flow effects become apparent. These, in particular, manifest themselves in the phenomenon called **stall**. Consider the flow depicted on Figure 5.12. Here the flow impinges on the blade in such a direction as to provide a '*lift*' force which will move the blade in the required direction. The angle between the flow and the chord of the aerofoil is termed the **angle of attack**. As the angle of attack increases, the lift increases but so does the drag, which is the force normal to the lift. These forces are usually expressed as lift and drag coefficients, C_L and C_D, respectively:

$$C_L = \frac{F_L}{\frac{1}{2}\rho v^2}, \quad C_D = \frac{F_D}{\frac{1}{2}\rho v^2}$$

Here F_L and F_D are the lift and drag forces per unit area respectively and ρ and v are the flow density and velocity.

As the angle of attack increases, so does both the lift and drag coefficient as would be expected from the greater change in direction imposed on the fluid. However, at a limiting value, the separation of the flow behind the aerofoil causes the coefficient of lift to fall while that of drag still increases. At this point, the aerofoil is said to be stalled. A typical diagram showing this trend is given on Figure 5.11.

Drag is due to viscosity whether directly as in skin friction drag or indirectly in that it causes separation of the boundary layer with consequent insufficient momentum for the pressure to be regained behind the object. It should be noted that, as the flow in a compressor is from a low to a high pressure region, it is slowed by the adverse pressure gradient and the boundary layer is more prone to separate than in a turbine where the pressure gradient aids the flow. The effects are best determined experimentally. For the flow over such things as aircraft wings, isolated aerofoils can be tested and a great deal of data is available, for example from NACA (USA), British C series and German Gottingen sections. However, for turbomachinery, the blades are in close proximity to each other and therefore interact with noticeably different results. Hence, in general isolated aerofoil data cannot be used with any great accuracy and other test procedures have to be adopted. This is best determined in terms of a **solidity ratio** defined as

$$\text{Solidity} = \frac{\text{Chord length}}{\text{Pitch}}$$

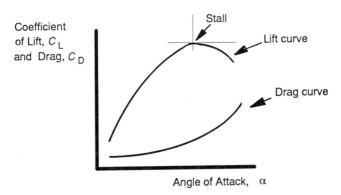

Figure 5.11 Effect of stall on the lift and drag coefficients of an aerofoil.

where chord length and pitch are as designated on Figure 5.12. For different solidity values, different data or approaches are possible as shown in Table 5.3.

Isolated aerofoil data is obtained by testing a single, representative aerofoil in a wind tunnel. Cascades are a set of similar blades, usually at least six, which are set up on a fixed frame at the appropriate pitch. This cascade is also tested in a wind tunnel and changes from the isolated aerofoil data due to the interference effects are detectable. Simple cascades as shown on Figure 5.13 are most commonly used and represent a small section of the total blade system. However, they can be arranged symmetrically around the periphery of a circle if desired. Systems with very high solidity are very rare and test data is generally not available. However, as the high solidity tends to isolate the passageways between the blades, theoretical approaches based on viscous channel flow can be used.

In real flows, losses other than those due to the blade profile exist and should be noted. These are illustrated on Figure 5.14. It can be seen that a boundary layer exists on both sides of the passageway due to the base annulus to which the blades are fixed and to the outer casing. This boundary layer can extend considerably into the flow, particularly in the later turbomachine stages. The effect from it is referred to as **annulus drag**. Three-dimensional flow losses also exist due to a vortex shedding at the exit of the

Table 5.3

Approximate solidity range	Appropriate data or theory
0.5 – 1.0	Isolated aerofoil data
1.0 – 15	Cascade data
15 and higher	Channel theory

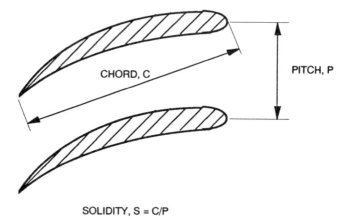

SOLIDITY, S = C/P

Figure 5.12 Geometry of a group of aerofoils.

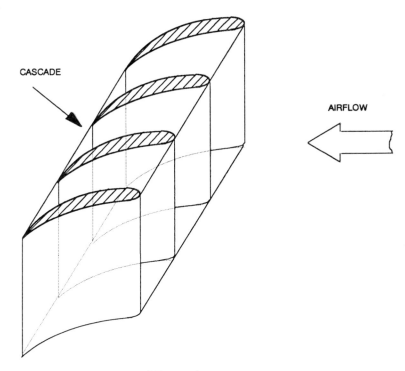

CASCADE

AIRFLOW

Figure 5.13 Rectilinear aerofoil cascade.

blades and secondary flow can exist in the cavity between blades introducing additional frictional effects at the blade wall. At the blade tips, the
sealing is never perfect and some leakage occurs from the high to the low
pressure side of the blade. All these combine to make the design, in practice,

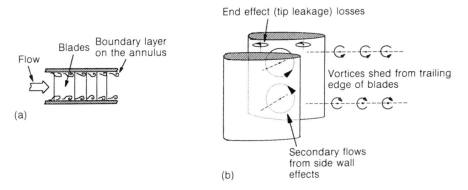

End effect (tip leakage) losses

Blades Boundary layer on the annulus

Flow

Vortices shed from trailing edge of blades

(a)

Secondary flows from side wall effects

(b)

Figure 5.14 Losses in blade rows in cascades or turbomachinery (a) annulus drag,
(b) secondary flow effects.

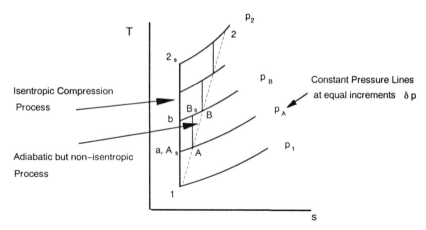

Figure 5.15 The effect of a number of 'small' stages on the isentropic efficiency of a compressor.

of turbomachinery much more complex than a simple, loss-free theory would indicate.

5.2 TURBOMACHINERY EFFICIENCY

5.2.1 Overall isentropic efficiency of the turbomachine

For analysis, turbomachines are usually assumed to be steady flow devices with the steady flow energy equation (SFEE) applying. As the stage length is short, any heat transfer is small compared to the energy of the flow and is neglected, as are changes in potential energy. Hence

$$w = (h_1 + v_1^2/2) - (h_2 + v_2^2/2) = h_{01} - h_{02} = \Delta h_0 \qquad (5.8)$$

which is the change in stagnation enthalpy. The change in kinetic energy per unit mass is also usually small and the above is often written simply as change in enthalpy, Δh.

It is usual to assume ideal gas behaviour and to neglect the kinetic energy terms to give a simple analysis in terms of pressure ratios although the latter may readily be included if so desired. That is, the following temperatures and the pressure ratio may be used to represent the stagnation values. Following this procedure, the equations for work for a turbine w_T and for a compressor w_c become

$$w_T = c_p T_1 (1 - T_2/T_1)$$

$$= c_p T_1 \left[1 - \frac{1}{r_p^{(\gamma - 1)/\gamma}} \right] \qquad (5.9)$$

and

$$w_c = c_p T_1 (r_p^{(\gamma-1)/\gamma} - 1) \qquad (5.10)$$

Note that the pressure ratio, r_p is p_1/p_2 for the turbine and p_2/p_1 for the compressor in the above cases. Once included in a cycle analysis, the nomenclature for the subscripts will be different. A general way of treating adiabatic turbomachine stages is to assume an isentropic efficiency which, as defined previously, is for a compressor

$$\left. \begin{aligned} \eta_c &= \frac{\text{Work for the ideal isentropic process}}{\text{Work for the actual process}} \\ &= \frac{h_{2s} - h_1}{h_2 - h_1} \\ &= \frac{T_{2s} - T_1}{T_2 - T_1} \quad \text{for an ideal gas} \end{aligned} \right\} \qquad (5.11)$$

The turbine isentropic efficiency is

$$\left. \begin{aligned} \eta_T &= \frac{\text{Work for the actual process}}{\text{Work for the ideal process}} \\ &= \frac{h_1 - h_2}{h_1 - h_{2s}} \\ &= \frac{T_1 - T_2}{T_1 - T_{2s}} \quad \text{for an ideal gas} \end{aligned} \right\} \qquad (5.12)$$

Note that both the ideal and actual processes are adiabatic. The subscripts 1 and 2 represent the beginning and end of the respective compressor and turbine processes respectively and s represents a change taking place isentropically.

5.2.2 Small stage efficiencies

In order to better understand these isentropic efficiencies, another efficiency is defined in exactly the same manner but for an incremental pressure rise δp instead of the overall change which is, say, p_2-p_1. That is, the overall compression (or expansion) process is made up of a large number of these stages each of pressure rise δp. These are usually termed 'small' stage and have an efficiency related to them. A compression process made up of several of these δp pressure rises is shown on the temperature–entropy diagram of Figure 5.15. It can be seen that the temperature at entry to each succeeding stage is higher than that for the equivalent isentropic process because of the inefficiency which manifests itself as an entropy increase in the previous stage. For each stage of the actual compression process, either more work is required than for the isentropic process or, for the same work, a lower pressure ratio is achieved. Assuming the working substance to be an ideal gas, the overall isentropic efficiency and the small stage efficiency can both be expressed in terms of temperature change, ΔT. That is, isentropic efficiency is

$$\eta_c = \frac{\Delta T_{(1-2s)}}{\Delta T_{(1-2)}}$$

$$= \left[\frac{(T_2 - T_1)_s}{(T_2 - T_1)}\right] \tag{5.13}$$

Now assume a number of small compression stages, $1 - A$, $A - B$, $B - C$, etc. of equal small stage efficiency. Here, the small stage efficiency is

$$\eta_s = \frac{\Delta T_{(1-As)}}{\Delta T_{(1-A)}}$$

$$= \frac{\Delta T_{(A-Bs)}}{\Delta T_{(A-B)}} \tag{5.14}$$

$$= \text{similar term for each stage}$$

The overall work is

$$w = c_P \Delta T_{(1-2)}$$

$$= c_P(\Delta T_{(1-A)} + \Delta T_{(A-B)} + \Delta T_{(B-C)}\ldots) \tag{5.15}$$

or

$$\Delta T_{(1-2)} = \frac{1}{\eta_s}(\Delta T_{(1-As)} + \Delta T_{(A-Bs)} + \Delta T_{(B-Cs)}\ldots) \tag{5.16}$$

On a T–s diagram, the lines of constant pressure diverge as they move to higher temperature and entropy regions giving larger changes of temperature for the isentropic process across that pressure change. Hence, for example,

$$\Delta T_{(A-Bs)} > \Delta T_{(a-b)s} \tag{5.17}$$

the s outside the bracket indicating that the change $a - b$ is isentropic. This gives

$$\Delta T_{(1-As)} + \Delta T_{(A-Bs)} + \cdots > \Delta T_{(1-a)s} + \Delta T_{(a-b)s} + \cdots$$

$$> \Delta T_{(1-2s)}$$

$$\eta_s \Delta T_{(1-2)} > \eta_c \Delta T_{(1-2)}$$

$$\eta_s > \eta_c \tag{5.18}$$

In summary, the overall isentropic efficiency over a finite pressure ratio must be less than the small stage efficiency of the pressure rise elements which constitute it. For a turbine, the opposite is the case and it can be shown, using similar reasoning, that

$$\eta_s < \eta_T \tag{5.19}$$

Essentially, this is because the isentropic inefficiency associated with the stage reduces the temperature drop for that stage giving a higher inlet enthalpy to the next stage. The possible work output attainable is therefore increased. The turbine effect is sometimes referred to as a reheat factor while the compressor effect is called a preheat factor.

The above gives a simple physical explanation of the phenomenon. Although turbomachines actually consist of a number of discrete rotating blade stages with stationary (stator) blade stages intervening, it can be assumed for a mathematical treatment that the compression or expansion processes proceed continuously through all stages. This leads to the idea of an 'infinitesimal' stage with a pressure rise, dp, and an efficiency associated with it. A mathematical relationship between the overall and infinitesimal stage efficiencies can then be obtained as follows.

The infinitesimal stage efficiency for a compressor can be defined similarly to the small stage efficiency as

$$\eta_P = \frac{dT_s}{dT} \tag{5.20}$$

the subscript s referring to the isentropic process while the unsubscripted value refers to the actual, adiabatic one. Substituting for dT_s from the normal isentropic relationship between temperature and pressure rearranging gives

$$\frac{dT}{T} = \frac{(\gamma - 1)}{\eta_P \gamma} \frac{dp}{p} \tag{5.21}$$

Integrating both sides now over a finite temperature and pressure range, T_1 to T_2 and p_1 to p_2, gives

$$\frac{T_2}{T_1} = \left(\frac{p_2}{p_1}\right)^{(\gamma-1)/(\eta_p \gamma)} \tag{5.22}$$

$$T_2 - T_1 = T_1 \left[\left(\frac{p_2}{p_1}\right)^{(\gamma-1)/(\eta_p \gamma)} - 1 \right] \tag{5.23}$$

But, for the overall process, the temperature difference can, from (5.11) be expressed in terms of the isentropic efficiency, η_c. That is

$$T_2 - T_1 = \frac{T_{2s} - T_1}{\eta_c}$$

$$= \frac{T_1}{\eta_c} \left[\left(\frac{p_2}{p_1}\right)^{(\gamma-1)/\gamma} - 1 \right] \tag{5.24}$$

Comparing $T_2 - T_1$ from (5.23) and (5.24) and rearranging then gives

$$\eta_c = \frac{\left(\dfrac{p_2}{p_1}\right)^{(\gamma-1)/\gamma} - 1}{\left(\dfrac{p_2}{p_1}\right)^{(\gamma-1)/(\eta_p \gamma)} - 1} \tag{5.25}$$

That is, the isentropic efficiency can be calculated from the infinitesimal stage efficiency for a given pressure ratio and isentropic index. A similar result can be obtained for a turbine.

$$\eta_T = \frac{1 - \left(\dfrac{p_2}{p_1}\right)^{\eta_p(\gamma-1)/\gamma}}{1 - \left(\dfrac{p_2}{p_1}\right)^{(\gamma-1)/\gamma}} \tag{5.26}$$

Note that the pressure ratio, r_p, is p_2/p_1 for the compressor and p_1/p_2 for the turbine. That is, $p_2 > p_1$ in (5.25), $p_2 < p_1$ in (5.26).

Typical relationships are shown on Figure 5.16 and 5.17. The importance of these lies in demonstrating the way the overall isentropic efficiency of the turbomachine varies with the pressure ratio. In a compression process, moving to a higher pressure ratio reduces the isentropic efficiency while, for a turbine, the opposite occurs. This shows why compressor development has been regarded as critical in the development of gas turbines even though, as pointed out in Chapter 2, a turbine efficiency reduction has a larger effect on the overall thermal efficiency of the engine.

As the turbomachinery is generally broken into a number of finite stages, and each of these is likely to have a known stage efficiency, it is worthwhile considering the modification of the above equations for this purpose. If it is assumed that there are m stages with each stage having an equal pressure

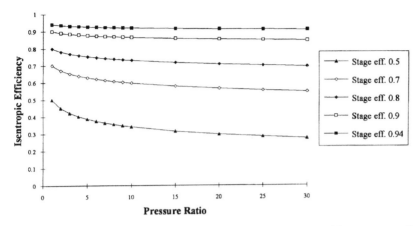

Figure 5.16 Variation of isentropic efficiency of a compressor with pressure ratio and infinitesimal stage efficiency. Isentropic index $\gamma = 1.4$.

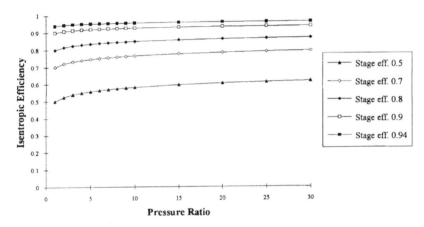

Figure 5.17 Variation of isentropic efficiency of a turbine with pressure ratio and infinitesimal stage efficiency. Isentropic index $\gamma = 1.4$.

ratio given by r_{ps} and a stage efficiency of η_s, the relationships (for $r_{ps} = r_p^{1/m}$ are

$$\eta_c = \frac{r_{ps}^{m(\gamma-1)/\gamma} - 1}{\left[1 + \dfrac{1}{\eta_s}(r_{ps}^{(\gamma-1)/\gamma} - 1)\right]^m - 1} \tag{5.27}$$

$$\eta_T = \frac{1 - \left[1 - \eta_s(1 - 1/r_{ps}^{(\gamma-1)/\gamma})\right]^m}{1 - 1/r_{ps}^{m(\gamma-1)/\gamma}} \tag{5.28}$$

As the number of stages increases, this stage efficiency will approach the infinitesimal stage value. Typically, for $m > 6$, the difference in the two is small enough to be neglected.

5.3 TURBOMACHINE AND ENGINE CONFIGURATIONS

The effect of turbomachinery on engine cycles has been dealt with briefly in Chapter 2. However, the overall relationship between the turbomachinery, combustor and work producing elements needs to be considered in more detail. That is, the turbomachine/engine needs to be examined as a system.

A great many configurations are possible, the conventional gas turbine and turbocharged reciprocating engine providing the opposite ends of a spectrum. That is, the gas turbine is a system where the combustor is a hot gas generator only with the expansion in the turbine (or jet nozzle) producing both the work to drive the compressor and the net output. With the turbocharged reciprocating engine, either SI or CI, the turbine and compressor merely enhance the mass flow rate through the combustion chamber which is itself integral with the element (i.e. the reciprocator) which provides the net work output. In these devices, no work is available from the turbine other than that required to drive the compressor. It is immediately obvious that an intermediate system is possible, this being a compound type where the exhaust gas energy from the reciprocator is in excess of the compressor requirements and is sufficient to justify an additional turbine to enhance the useful work output of the reciprocator. A number of configurations are possible with the compressor work and net work coming from some combination of either the reciprocator, turbine or both. This type of configuration may in fact be considered to be the general configuration. That is, it tends towards the gas turbine as the work extracted directly in the combustor via a reciprocator becomes smaller and the turbine work larger and to the normal turbocharged reciprocating engine when the opposite occurs. The choice is usually one or other end of this spectrum for reasons of both cost and simplicity of construction and control. As a general proposition, the best balance of extraction of useful work between the two devices, the reciprocator or the turbine, will depend on the size and the operating conditions for which it is designed and, in theory, some sort of compound system would be optimal in many cases. However, one important factor which works against a compound system is that the optimal rotational speeds of the piston/crank reciprocating device is much less than those of turbines. Integrating these into a suitable, single output shaft is therefore difficult and can only be achieved with the use of high ratio gearboxes which are heavy, costly and reduce efficiency. An alternative is to use the reciprocator to only drive the compressor and the turbine only for net work output but, while this has been considered, it has not been used in practical applications.

Three systems will therefore be discussed, these being the gas turbine at one end of the range and the SI and CI engines at the other.

5.3.1 Gas turbines

As discussed in Chapter 2, the basic limitations on the pressure ratios which can be used are those related to the effect on the overall efficiency of the individual turbomachine component efficiencies and on the maximum turbine entry temperature allowable because of metallurgical considerations. With modern aircraft gas turbines, pressure ratios of up to about 35:1 or even higher are possible. To achieve these, axial flow compressors with about fifteen or twenty blade rows in three stages are used driven from two or three shafts each rotating at different speeds. Turbine units are also axial flow with about six blade rows in the two or three (i.e. one per shaft) stages. Typical rotational speeds for the very large engines of around 300 000 N thrust are 6000 rev min^{-1} for the highest speed shaft, 2000 rev min^{-1} for the lowest. To allow for operating conditions throughout the range, various bleed valves are incorporated between compressor stages. Two-shaft engines require more flow adjustment of this type than do three-shaft but are mechanically simpler and likely to be lighter. While combustor design is always a difficult task in a gas turbine, the high inlet pressure and temperature present no difficulties other than mechanical strength problems. In fact, they aid the combustion process.

In smaller gas turbines, the centrifugal (radial) compressors which were used on Whittle's original machines are still occasionally considered for special purposes. However, the difficulty in multistaging limits their applicability to high pressure ratios.

5.3.2 CI engines

Turbocharging is very compatible with CI engines, rotational speeds for the compressor commonly being in the range 40 000 to 80 000 rpm. Compressors normally used in this application are radial types with some initial axial flow component. Peak pressure ratios up to about 3:1 are used in high output varieties although around 2:1 is a more common value. The blades are often straight (i.e. without sweep) because of the greater strength of this pattern although, at very high pressure ratios, some backsweep may be provided. The radial part of the blade is usually preceded by a short axial section, normally termed the **inducer**, but the pressure rise takes place predominantly due to the centrifugal compression in the impeller. The inducer and impeller are normally made as a single unit in small-to-medium size compressors but, in large marine or stationary engines, are often separate sections. In general, inlet guide vanes are not used but there are some examples of this at the large, high-output end of the range. Although vanes in the diffuser section improve the diffuser efficiency, vaneless types are normally used except on large, high-output units where the greater initial cost is justifiable. The turbines are radial inflow types for small (road vehicle) engines as these provide a simple low-cost robust solution but are commonly axial flow for larger types.

The most useful operating range for a heavy transport vehicle engine operating at high load is between the points of maximum torque and maximum power. Consider the case of a heavily loaded vehicle approaching an increasing grade with engine speed falling from an original value near to that for its maximum power. When a turbocharged engine is compared to a normally aspirated type of the same power output, the characteristics of the power and torque curves will be quite different. With the latter, the shape of the torque curve, and hence the power curve, is dictated principally by the rate at which the volumetric efficiency falls off at speeds above maximum torque and by the longer time available per cycle for heat transfer at the lower end of the range. There are additional effects with the turbocharged engine. It is clear that it will have a reducing turbine speed due to the falling exhaust gas mass flow rate as the engine speed is lowered and therefore a lower pressure ratio will be available from the compressor. That is, the mass flow rate through the engine falls due to the reducing engine volumetric flow rate which always occurs with engine speed reduction and is the same as with the normally aspirated engine. Additionally, it falls with the reduced density. The fall-off in power of the turbocharged engine is therefore more noticeable, the power and torque curves being peakier and the engine less flexible. This can be compensated for by the use of a gearbox with more speeds which allows engine speed in a more limited range to match road speed. It does, however, add to the expense of the turbocharging option. An alternative is to somehow maintain a high turbine speed as the gas throughput reduces thereby keeping the compressor speed and hence pressure ratio high. Turbine speed can be increased for a given exhaust gas flow rate by changing the inlet guide vane angles or reducing the nozzle area of the turbine thereby giving a high velocity head at impeller inlet. This increases the turbocharger speed, producing the desired result. However, too small a turbine nozzle results in choking at higher flow rates and hence limits the maximum turbine/compressor speed and inlet density at the high end of the range. The only way both the high and low flow rate changes can be properly optimized with a single device is by use of a variable nozzle area. While the theory is clear, producing a workable device is a major challenge. Some work has been carried out on variable geometry radial turbines but they are essentially still in the developmental stage.

The combustion in a CI engine is enhanced by the compression of the incoming charge. This is because the higher pressures and temperatures shorten the ignition delay time and reduce the tendency for diesel knock to occur. The engine compression ratio can therefore be maintained at its original level and the basic cycle thermal efficiency is unchanged. The improved combustion characteristics may, in fact, increase the burning rate reducing the thermodynamic loss in the cycle associated with slow heat addition. Because the diesel engine draws in air only, an intake flow in excess of the trapped air requirement may be used to ensure complete scavenging of the exhaust gases. This reduces their effect as a diluent, a factor which also slows combustion, and is particularly beneficial to two-stroke engines where good scavenging is difficult to obtain with normally aspirated engines. Additionally, the higher mean effective pressure ensures that friction becomes a lower proportion of the indicated work. In other words, there are many advantages in turbocharging CI engines and the process has become extremely widespread. In many cases, an aftercooler (sometimes called an intercooler), which is a heat exchanger using either air or liquid coolant as the medium, is installed between the compressor and engine. This further increases the intake density for a given compressor pressure ratio and gives an additional potential power gain. Typical plots of density ratio versus pressure ratio are shown on Figure 5.18. Many manufacturers produce several varieties of the one base engine with increasing power outputs, these being a normally aspirated, a turbocharged and a turbocharged/aftercooled version.

Improved efficiency occurs for operation at both full and part throttle because the engine is controlled only by changing the flow rate of fuel. While, as pointed out above, the reduced exhaust gas energy at part-throttle will cause some reduction in compressor pressure ratio, the overall engine cycle continues to operate at high pressures. Thus turbocharging is beneficial over a wide range of conditions. There are, however, some disadvantages in turbocharging CI engines, these being predominantly the increased

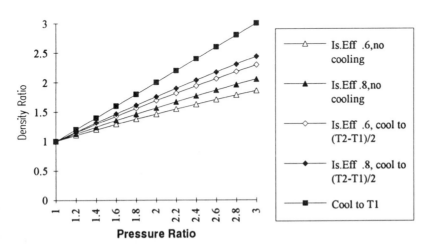

Figure 5.18 Effect of isentropic efficiency and intercooling on the density ratio obtainable across a turbocharger compressor. Isentropic index $\gamma = 1.4$.

mechanical and thermal loadings due to the higher pressures and the greater energy throughput for the system.

5.3.3 SI engines

SI engines are more difficult than CI types to turbocharge for a number of reasons, the principal being that the increased pressures and temperatures increase the tendency for end-gas autoignition (SI knock) to occur. This can, to some extent, be offset by use of an aftercooler but is usually controlled by a reduction in compression ratio (of several points from, say from 9 or 10:1 to perhaps 7 to 8:1), by spark retard or both. All reduce the engine efficiency. The latter allows different departures from the desired timing to be used at different operating points for the one engine and, at some conditions, the optimum timing may be possible. Timing can be handled by the electronic engine management system of the engine and efficiency reduction can be minimized. It is therefore the preferred option where possible. However, in most cases, some reduction in compression ratio from the equivalent, normally aspirated version is necessary and a fall in efficiency of the basic thermodynamic engine cycle must result. This is usually offset at full throttle because of the higher pressures throughout the cycle and therefore the proportionally lower friction losses. However, unlike the diesel (CI) engine, this advantage is not maintained at part load. Here, the engine control requires a reduced flow rate of both fuel and air by throttling. Throttling reduces the air density which is the opposite of turbocharging. There is the same power demand as long as it is within the normally aspirated engine range. Therefore, if a normally aspirated engine used in a particular vehicle is turbocharged, additional throttling is required to offset the turbocharging density increase at these part load operating points. This additional throttling, coupled with likely compression ratio reduction, means that the part load efficiency will always be lower than that for the original engine. The advantage of turbocharging is that maximum power and the related efficiency will be increased but not the efficiency throughout the range. However, an overall efficiency advantage over a normally aspirated engine is likely if the original normally aspirated engine is replaced by a turbocharged one of smaller capacity but the same maximum power. Here the advantages of the high engine loading will be maintained at part load.

The turbochargers used in SI engines are small very-high-speed devices. Typically, the desired pressure ratio will be about 1.5:1 maximum with 2:1 being used only on very high performance engines. Rotational speeds of 60 000 to 100 000 rev min^{-1} are common with considerably higher speeds being used on small units. The centrifugal compressor and radial inflow turbine are universally used. With an SI engine, the exhaust gas enthalpy is very high because the equivalence ratios in the combustion are around unity. Hence, there is more than adequate energy in the exhaust gas even at low engine speeds to provide the relatively low pressure ratio required from the compressor. This means that, in the medium to high speed range, the pressure and temperature of the intake charge will be excessive. The solution is to allow the unit to come up to its maximum pressure ratio at a relatively low speed and to bypass the additional exhaust gas once this condition is reached. This is usually achieved by a wastegate, which is a valve just upstream of the turbine which opens automatically to control the pressure ratios. While wastegates are sometimes used in CI engines, they or a similar system are essential for turbocharged SI engines.

5.4 TURBOMACHINE INDUCED LAG IN ENGINE RESPONSE

In all engines which use turbomachinery, response problems exist. For example, large aircraft gas turbines may take up to about seven seconds to respond from a low power setting (idle or close to it) to full power. This can cause difficulties on approach to landing when adjustment has to be made for windshear, an incorrect glide path or problems on the ground necessitating a go-around. Problems of slow response also exist with reciprocating engines. While these are not usually as critical as those in aircraft engines, they range from causing minor inconvenience to potential accidents.

The slow response of an engine is termed lag. It is due essentially to the fact that the control system is based on supplying extra fuel to the combustor but this is only a first step in the change of engine conditions. In a normally aspirated reciprocating engine, the response is rapid because the combustion pressures rise almost immediately once the air and fuel flow rates have been adjusted and only the inertia of the engine is responsible for any further lag to the final accommodation to the new conditions. Consider now a turbocharged engine or a gas turbine undergoing a rapid increase in throttle position. Note that throttle here indicates the main control lever and does not necessarily refer to any specific thermodynamic process. The engine cannot respond until the higher enthalpy gas reaches the turbine unit and brings it up to a higher speed. That is, a further time delay exists, made up of two parts which are the time from injection of the additional fuel to when the ensuing exhaust gas reaches the turbine and the turbomachinery lag itself which is quite considerable due to the inertia of the rotating components and the high rotational speed change which is often required. Further, a full response is not achieved until the higher density airflow now generated from the compressor reaches, first the combustion chamber and then the turbine. Thus there is a progressive change and, ideally, the fuelling scheduling should be adjusted progressively with the airflow. This is not always possible as is sometimes evidenced by heavy exhaust smoke from an accelerating turbocharged CI engine.

It is very necessary to overcome this lag. This is achieved by units with low rotational inertia (low mass and small diameter) and by use of wastegates where the turbocharger speed is brought rapidly to its maximum at low engine speeds. With SI engines, the fuel may be injected upstream or downstream of the turbocharger. With upstream injection (i.e. using a normal atmospheric pressure carburettor or throttle body injector), additional wall wetting and hence fuel collection on the extensive turbocharger surfaces can temporarily create over-lean mixtures at the cylinders with noticeable engine stumbling. This can be followed by a later over-enrichment at the end of the acceleration period. It should be noted, however, that there are some advantages in upstream injection because the air is cooled by the, at least partial, fuel evaporation before it enters the compressor. Simple calculations show that this temperature drop is increased in magnitude during the compression process giving a more dense charge. Nevertheless, the considerable advantages of downstream injection outweigh this and turbocharging in SI engines is therefore mostly now confined to port injection systems with injection after the compressor.

5.5 MATCHING OF TURBOMACHINERY IN ENGINES

The individual elements of a turbomachinery system used in an engine can be readily analysed by the above methods, at least under steady-flow conditions. However, an important part of the total design of an engine which contains turbomachinery is the determination of the way the system operates as a whole. For example, a typical system consists of a combustor, a turbine and a compressor. The net usable work may be obtained from reciprocating pistons directly associated with the combustor, by downstream jet expansion or from part or all of the turbine output but, in each case, the general principle that the components must relate to each other in a specific way under each set of flow conditions applies. For example, consider a system at one operating point. The combustor is provided with a suitable air flow rate at an appropriate pressure from the compressor. The exhaust reaching the turbine will consist of a related mass flow rate of gas with an energy determined from both the fuelling rate and the work extracted upstream. This gas provides the driving fluid for the turbine and will therefore fix its operating point which can be ascertained from the turbine map. The turbine in turn provides the power for the compressor whose operating point can be similarly determined. It is obvious that the mass flow rates (of gas plus the relatively small quantity of fuel from the point at which it is injected) must be the same through the three units. Assuming, in the configuration being examined, that the work output from the turbine is soley used to provide the work input for the compressor, these must also be identical when adjusted for the mechanical efficiency of each component. Further, a direct drive linking turbine and compressor means that their rotational speeds are identical. That is, the particular match of components will provide an operating point for the total system with a specified fluid flow throughput, compressor and turbine speeds and pressure ratio, engine work output and engine rotational speed (in the case of a reciprocating engine).

Components may be badly matched. For example, a particular engine operating point may be desired but a turbine may have been selected with a swallowing capacity which is too great, making it rotate too slowly with insufficient power to drive the compressor at the appropriate speed. Thus, the desired mass flow rate and pressure ratio of the compressor cannot be reached and the operating point will not be achieved. This type of mis-matching can occur in a number of ways and appropriate matching studies are essential for all engines containing turbomachinery so that the correct choice of components is achieved. The matching procedure can be carried out by a series of individual hand calculations in conjuction with the compressor and turbine maps but computer programs are now available which make this process much simpler.

The matching procedure requires the mass flow rate balance equation

$$\dot{m}_{\text{compressor air}} = \dot{m}_{\text{combustion air}} \tag{5.29}$$

$$\dot{m}_{\text{combustion air}} + \dot{m}_{\text{fuel}} = \dot{m}_{\text{exhaust}} \tag{5.30}$$

and the work (power) balance equation

$$\eta_{\text{MT}}\dot{W}_{\text{T}} = \dot{W}_{\text{c}}/\eta_{\text{MC}} \tag{5.31}$$

where η_{MT} and η_{MC} are the mechanical efficiencies and $\dot{W}_{\text{T}}$ and $\dot{W}_{\text{c}}$ are the actual (not isentropic) work output and input for the turbine and compressor respectively. The final equation is the rotational speed balance

$$N_{\text{T}} = N_{\text{c}} \tag{5.32}$$

These equations must be used with either compressor and turbine maps or with equations or tabulated data points which represent the appropriate contours. In addition, an engine or combustor model is required.

Initially, quantitative values of the above variables are generally not known. For example, if a reciprocating SI engine is to be examined when operating at a certain speed and, say, throttle position, there is no immediate knowledge of the turbomachinery speed or pressure ratios and hence the density of the intake air which will allow a direct calculation of all the operating parameters. That is, the mass flow rate of air to the engine can only be calculated from the operating point data by assuming a compressor pressure ratio. The engine model can then be used to calculate the mass flow rate of air (from the swept volume, volumetric efficiency and engine speed) and the exhaust pressure. This allows the turbine pressure ratio to be calculated giving, together with the mass flow rate, values on each of the axes of the turbine map. By either reading directly from this map or by use of the turbine equations or data, it is then possible to determine the isentropic turbine efficiency and turbine speed. Hence the power output can now be calculated. Compressor power follows and a similar use of its map, equations or data provides new values for air mass flow rate and compressor pressure ratio. These are unlikely to immediately match the originally selected ones and the process must be repeated systematically until matching occurs. It is then possible to calculate engine power output at the operating speed. This process is then repeated for other operating conditions.

For any given combination of components, the appropriate matching study will specify the points at which operation is possible and this can be easily obtained for a series of steady state conditions. The appropriate points are often plotted on the compressor map forming a full-load operating line. A typical example is shown on Figure 5.19. For part throttle with SI engines or lower fuelling scheduling for CI or gas turbines, different operating lines will result.

5.6 TURBOMACHINERY IN RECIPROCATING ENGINES

5.6.1 Unsteady flow effects

When used in conjunction with a reciprocating engine, the flow induced through the turbomachinery will usually experience pulsations emanating from the piston movement. This applies to both compressor and turbine. It is generally assumed that the steady flow analysis is sufficient for the turbomachinery but that the isentropic component efficiencies are reduced. This is because the unsteadiness will disturb the boundary layers increasing separation across the blades and the losses at the annulus. Nevertheless, most isentropic efficiencies are determined from steady flow tests. More work is required on pulsating flows.

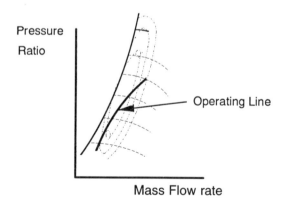

Figure 5.19 Compressor map showing a typical operating line. Note that the system is matched so that it passes as near as possible through the peak efficiency region.

5.6.2 Constant pressure and pulsed turbines

The turbine of a reciprocating engine can be connected to the exhaust system either by using a large manifold which essentially damps out the major pulsations from the blowdown process or via a small manifold that allows these pulsations to directly affect the turbine rotor. The former is known as constant pressure turbocharging. Here the high-pressure blow-down gas exhausts into the large manifold acting like a reservoir which is fed by the exhaust gases from all cylinders and is at a mean pressure that is much lower than the blowdown value. The gas from an individual cylinder enters this large volume behind a compression wave and mixes in a type of turbulent jet action, which is an irreversible process. The actual losses will depend on the insulation around this manifold, an adiabatic manifold retaining the total energy but converting it giving an increased enthalpy (i.e. increasing the temperature) at the expense of a pressure drop. In practice, however, it is not possible to retain all the initial exhaust energy in a usable form. The large manifold required by this system is not only bulky but prevents rapid engine response because the total reservoir must adjust to the new conditions before the system again stabilizes. Its advantages are the high steady-flow efficiency of the turbine and the ability on large engines to use a single large turbine unit. The disadvantages generally outweigh these and it is therefore not commonly used except on very large engines designed for operations where frequent speed changes and rapid response are not required.

The alternative pulsed system makes use of the blowdown process direct-ly by keeping the manifold volume small. On larger engines, this may require twin entry turbine units or even multiple turbines. The comparison of the energy available to the turbine can be seen on Figure 5.20. For the

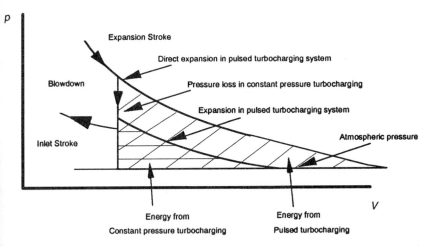

Figure 5.20 Comparison of the expansion process and possible energy obtainable from a constant pressure and a pulsed turbocharging system.

pulsed unit, the turbine receives the exhaust gas at near to the initial blow-down pressure and expands it directly to the final exhaust pressure. While in practice, this will not be an ideal process, isentropic conditions may be assumed as a first approximation. The ideal process is therefore a continuation of the engine isentropic expansion down to a minimum exhaustible pressure. Most turbochargers use this system. While the isentropic efficiencies of pulsed turbocharging are lower, its greater use of the available exhaust energy makes it more attractive than constant pressure turbocharging. The details of both systems are complex and cannot be dealt with here. The reader is referred to Watson and Janota for further information.

5.7 REFERENCES

Watson, N. and Janota, M. S. (1982) *Turbocharging the Internal Combustion Engine*, Macmillan, London.

PROBLEMS

5.1 A centrifugal compressor is to be designed for use in an engine turbocharger. It is to produce a pressure ratio of 2:1 and a mass flow rate of $0.14\,\mathrm{kg\,s^{-1}}$ of air at a turbocharger speed of $40\,000\,\mathrm{rev\,min^{-1}}$. The flow is radially outwards and the blades have an inner diameter of 50 mm and an outer diameter of 140 mm, with corresponding widths of 15 and 4 mm. It is assumed that there is zero swirl at inlet while the blades at outlet are to be backward swept at an angle of 30° to the tangent. Inlet conditions are 101 kPa, 300 K. Draw the velocity triangles at inlet and outlet and determine the inlet blade angle, the theoretical head (of air) developed and the torque and power required assuming that the flow is at constant density.

5.2 Re-examine the torque and power requirements of Problem 5.1 assuming now that (a) the compression takes place isentropically (with isentropic index of 1.4), (b) the compression process has an isentropic efficiency of 70%. Assume the only change from Problem 5.1 is in the outlet flow valves.

5.3 The situation of Problem 5.2(b) is assumed to apply. However, it is estimated that the spinning compressor rotor generates a whirl at inlet in the direction of rotation equal to 10% of the inlet tangential velocity. Recalculate the inlet blade angle, torque and power required. Is the inlet blade angle and/or the power calculated here or in Problem 5.2(b) likely to be that required in practice?

5.4 A turbine unit is to be designed to run the compressor of Problem 5.3. The mechanical efficiency of the compressor and turbine are both expected to be 85% and the turbine is to be directly coupled. The power requirement to compress the air in the compressor may be assumed to be 8.9 kW as calculated in Problem 5.2. The flow through the turbine may be considered to be radially inward and a design criterion is zero whirl at outlet. The dimensions are: outer diameter, 140 mm, inner diameter, 70 mm, blades parallel sided with blade width of 15 mm.

The mass flow rate will be 7% greater than the air flow rate through the compressor due to the addition of fuel. The exhaust gas may be considered to have an average isentropic index of 1.35, a molecular mass of 29 and is at conditions of 800 K, 350 kPa at the turbine entrance. If the expansion (pressure) ratio is to be 2.75 and the expected isentropic efficiency 75%, calculate the blade angles at inlet and outlet. If consideration is to be given to adding inlet guide vanes to improve the flow under these conditions, nominate their angle.

5.5 An axial turbine stage is designed as a fully impulse row with the blade angle at inlet and exit equal but in opposing directions. The exit velocity is solely in the axial direction. If the tangential velocity of the blade at its mean radius is $300 \, \mathrm{m\,s^{-1}}$ and the mean axial velocity is $150 \, \mathrm{m\,s^{-1}}$, determine the power output per unit of mass flow rate, the angle of the blade at inlet and outlet, the value of the absolute velocity at inlet and the outlet angle of the stator blade preceding it to the tangential direction.

5.6 A bypass fan for an aircraft is to operate at the following conditions. At 0.5 m from the centreline, the air following the diffuser section approaches an axial flow rotor at 190 m/s with conditions of 230 K, 35 kPa. The rotor spins at $2880 \, \mathrm{rev\,min^{-1}}$, and the pressure ratio across the fan is to be 1.5:1 with an isentropic efficiency of 94%. Base all calculations on a mean radius of 0.5 m from the centreline of the fan. The mechanical efficiency from the drive shaft from the turbine is 98%. Determine the work input to the gas and the work supplied from the turbine per unit of mass flow rate, the blade angle at inlet and outlet and the velocity at outlet. Assume zero whirl at inlet.

5.7 For the turbine of Problem 5.5, the inner and outer diameters of the stage are 0.3 m and 0.6 m respectively. If the blades are designed to maintain an even whirl and axial flow velocity over the entire cross section (ignoring boundary layer effects) determine the inner and outer blade and the stator angles. Also determine the net power output if the inlet is assumed to be exhaust gas at 600 kPa, 1000 K with a gas constant of 29 and the blades provide a 10% blockage of the cross-sectional area.

5.8 An air compressor for a small gas turbine with an overall pressure ratio of 10:1 consists of four stages each having an identical pressure ratio and a small stage efficiency of 90%. Calculate the overall compressor isentropic efficiency based on the infinitesimal and the small stage concepts. The isentropic index for air under these conditions is 1.4.

5.9 Show that, for two successive compressor sections with different isentropic efficiencies and different pressure ratios, the overall isentropic efficiency is given by

$$\eta_c = \frac{r_p^{(\gamma-1)/\gamma} - 1}{\left[1 + \dfrac{1}{\eta_{c1}}(r_{p1}^{(\gamma-1)/\gamma} - 1)\right]\left[1 + \dfrac{1}{\eta_{c2}}(r_{p2}^{(\gamma-1)/\gamma} - 1)\right] - 1}$$

5.10 A small stage compressor efficiency is measured in a test rig for a single stage (stator and rotor) as 92%. The compressor is to have 16 stages and is run on two shafts, the first (upstream) carrying 10 stages while the second (downstream) carries the remaining 6. The overall pressure ratio is to be 35:1. Calculate the overall isentropic compressor efficiency if (a) the calculation is based on an infinitesimal stage efficiency equivalent to that measured, (b) all stages have an identical pressure rise, (c) the second section stages provide a pressure rise per stage of 10% less than the first section. In this part, determine the isentropic efficiency of each compressor section as well as the overall value.

5.11 A compressor for a gas turbine consists of 15 equal pressure ratio stages to obtain an overall pressure ratio of 30:1 while the turbine has three equal stages only. Assuming that both compressor and turbine have the same stage efficiency of 90%, compare their overall isentropic efficiencies. For the compressor, the isentropic index is 1.4 while for the turbine it is 1.35.

Introduction to fuels and combustion

6

6.1 INTRODUCTORY COMMENTS

Human society in its modern form is heavily dependent on the provision, conversion and use of nonbiological forms of energy and it is reasonable to assert that current world population levels would not be sustainable without this input. Energy is therefore one of the most critical of the world's resources. Because the vast majority of our energy utilization emanates from nonrenewable resources, it is imperative that these be used to their maximum advantage in extraction, conversion and end-use efficiency and in maintaining the lowest possible levels of related pollution. Machines in current use rely predominantly on the conversion of the chemical energy of the fuel to first raise the temperature of the working substance to a high level before generating their work output and may therefore be regarded as forms of heat engines, subject, at the very best, to the limitations imposed by the Carnot efficiency. Biological systems, as pointed out in Chapter 3, operate by directly converting the chemical energy to work at low temperatures and can exceed these limits. While some progress has been made in developing direct conversion systems which operate on similar lines to biological systems (e.g. the fuel cell), it is likely to be a very long time before these types have the high work output rates (power), compact size, response and versatility of combustion engines, a prime requirement for their widespread introduction. Combustion engines are likely to dominate for a long time. The use of fuels to produce high-temperature working substances by combustion should therefore be carefully studied.

Although energy for heating may be obtained by methods other than combustion, for example by nuclear reactions, combustion is by far the most common method by which a stored form of energy is converted for this purpose. Combustion is essentially a reaction of molecular breakdown or formation which releases bond energy (i.e. is exothermic) and the energy so liberated may be used for a variety of tasks from simple space heating to work-producing devices (heat engines). To be self-sustaining, the rate of energy release must be sufficient to maintain the combustion temperature once the related work and heat transfers from the region take place. Thus, the combustion and thermodynamics must be examined simultaneously. In this chapter, some of the basic concepts related to combustible fuels and combustion systems are examined in a general context although, in keeping with the theme of this text, the applications will be directed towards internal combustion engines. Details of the peculiarities of internal combustion engine processes will be dealt in Chapter 7.

6.2 COMBUSTIBLE FUELS

6.2.1 Fuel types and resources

The combustible fuels in common use are predominatly from fossilized resources containing carbon and hydrogen which react with oxygen (usually in the form of air) to form major products of carbon dioxide, carbon mon-oxide and water. Pure carbon or hydrogen are themselves combustible but common fuels are likely to be some compounds of the two generally and are then designated as hydrocarbons, C_xH_y where x and y commonly cover a range from one to about twenty. Of these, coal provides the most extensive resource but, as well as being more difficult to extract and transport than liquid or gaseous types, is not suited to modern engine combustion and is used mainly for the production of steam for electrical power production. Petroleum-based oils resources are the most limited but their fuels are the most useful for internal combustion engines because of their high energy density and ease of fuel delivery. Gaseous fuels are similar or even better on this last point but considerably worse on energy density. However, their reserves are reasonable. It should be noted that the rate at which a reserve is being depleted is of at least as great an importance as the total reserve. Most projections indicate that, at current usage rates, oil supplies will be depleted before the middle of the twenty-first century, gas (i.e. natural gas, NG) should last through much of that century, while coal is a long-term resource with potential supplies for several hundred years. Obviously, all these projections are somewhat speculative as they depend on many factors which have the potential to change. These factors include new finds, the introduction of alternatives, the destruction of fields due to hostile action and the change in energy usage rates due to increased efficiencies, economic and population growth.

6.3 FUEL CLASSIFICATIONS

For an understanding of the properties of fuels and for their analysis, it is important to classify fuels in several different ways. Here, the classifications are based on the chemical structure of the individual compounds which make up the fuel and on the overall fuel properties.

6.3.1 Classification of hydrocarbons by chemical structure

Carbon is tetravalent which means that it has four valence electrons while hydrogen has a valency of one. The simplest hydrocarbon molecule is therefore CH_4, methane. However, carbon combines with itself with either single, double or triple bonds. In addition, hydrocarbons form either straight or branched chain structures and cyclic molecules and so a very large number of HC molecule types exist. To date, somewhere about 2×10^6 have been identified. These are classified into two different types and several different series. The two types of HC molecule are designated as **saturated** and **unsaturated,** these representing a full complement of hydrogen atoms or the ability to take up additional hydrogen atoms respectively. These extra atoms may be accomodated by reducing a double or triple bond or by breaking a cyclic molecule into a chain. Some of the more important series of hydrocarbons are as follows.

Alkanes $C_n H_{2n-2}$

These are usually better known under the designation of paraffins. They are open chain saturated types and may exist as a straight chain, called normal, n- types or branched chain, called iso- types. Examples are n-hexane $C_6 H_{14}$ and iso-octane $C_8 H_{18}$ the octane referring to the eight carbon atoms. The latter is, in fact, a pentane chain with three methyl groups (CH_3) added at branches, two at the second C atom and one at the fourth. The more accurate description is therefore 2, 2, 4 trimethylpentane. Alkanes have a high hydrogen content and high heating values. Branched chain types have a more compact molecular structure and have greater resistance to auto-ignition in engine combustion which is an important factor in reciprocating engines.

Cyclanes $C_n H_{2n}$

These are also often called either cycloparaffins or naphthenes. They have closed chain structure and are essentially straight-chain saturated hydrocarbons turned into a circular structure. The hydrogen atom at either end is therefore deleted to allow the final C to C bond and can be replaced if the ring is broken. Thus they are unsaturated themselves. An example is cyclopentane $C_5 H_{10}$.

Alkenes $C_n H_{2n}$

These are also called olefins. They contain one double bond and are open chain unsaturated hydrocarbons. The double bond may occur between different carbon atoms in large molecules giving a series of different compounds with the same identity. An example is butene $C_4 H_8$.

Alkynes C_nH_{2n-2}

These are sometimes called acetylenes because they are based on the structure of the compound acetylene which is the simplest molecule of the series. To expand the series, additional methyl groups replace the hydrogen atoms. That is, they all contain one triple bond and are open chain unsaturated hydrocarbons. An example is acetylene C_2H_2.

Aromatic hydrocarbons C_nH_{2n-6}

These are again built up on a basic molecule which is the benzene ring C_6H_6 which contains three double and three single bonds and is a closed-chain unsaturated hydrocarbon. To expand the series, methyl groups replace the hydrogen atom. The aromatics have a strong odour, hence the name and some are the basis for pleasant smelling oils. They are stable molecules with high knock resistance but are targeted for reduction in fuels because of their carcinogenic potential and high relative reactivity in photochemical smog formation.

Monohydric alcohols $C_nH_{2n-1}OH$

These are essentially alkanes with one H atom replaced by an OH radical and are one form of a group of fuel compounds known as oxygenates because they contain oxygen. They are not fossil fuels but may be obtained from them or by fermentation of vegetable matter. The common types are ethyl alcohol C_2H_5OH and methyl alcohol CH_3OH.

Representative diagrams of the different molecular structures are shown on Figure 6.1.

6.3.2 Classification by properties of petroleum-based fuels

Petroleum-based fuels are by far the most common internal combustion engine fuels and are a blend of many hydrocarbons distilled and reformed in the refining process to give the appropriate properties. A typical sample might contain around fifty compounds of which fifteen to twenty might be in significant enough proportions to affect the properties. Samples will differ from time to time from the one refinery depending on its crude oil input, the overall output and the type of fuel required. For example, winter grade gasolines can contain more low boiling point components than summer grade. The most common groups are as follows.

Gasoline (petrol)

This is a blend of colourless, volatile liquid petroleum fractions which boil between about 30 °C and 200 °C, a typical relative density being about 0.73.

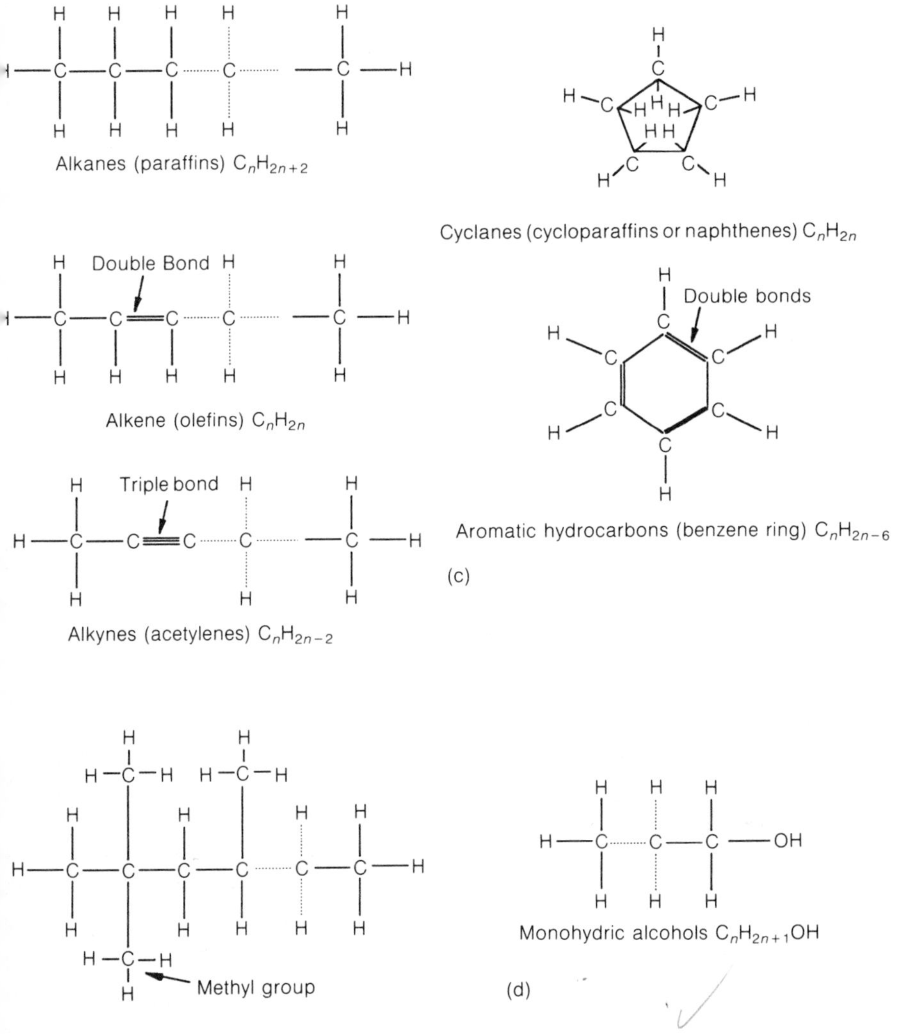
Alkanes (paraffins) C_nH_{2n+2}
Cyclanes (cycloparaffins or naphthenes) C_nH_{2n}
Double Bond
Alkene (olefins) C_nH_{2n}
Double bonds
Triple bond
Aromatic hydrocarbons (benzene ring) C_nH_{2n-6}
(c)
Alkynes (acetylenes) C_nH_{2n-2}
Methyl group
Monohydric alcohols $C_nH_{2n+1}OH$
(d)

········ Indicates possible extension of chain in multiples of this group of larger molecules

Figure 6.1 Some basic structures of the common hydrocarbon fuels (a) straight chain (n-) molecules, (b) branched chain (iso-) molecules, (c) cyclo molecules, (d) oxygenates (alcohols).

Motor fuels cover the whole of this boiling point range, aero fuels (for SI reciprocating aero engines) usually only 50 °C to 170 °C. For overall average properties, gasolines are sometimes considered to be the single compound, iso-octane C_8H_{18} but this is substantially different, particularly in volatility. Actual components range from C_4 to C_9 (sometimes to C_{12}) and cover many types of hydrocarbons depending on the crude oil source and the refining process. A more typical average composition is $(CH_{1.85})_n$ where $n \simeq 7$. The fuel is refined to give appropriate octane numbers, volatility, etc., but the composition of the final product may still vary considerably from one source to another. Net heating values (lower) are about $44\,MJ\,kg^{-1}$ $(32\,MJ\,l^{-1})$.

Because gasoline is a spark ignition engine fuel, one of its most important characteristics is the antiknock property called **octane number**. There are several scales, the most commonly used being the **research octane number** (RON) which is a light duty test and the **motor octane number** (MON), a heavy duty test. On-road tests indicate that something between the two provides the best indication, the midpoint being commonly used. The tests compare the antiknock characteristics of the fuel with a blend of iso-octane which has good antiknock properties and normal heptane which has poor ones. The percentage of iso-octane in the blend which produces the same knock intensity as that of the fuel under consideration is termed that fuels octane number. For most commercially available petrols, the RON rating is around 90 to 98 which is about 8 to 10 points higher than that of the MON rating. The difference, RON minus MON is called the **fuel sensitivity**, i.e. sensitivity to increased knock under heavy load conditions. There may, of course be very little difference between the scales. For example, iso-octane itself must have a value of 100 on both scales. Various additives can improve the octane rating, the most notable being tetra-ethyl lead (TEL). Addition of only small quantities, from 0.64 to $1.28\,gml^{-1}$ improves the octane rating by 5 to 10 points, but, due to the adverse health effects of lead and its effect on catalytic converters for emission control, it is now phased out in many countries. More compact HC molecules improve the knock rating as does alcohol and other (ether) oxygenates but much larger quantities than TEL are required.

Kerosene

This is a relatively involatile, colourless fraction which boils between $150\,°C$ and $250\,°C$. Relative density is about 0.8. Its net heating value (lower) is about $43\,MJ\,kg^{-1}(32\,MJ\,l^{-1})$. Overall properties are sometimes taken as those of the single component substance dodecane $C_{12}H_{26}$ but similar comments to those which were made for gasoline apply and a formula, $(CH_{1.85})_n$ where $n \simeq 10$ may be more appropriate. Kerosene is most widely used as a gas turbine fuel, and can be separately classified as Jet A, Jet B, etc. The latter is a wider-cut fuel and, as it therefore contains more volatile components, it is more dangerous and is generally prohibited for commercial, though not always from military use.

Diesel fuels ✓

These compete with aviation kerosene for the same sector of the crude oil barrel and, as both applications are increasing rapidly, they are one of the fuels for which supply in the future may be potentially difficult. The term **diesel oil** covers a fairly broad spectrum from

1. Light oils for automotive use. These are brown oils boiling between 180 °C and 360 °C with a relative density of about 0.86. Lower heating value is about $42 \, \text{MJ} \, \text{kg}^{-1} (35 \, \text{MJl}^{-1})$.
2. Heavy oils for stationary or marine engines. These are darker brown, heavy oils of relative density of about 0.87 with a similar boiling range to light oils. Their lower heating value is about $41.5 \, \text{MJ} \, \text{kg}^{-1} (36 \, \text{MJl}^{-1})$.

Again diesel fuels consist of many compounds and while the single component, dodecane, is sometimes used to simulate them, a more appropriate estimate is $(CH_{1.85})_n$ where $n \simeq 11$. For their knock quality, diesel fuels are given a **cetane number** which is their equivalent to the antiknock octane scale used with gasoline. It should be noted that the knock in CI engines has an entirely different cause to that in SI engines and the scales are not equivalent. Fuels with a high octane number generally have a low cetane number and vice versa but one cannot be inferred from the other. The cetane number is again a comparative test with originally the test fuel being compared with a mixture of normal cetane, $(C_{16}H_{34})$ rating 100 and α-methylnapthalene $(C_{11}H_{10})$ rating 0. The proportion of cetane in the comparative fuel gives the cetane number of the fuel under test. In fact, the low end test fuel has now been replaced with heptamethylnonane with a rating of 15 and the proportioning is modified accordingly to give the same result. Values which are typical of a good quality fuel are in the range 45 to 50.

Fuel oils ✓

These, in the past were predominantly used in industrial furnaces although they are now sometimes used in large marine diesels which have to fuel at a wide range of ports. This fuel is very viscous and requires heating so that it will flow adequately. It is a brownish, black oil, of relative density about 0.95. Lower heating value is about $40 \, \text{MJ} \, \text{kg}^{-1} (38 \, \text{MJl}^{-1})$.

STol

6.3.3 Classification by properties of other fossil fuels

Coals ✓

Coal is classified from the oldest in order in fossilization, which is black coal, anthracite, to the youngest, which is lignite or brown coal, only one step further in the process than peat. Coal contains carbon, hydrogen, oxygen, sulphur and ash in different proportions depending on its source. Although considerable local variations might be found a typical analysis by weight could be ash, sulphur, hydrogen, carbon and oxygen in proportions by mass of 9%,1%,3%,83% and 4%. In general, much of the carbon is unattached to hydrogen atoms. The oxygen is in the form of water vapour and hence, some of the 3% hydrogen is bound with oxygen rather than carbon. The dry coal would have a composition somewhere around $C_{18}H_5$.

Coal is used either directly in solid form where it is normally pulverised (PF) but is sometimes gasified. Coal gasification is a very old process dating to about the end of the eighteenth century with methods giving low-energy gases such as producer gas and water gas. These have been important in the supply of gas to towns (hence the name town gas in Britain) but are now being superseded in many places by natural gas and LPG. Producer gas is obtained by passing insufficient air for total combustion over coal or coke. Water gas in various forms adds steam to the process which is reduced to hydrogen. Both producer and water gas are high in carbon monoxide. Modern processes are available from which gases with greater heating values are formed and are generally classified as medium and high-energy gas. These production processes occur at high pressures and use oxygen rather than air so that the nitrogen content of the gas is low. In general, medium energy gas is practicable at present and is produced by either the Lurgi, Koffers–Totzek or Winkler processes.

Gaseous fuels

Other types of gaseous fuels that are not derived from coal gas are in widespread use. These are, in particular, natural gas (NG) and liquefied petroleum gas, (LPG). The latter is currently widely used in engines as well as having domestic and industrial usage. The former has, as yet, only small internal combustion engine use although it is under consideration for the future because of its wide supply and low emission levels.

Natural gas

Natural gas is predominantly methane (CH_4) but it also contains other gases which may include ethane, propane, butane and heavier hydrocarbons in the form of condensate. Typically at the well-head 75% to 95% by volume is methane while 2% to 8% is ethane. Methane and ethane boil at atmospheric pressure at temperatures of $-162\,°C$ and $-87\,°C$ respectively and therefore can only be liquefied at very low temperatures, pressure alone being insufficient. Generally liquefaction (to LNG) is impractical for many purposes at present and the gases may be stored for use as compressed natural gas (CNG) at 15 MPa to 25 MPa. This means that the weight of container to weight of stored fuel is high which is a severe impediment in transport applications. Current improvements in LNG tanks with vastly improved insulation may improve its prospects in the near future. There are, however many advantages of natural gas in that methane is a clean burning (it has a high H/C ratio), safe, easily handled fuel. It has a low laminar flame speed and therefore does not ignite easily. Also it is lighter than air and disperses readily. It is in wide supply throughout the world and is regarded by some as the essential bridging fuel from when current petroleum stocks are exhausted to the introduction of liquid alternative fuels. It has a lower heating value of about $47\,\mathrm{MJ\,kg^{-1}}$.

Liquefied petroleum gas

The chief components of LPG are propane C_3H_8 and butane C_4H_{10} and under moderate pressures of around ten atmospheres (1 MPa), these become liquid at atmospheric temperature which facilitates storing and handling in comparison with LNG. Boiling points at atmospheric pressure are propane $-42\,°C$, butane $-0.5\,°C$. The majority of LPG (about 70%) occurs as well gas often alongside natural gas from which it is separated and contains only paraffin hydrocarbons. The remainder is a refinery gas and contains a small proportion of olefins. Most natural gas is odourless and a malodorant is added so that it can be easily detected. It is not inherently a safe fuel because of its easy evaporation and density greater than air. It can therefore collect as a mixed air/fuel vapour in low areas with explosive potential. Several major disasters have ensued from LPG explosions of this type. Relative density of the liquid is about 0.51 (propane) to 0.58 (butane). The lower heating values for both are about $46\,\mathrm{MJ\,kg^{-1}}$.

6.3.4 Classification by properties of 'alternatives' fuels

A number of other fuels exist and, although they are not in widespread use at the moment, are potential alternatives for the future. Some of these are considered below.

Alcohols (methanol, ethanol, propanol, butanol)

Of these, the most commonly available are methanol and ethanol and, while there has been some consideration of their use for general combustion processes, they are predominantly considered as an alternative engine fuel. They can be used either as fuels in their own right or for blending with petroleum-based compounds either as a fuel extender or octane improver as they have good antiknock properties, particularly on the RON scale. It should be noted that the octane number, however, is not directly proportional to the constituents in blends. Proportions up to 15% to 20% with petroleum require little engine modification for spark ignition engines although high methanol content fuels are under consideration and these will require substantial changes in order to optimize engine performance. They have high latent heat, high boiling point and flash point and therefore greater handling safety than petrol. Cold starting of engines and high alcohol content fuels is difficult due to the high latent heat. Production is, for ethanol, by processing (fermenting) vegetable matter and, for methanol, by wood distillation, or by chemical formation from natural gas or coal.

Hydrogen

This is a most attractive fuel for advanced applications with a high heating value on a mass basis although the density is very low. It is very abundant in nature although, because it is highly reactive, it is in the form of compounds. It vaporizes at $-253\,°C$ at atmospheric pressure, $-240\,°C$ at 12 atmospheres which, even more so than NG, makes it difficult to store. It has a very high flame velocity, wide flammability limits and low ignition energy and must be handled with care although its very buoyant flame tends, in some circumstances, to minimize damage. It also escapes through seals very readily but its lightness and high diffusivity allow it to disperse readily. Its characteristics give it some advantages and disadvantages as an engine fuel. The wide flammability limits allow quality control (by varying the fuel/air ratio) which is more efficient than quantity (throttling) control. However, the low ignition energy may cause abnormal combustion, for example, pre-ignition from surfaces or particles.

Biogas

This, like NG, is mainly methane, CH_6. It is a major decomposition product and is, for example, the principal component of sewerage gas or of farm manure, the former being about 65% methane, 35% CO.

Synthetic fuels

Synthetic oxygenates such as MTBE, ETBE (methyl, ethyl tertiary butyl ether) which are derived from methyl and ethyl alcohol respectively via ethers. Consequently they are more expensive than the alcohols but have better characteristics for use in IC engines. Other synthetic petrols may be derived from coal or NG either via methanol or by hydrogenation of coal. Various techniques for the production of olefins from coal are under investigation but are essentially at an experimental level at present.

Nitrogen hydrides (ammonia NH_3, hydrozine N_2H_4)

These are synthesized from nitrogen and hydrogen, the latter usually being produced by electrolysis. Ammonia is toxic but is readily identifiable through its distinct pungent smell. Nitrogen hydrides have good antiknock properties which allow compression ratios of up to about 12:1 in otherwise conventional SI engines. The energy densities are low making them relatively expensive alternatives.

Benzol

This is a byproduct of coal tar and is an aromatic hydrocarbon blend of approximately 70% benzene, 18% toluene. It has high antiknock properties and a high initial boiling point and has sometimes been added to petrol to improve its octane rating. Doubts about its safety in relation to carcinogenic effects and problems in smog formation mean that it is now eliminated as an alternative.

6.4 COMBUSTION, HEAT RELEASE AND CONTROL

The essential points in a study of combustion are:

- to be able to calculate the quantity of heat that a fuel can release;
- to understand the combustion process, and control it so that this heat will be released in the appropriate quantities at the appropriate rate;
- to operate in such a way that undesirable emission products will be minimized.

Each different combustion device has its own characteristics and so each is controlled in a different way. While specific situations related to engines will be considered later, a general overview is given here.

6.4.1 Reactions and reaction chains

Reactions occur when the molecules of the various reactants collide with a sufficient energy so that molecular modification rather than simple rebound occurs. The minimum energy requirement is called the **activation energy** of which more will be discussed later. If, however, the reaction had to wait until the correct species collided in such a way, the time during which it occurred would be substantially longer than observed in practice. What in fact occurs is that a series of 'reaction chains' form where highly active intermediate radicals are formed. Fast reactions occur when the chains branch and the active radicals proliferate. Alternatively, some reactions may break the chain and slow the reaction and it may then cease entirely.

To understand this mechanism, consider, for example, the simple H_2, O_2 reaction to form water. It should be noted that the following is not in any way intended to be a complete description but one route only to the completed compound.

Assume that an OH radical and a H_2 molecule are available and collide with the appropriate energy.

$$H_2 + OH \longrightarrow H_2O + H \qquad \text{Reaction 1}$$

A highly active hydrogen radical is now available and may react with an oxygen molecule.

$$H + O_2 \longrightarrow OH + O \qquad \text{Reaction 2}$$

The original OH radical has been replaced and Reactions 1 and 2 can be repeated. That is, a chain exists whereby by formation of water creates an active radical which can help the process repeat itself. However, in addition, there is still the O radical left over from the second reaction. This can combine with a H_2 molecule to form two more radicals.

$$O + H_2 \longrightarrow OH + H \qquad \text{Reaction 3}$$

This frees an additional OH to start a new chain via Reaction 1 and a new H to start a third chain through the mechanism of Reaction 2. Thus within a very short time, the number of active species and chains has increased and this continues essentially as a geometrical progression at each further step. The chain branching mechanism, which may be extremely complex, is therefore an integral part of a fast reaction.

Chain breaking, as mentioned above, is the reverse phenomenon. Here a reaction may remove active radicals and form stable compounds as for example in a methyl CH_3, H reaction.

$$CH_3 + H \longrightarrow CH_4$$

Alternatively, collision with an inert substance, which may be either the wall of the vessel or an inert molecule such as argon, may inhibit a reaction by removing energy. However, this latter collision sometimes acts to increase the reaction energy and provides a catalyst for its continuation.

6.4.2 Homogeneous mixtures of fuel and oxidant

While in practice homogeneous mixtures are not always achieved, it is the mode in which most spark ignition engines are designed to operate. Several situations are possible. Consider a closed, fixed volume vessel which contains a homogeneous mixture. If the temperature of the mixture is raised completely uniformly, as shown on Figure 6.2(a), until the ignition temperature of the fuel at the then-prevailing vessel pressure is reached, the fuel will theoretically autoignite simultaneously everywhere. Here the only thing that controls the energy release and hence the pressure and temperature rise is the rate at which the chemical processes occur. This is called the reaction rate and is studied as a branch of science called **chemical kinetics**. Even for relatively simple fuels, there are a series of reaction chains through intermediate species which must be followed for the reaction to proceed to its conclusion. Thus chemical kinetics is a study of the rates at which a series of many relatively simple reactions occur. The total reaction time is typically very short, of the order of milliseconds.

A second possibility is that the mixture temperature is well below the autoignition temperature and a point energy source, such as a spark, is required as shown on Figure 6.2(b). The flame front now progresses through the mixture from the point of ignition. This occurs because of the heat and mass transfer from the reacting gases to the mixture immediately ahead of it raising it to the ignition temperature. As this reacts, it now affects the unburned gas ahead of it in the same way. The flame velocity

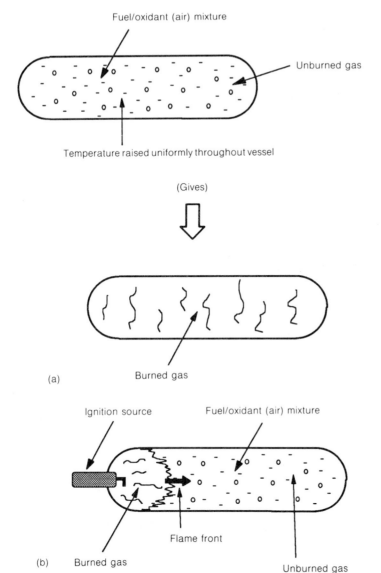

Figure 6.2 Some methods by which reactions could occur in closed vessels. (a) reaction under uniform ignition throughout, (b) reaction from a point source ignition.

depends on many things, such as the properties of the mixture both burnt and unburnt and the reaction rate itself. Two types of flame fronts are readily distinguishable. These are a laminar flame front which progresses at an experimentally very repeatable speed at each set of conditions. There are many correlations available for simple fuels giving the laminar flame front velocity, S_l, as a function of pressure and temperature. When the mixture becomes turbulent, the flame velocity increases by about an order of magnitude. A turbulent flame front is much less predictable as the level of turbulence has to be defined. Generally, most modern theories show the turbulent flame front velocity, S_t, as a function of S_l and a characteristic associated with the turbulence. A turbulent flame front has a much higher energy release rate than a laminar one and is necessary for high output combustors. Note that the phenomena in steady flow burners are similar with the flame burning back against the outflowing unburnt mixture.

6.4.3 One-dimensional homogeneous mixtures

A special case of flame fronts in homogeneous mixtures is when the combustion occurs in a long parallel tube closed at the end near the ignition point with the gas initially settled. Because of its simplicity, this case is most amenable to mathematical analysis and it has been extensively studied. Two types of combustion are identifiable and are shown on Figure 6.3. These are deflagrations and detonations. The burning gas causes a pressure wave to move ahead of it down the tube setting the unburned gas in motion away from the flame front. In a deflagration, the flame front is very much slower than the pressure wave and it progresses in the same way as above, i.e. by heat transfer to the gas ahead. A detonation occurs under special conditions when the pressure (i.e. shock) wave is of sufficient strength to raise the gas behind it to the ignition temperature. Hence, the flame front becomes attached to the shock and moves at supersonic velocity with very high pressures and temperatures behind it.

6.4.4 Injection of fuels

This is a very common combustion problem. Here the fuel must mix or diffuse into the air to form appropriate air/fuel ratios before combustion can occur. The fluid mechanics of the process are therefore extremely

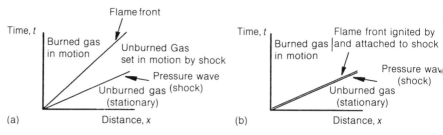

Figure 6.3 Modes of burning in a one-dimensional situation (deflagration and detonation from the closed end of a pipe) (a) deflagration, (b) detonation.

important and include entrainment of the surrounding gas, mixing, diffusion and, possibly, formation of fuel layers (wall jets) on the surfaces when impact with them occurs. With gaseous fuels, the diffusion is usually very rapid. With liquids the fuel must evaporate before the diffusion can take place. Hence, atomization into drops, drop size and fuel volatility are important as well as the mixing and diffusion processes. For solid fuels, similar processes to drive off the volatile substances for later mixing play a role in controlling the reaction. For diesel engines, gas turbines, industrial oil burners, the droplet burning processes are important. Figure 6.4 illustrates these situations.

6.4.5 Flame stability

The stability of a flame is an important criterion in steady-flow burners such as gas turbine combustors. Consider burning in a duct in which a fuel/air is flowing as shown in Figure 6.5. For the flame to be stable, i.e. for it to remain in a fixed location, the mixture flow must be adjusted so that, at the flame front, its velocity is exactly equal and opposite to that of the flame front. Hence, the absolute velocity which is the velocity of the flow plus the velocity of the flame relative to the flow is zero. When this condition breaks down, the flame may either '*flash back*' or '*blow out*'. In the former, the mixture velocity is too low and the flame progresses upstream. In the latter, the mixture velocity is too great forcing the flame downstream. To stabilize flames over a range of conditions in burners, baffles are used to create recirculation zones and swirl. This is called **flame stabilization** and is inherent in good combustor and burner designs. If the flame can adjust so

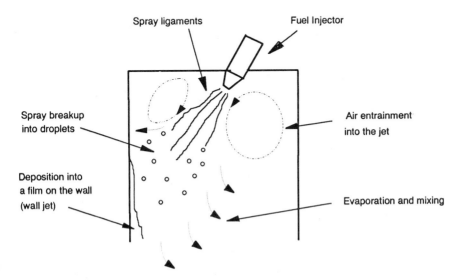

Figure 6.4 Mixing processes following fuel injection.

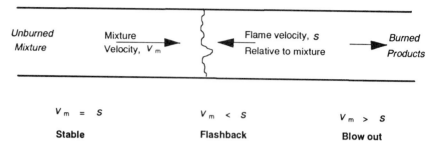

Figure 6.5 Stability of a plane flame front moving in one-dimensional flow relative to the unburned mixture.

that the angle of the cone that it forms can vary, a reasonable range of gas flow velocities can be tolerated. This is because the flame front velocity is a component of the progression against the flow as shown in Figure 6.6. That is, the flow velocity is $S/\sin \theta$ where θ is the half cone angle. The change of cone angle with flow adjustment is observable in most open burners such as the well-known Bunsen burner.

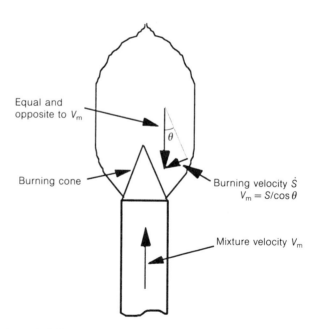

Figure 6.6 A conical flame showing the ability to adjust to different mixture velocities by change of cone angles.

6.5 BASIC COMBUSTION DEFINITIONS AND LIMITS

A number of basic definitions are required before a study of combustion can be undertaken. Some of the necessary ones related to engine operation follow.

6.5.1 Fuel/oxygen relations

Stoichiometric

This term applies where there is just sufficient oxygen to complete combustion of fuel to CO_2 and H_2O only, e.g. taking butane, C_4H_{10} as an example

$$2C_4H_{10} + 13O_2 \longrightarrow 8CO_2 + 10H_2O$$

Insufficient oxygen (i.e. 'rich' mixtures)

In the above example less than 13 molecules of O_2 would be available and CO, CO_2 and H_2O would be formed.

Excess oxygen (i.e. 'lean' mixtures)

Also, for the above example, more than 13 molecules would be available and CO_2, H_2O and O_2 would exist on the right hand side at the end of the reaction.

Use of air

The oxygen-carrying medium is usually air and similar terms apply. The mean molecular weight of air is 28.97. Appropriate values, both actual and reasonable approximations for calculations, are given on Table 6.1.

Ratios for calculations are 3.76 of N_2:1.0 of O_2 by volume, 3.31 of N_2:1.0 of O_2 by mass. Nitrogen and oxygen combine at high temperatures to form NO_x. This is less than 1% of the combustion products but helps form

Table 6.1

Air	Proportion by volume (%)		Proportion by mass (%)	
	Actual	*Use*	*Actual*	*Use*
Nitrogen	78.03	79	75.45	76.8
Oxygen	20.99	21	23.20	23.2
Argon	0.94	0	1.30	0
CO	0.03	0	0.05	0
Other	0.01	0	negligible	0

photochemical smog. In general, nitrogen acts as a moderator to reduce combustion temperatures and alter equilibrium composition.

Air/fuel (or fuel/air)

These ratios are expressed as mass ratios unless otherwise noted.

6.5.2 Combustion limitations

Flash point

This is important in determining the safety with which a fuel can be handled and is determined by standard tests. If a quantity of liquid fuel in a '*cup*' is heated, it will vaporize, mix with the air above it and eventually will ignite if the temperature becomes high enough. The temperature at which ignition occurs but after which the flame immediately extinguishes, is called the **flash point.**

Fire point

This is similar to the flash point except the flame now continues to burn. The fire point is the lowest temperature at which this occurs.

Explosion limits

These are the thermodynamic conditions (pressure, temperature) within which, for a given air/fuel mixture ratio, an explosive reaction must occur spontaneously. That is, raising the pressure at a fixed temperature and equivalence ratio will eventually lead to an explosion. This is the **explosion limit**. It essentially occurs because the molecules are now more densely packed and collision probabilities will increase. The explosion also occurs if the temperature is raised at lower pressures.

Some substances (e.g. CO, H) have two or more pressure limits, a lower and a higher one, where an explosion occurs. While the '*no explosion*' region at very low pressures occurs because of the increasing proportion of wall collisions with their effect of breaking the reaction chain is clear, the actual mechanism for the higher limit is not so obvious. It is related to the formation of intermediate species which end the reaction. For example, hydrogen has three explosive limits at temperatures between $400\,°C$ and $570\,°C$ due to the formation of an intermediate species. With higher order hydrocarbons, (above propane, CH_4), there is a region where '*cool flames*' occur. These are progressive reactions from the wall and usually occur as a low temperature blue flame which is a partial reaction but they are not explosions.

Flammability limits

Flames do not necessarily propagate through a fuel/air mixture if an ignition source is introduced. That is, there are rich and lean mixture ratios outside of which the mixture will not continue to support combustion. Although wall quenching effects may appear to alter the flammability limits in some experiments, when these effects are removed, the following general conclusions can be reached.

1. Pressure change below atmospheric does not affect the range except at very low pressures.
2. In general, the lean limit is unaffected at higher than atmospheric pressure although the rich limit is increased.
3. When burning in oxygen rather than air, the lean limit is unaffected although the rich limit increases.

Some flammability limits of commonly used fuels are shown on Table 6.2. Table 6.3 gives a comparison of oxygen and air flammability limits.

6.6 COMBUSTION FUNDAMENTALS

A basic understanding of the chemical reactions during combustion and the way the appropriate heat release can be calculated is required. The following outlines the appropriate procedures.

Table 6.2 Air flammability limits

	Lean	*Rich*	*Stoichiometric**
Methane	5	14	9.47
Propane			
Heptane	1	6	1.87
Hydrogen	4	76.2	29.27
Carbon monoxide	12.5	76.2	29.5
Acetyladehyde	3.97	57	7.7
Acetylene	2.5	80	7.7

Table 6.3 Comparison of oxygen and air flammability limits

	Lean		*Rich*	
	Air	O_2	*Air*	O_2
H_2	4	4	74	94
CO	12	16	75	94
NH_3	15	15	28	79
CH_4	5	5	14	61
C_3H_8	2	2	10	55

6.6.1 Energy liberated during combustion

Constant volume combustion

The energy liberated as heat during a combustion process is normally determined from measurements in constant volume vessels called *combustion bombs*. Here the fuel and air at known initial conditions is ignited and burned with subsequent cooling to allow the products to return to a low temperature. The heat released is collected in a calorimetric container surrounding the bomb and so can be precisely measured. Conditions are specified as a standard state of 1 atmosphere, 25 °C before and after the reaction. If the test is not carried out to these conditions, corrections must be made for the internal energy variation with temperature both for the reactants and the products in order to obtain the correct value of internal energy of combustion which is essentially the chemical energy available from the fuel.

For a constant volume vessel, there is no work as the $\int p\,dv$ is zero. Hence, from the first law of thermodynamics the heat transfer is equal to the final minus the initial internal energy.

$$Q = U_{P,T_2} - U_{R,T_1} \qquad (6.1)$$

where subscripts P and R stand for products and reactants respectively and T_2 and T_1 are the temperatures at the end, following the transfer of the heat Q, and beginning of the combustion process.

Now at the standard state, T_1 and T_2, are both 298 K (25 °C). The atmosphere condition is represented by a superscript O. The **internal energy of combustion** is defined as being equal to

$$\Delta U_c^0 = U_{P,298}^0 - U_{R,298}^0 \qquad (6.2)$$

This is a standardized value and is available in tabulations as a property of the fuel. It should be noted that, in general, the number of kmoles of products and reactants are not equal and that it is therefore not possible to reach this pressure condition experimentally. For general use, a correction must be used to relate the standard state values of the products and reactants to those that actually exist. The first law of thermodynamics expression must now be altered to include the term ΔU_c. That is

$$Q = U_{P,T_2} - U_{R,T_1} = (U_{P,298}^0 + \Delta U_{P,T_2}) - (U_{R,298}^0 + \Delta U_{R,T_1}) \qquad (6.3)$$

where $\Delta U_{P,T_2}$ and U_{R,T_1} are the changes in sensible internal energy from the standard state to the required values. The above equation can be modified to

$$Q = \Delta U_c^0 + \Delta U_{P,T_2} - \Delta U_{R,T_1} \qquad (6.4)$$

If the products and reactants are ideal gases the correction terms are a function of temperature only. For the general case, the value ΔU_c^0 or its equivalent is called the **internal energy of reaction**. This is because many reactions are endothermic (i.e. require heat) rather than exothermic (i.e. liberating heat) as in the combustion case. As this text is concerned only with combustion, the term internal energy of combustion will be used. The values are normally expressed per kmole of the fuel and are therefore given an intensive symbol, Δu_c. Heat transfer Q (or q when expressed per kmole of fuel) is from the substances in the reaction and is therefore negative. Hence, the internal energy of combustion should also be expressed as a negative value but this is not always adhered to in tabulations. Finally, it needs to be noted that (6.4) is a general expression and relates the heat transfer for any constant volume, combusting system even when temperatures T_1 and T_2 differ substantially from the standard state value of 298 K.

Constant pressure and steady flow combustion

Most combustion processes, even in closed systems, are not constant volume while a great many are open systems. In each case, the correct First Law relationship must be used. For example, in both closed system, constant pressure and open system, steady flow systems (with negligible kinetic energy and potential energy changes), the First Law gives the heat transfer as the change in enthalpy.

$$Q = H_{P,T_2} - H_{R,T_1} \qquad (6.5)$$

Using the same logic as previously, the relation becomes

$$Q = H_c^0 + \Delta H_{P,T_2} - \Delta H_{P,T_1} \qquad (6.6)$$

where ΔH_c^0 is called the **enthalpy of combustion**. It is enthalpy of combustion rather than Internal Energy of Combustion values that are most usually available. Again, tabulations are usually per kmole of the fuel, are given as Δh_c^0 and are negative. The value of ΔH_c^0 and ΔU_c^0 can be related in the usual way. That is

$$\Delta H_c^0 = \Delta U_c^0 + \Delta(pV) = \Delta U_c^0 + \Delta(NR_0 T) \qquad (6.7)$$

for ideal gases, assuming both products and reactants are in the gas phase. Here N is the number of kmoles and R_0 is the universal gas constant. But in this case, the temperature is fixed at 298 K. Hence

$$\Delta H_c^0 = \Delta U_c^0 + RT_{298}\Delta N = \Delta U_c^0 + R_0 T_{298}(N_P - N_R) \qquad (6.8)$$

Typical values of enthalpy of combustion are high. For example, the reaction of CO to form CO_2 has an enthalpy of combustion of about $-283\,MJ\,kmol^{-1}$. The values of $R_0 T\Delta N$ are, however, relatively small. For the oxidation of CO, it is about $-1.24\,MJ\,kmol^{-1}$ or 0.43% of the enthalpy of combustion. This is less than the normal uncertainty in the calculation and it is therefore usually ignored. That is, internal energy of combustion values are taken as identical to enthalpy of combustion in most cases. It must be stressed, however, that this does not apply to the correction terms, $\Delta H_{\mathrm{P},T_2}$ and $\Delta H_{\mathrm{P},T_1}$ (or $\Delta U_{\mathrm{P},T_2}$ and $\Delta H_{\mathrm{P},T_1}$) where the correct process relations must be used.

The following points should be noted.

1. Tabulated values are quite often given per kilogram of the initial substance instead of per kilomole of the fuel.
2. In such tabulations, the negative sign is often not used. It must then be remembered that the heat is being removed from the combusting substance.
3. When water is formed in the combustion process, more heat is removed if the water is condensed than if it remains as vapour. This is referred to as the higher heating (or enthalpy) value. The lower heating (or enthalpy) value retains water as vapour, e.g. combustion of methane, CH_4:

   higher heating value $55\,564\,\mathrm{kJ\,kg^{-1}}$
   lower heating value $50\,047\,\mathrm{kJ\,kg^{-1}}$

Enthalpy of formation

Enthalpy values are tabulated in two different ways – either as enthalpy of combustion or as enthalpy of formation, the latter now being the more common. Both are evaluated at the standard state of 101.3 kPa, 25 °C. Enthalpy of formation is designated here as $h^0_{\mathrm{f},298}$. It is important to examine the relation between these two. Note that several different nomenclatures are used in different sources. In some texts, the term $\Delta h^0_{\mathrm{f},298}$ is used instead, in others a bar is added over the h to indicate that the units are per kmole of the substance.

Take, for example, the hydrogen oxidation reaction

$$H_2 + \tfrac{1}{2}O_2 = H_2O$$

For this reaction, the enthalpy of combustion is $-285.83\,\mathrm{MJ\,kmol^{-1}}$ and is essentially the heat which will be extracted in starting with the reactants at 298 K, burning the hydrogen and returning the products to the same temperature. However, instead of considering the combustion of hydrogen as the base, the formation of water may be used. Assigning zero enthalpy of formation to the stable elements on the left hand side of the reaction allows the value of $-285.83\,\mathrm{MJ\,kmol^{-1}}$ to be considered as the energy release from them in forming water. That is, the enthalpy of formation of water is $-285.83\,\mathrm{MJ\,kmol^{-1}}$.

This may seem a small and perhaps unnecessary change. However, in practice, it allows a more concise tabulation and simple calculation. Consider, for example, the oxidation of carbon. There are several possibilities, C to CO, C to CO_2, CO to CO_2. Enthalpy of combustion values would therefore be: (i) Carbon to CO, $-110.53\,\mathrm{MJ\,kmol^{-1}}$; (ii) Carbon to CO_2, $-393.52\,\mathrm{MJ\,kmol^{-1}}$; (iii) CO to CO_2, $-282.99\,\mathrm{MJ\,kmol^{-1}}$. The value in

(iii) may be determined by subtracting (i) from (ii). Intermediate values where some CO and some CO_2 are formed also occur.

In cases (i) and (ii) the values are also enthalpies of formation of CO and CO_2. If enthalpy of formation values are used, only these need to be considered even for reaction (iii).

$$\text{Enthalpy of formation} \quad \underbrace{CO}_{-110.53} \quad + \quad \underbrace{\tfrac{1}{2}O_2}_{0} \quad \longrightarrow \quad \underbrace{CO_2}_{-393.52}$$

$$Q = H_{P,T_2} - H_{R,T_2} = -393.52 + 110.53 = -282.99 \text{ MJ kg}^{-1}$$

That is, the enthalpy available is calculated directly as the difference between the enthalpies of formation of the products and reactants. In complex reactions, this provides a system less prone to error and ties in with the concept of using the difference in the total enthalpy (i.e. enthalpy of formation plus the correction for temperature) of each substance at the beginning and end of the reaction in order to find the energy available.

It should be reiterated that the enthalpy of formation of all stable elements is arbitrarily defined as zero at the standard state.

Using an internal energy or enthalpy of formation scheme, the first law statements becomes, for a constant volume reaction

$$Q = (U^0_{fP\,298} + \Delta U_{P,T_2}) - (U^0_{fR\,298} + \Delta U_{R,T_1}) \tag{6.9}$$

$$= \sum N_P(u^0_{fP\,298} + \Delta u_{P,T_2}) - \sum N_R(u^0_{fR\,298} + \Delta u_{R,T_1})$$

Noting the negligible difference between internal energy and enthalpy of formation, it can be rewritten as

$$Q = \sum N_P(h^0_{fP\,298} + \Delta u_{P,T_2}) - \sum N_R(h^0_{fR\,298} + \Delta u_{R,T_1}) \tag{6.10}$$

For a constant pressure or steady flow reaction

$$Q = (H^0_{fP\,298} + \Delta H_{P,T_2}) - (H^0_{fR\,298} + \Delta H_{R,T_1}) \tag{6.11}$$

$$= \sum N_P(h^0_{fP\,298} + \Delta h_{P,T_2}) - \sum N_R(h^0_{fR\,298} + \Delta h_{R,T_1})$$

Enthalpy of formation data for selected substances is given on Table 6.4.

6.6.2 Flame temperature

Excluding for the present the effects of dissociation which will be discussed later, the temperature at the end of combustion may be calculated by a simple energy balance. For example, if no heat is transferred during a process, the heat, Q, that would have reduced the enthalpy of the products to the standard state is now used in raising the temperature of the products. This limiting temperature for zero heat transfer gives what is called the **adiabatic flame temperature**. The first law is applied as before. For

Table 6.4 Entropy of formation of selected substances* all substances are gas phase unless indicated by s(solid) or ℓ(liquid)

Substance	Formula	$h_f^0(kJ\,kmol^{-1})$
Carbon	C (s)	0
Hydrogen	H_2	0
Nitrogen	N_2	0
Oxygen	O_2	0
Carbon monoxide	CO	$-110\,530$
Carbon dioxide	CO_2	$-393\,520$
Water	H_2O	$-241\,820$
Water	H_2O (ℓ)	$-285\,830$
Oxygen	O	$249\,170$
Hydrogen	H	$218\,000$
Nitrogen	N	$472\,680$
Hydroxyl	OH	$39\,040$
Methane	CH_4	$-74\,850$
Acetylene	C_2H_2	$-26\,730$
Ethene (Ethylene)	C_2H_4	$52\,280$
Ethane	C_2H_6	$-84\,680$
Propene (Propylene)	C_3H_6	$20\,410$
Propane	C_3H_8	$-103\,850$
n-Butane	C_4H_{10}	$-126\,150$
n-Pentane	C_5H_{12}	$-146\,440$
n-Octane	C_8H_{18}	$-208\,450$
n-Octane	C_8H_{18} (ℓ)	$-249\,910$
Benzene	C_6H_6	$82\,930$
Methyl Alcohol (Methanol)	CH_3OH	$-200\,890$
Methyl Alcohol (Methanol)	CH_3OH (ℓ)	$-238\,810$
Ethyl Alcohol (Ethanol)	C_3H_5OH	$-235\,310$
Ethyl Alcohol (Ethanol)	C_2H_5OH (ℓ)	$-277\,690$

* From Selected Values of Chemical Thermodynamic Properties JANAF Thermochemical Tables, 1971, Dow Chemical Co. additional data may be found in texts such as Wark, K., 1989, *Thermodynamics*, McGraw-Hill, New York and Sonntag, R. E., and Van Wylen, G. J., 1991, *Introduction to Thermodynamics*, Wiley, New York.

the constant volume case

$$Q = \sum N_P(h_{fP\,298}^0 + \Delta u_{P,\,T_2}) - \sum N_R(h_{fR\,298}^0 + \Delta u_{R,\,T_1}) = 0 \qquad (6.12)$$

That is

$$\sum N_P(h_{fP\,298}^0 + \Delta u_{P,\,T_2}) = \sum N_R(h_{fR\,298}^0 + \Delta u_{R,\,T_1}) \qquad (6.13)$$

and, for constant pressure, steady flow

$$\sum N_P(h_{fP\,298}^0 + \Delta h_{P,\,T_2}) = \sum N_R(h_{fR\,298}^0 + \Delta h_{R,\,T_1}) \qquad (6.14)$$

Here T_2 is the adiabatic flame temperature. In either case, the enthalpies of formation of both products and reactants can be found from standard data and the initial condition of the reactants is usually known, allowing temperature T_2 to be determined such that the equation balances. In cases where a standard value of the adiabatic flame temperature is quoted, this assumes that temperature T_1 is at 298 K.

It should be noted that tabulated values (Table 6.5) of enthalpy (or internal energy but usually the former) with temperature are normally used to obtain the corrections $\Delta h_{P, T_2}$ etc. and this requires a trial and error solution. These assume the substance is an ideal gas throughout the temperature range considered. Alternatively, rough estimates may be obtained directly by using approximate values of c_v or c_P for the temperature range of the correction. For example, if the reactants are H_2 and O_2 at an initial 75 °C, an approximate calculation can be made for each using $\Delta u = c_v \Delta T$, with, for a diatomic gas, $c_v \approx 21$ kJ kmol^{-1} K^{-1} in the 25 °C to 75 °C range. For the reactants, the error is usually small because reactant temperatures are mostly within a limited temperature range from 298 K but greater error is likely for very high adiabatic flame temperatures. To replace the trial and error method by a direct exact solution, the values of c_P and c_v must be expressed as functions of temperature and integrated over the temperature range. This is most desirable for analysis by computer.

In most real cases, an intermediate situation exists in relation to the heat transfer. That is, there is some heat transfer but it is insufficient to reduce the products to 298 K and the flame temperature rises to a lesser extent than in the adiabatic case. Here the actual heat transfer must be estimated and the general first law equation applied. The adiabatic case is, however, useful as it shows the limiting situation for a given reaction.

6.6.3 Pressure change in combustion processes

The final pressure, p, in a nonreacting case with heat addition is usually determined from the initial conditions and the final temperature by use of the intensive form of the ideal gas equation. For example, in the closed system, constant volume case, use of $pv = RT$ gives

$$p_2 = \frac{T_2}{T_1} p_1 \tag{6.15}$$

However, in a reacting case, the number of kmoles changes from the reactants to the products and the form of the ideal gas equation which is most conveniently used is that with the Universal gas constant R_0

$$pv = NR_0 T \tag{6.16}$$

For the constant volume example above, the final pressure is then

$$p_2 = \frac{N_P T_2}{N_R T_1} p_1 \tag{6.17}$$

Table 6.5 Selected values of ideal gas enthalpies (kJ/kmol) of gaseous substances involved in combustion*

Temperature (K)	0	289	500	1000	1500	2000	2500	3000	4000	5000
N_2	−8669	0	5912	21460	38405	56141	74312	92738	130076	167858
O_2	−8682	0	6088	22707	40610	59199	78375	98098	138913	180987
H_2	−8468	0	5883	20686	36267	52932	70492	88743	126846	166808
N	−6197	0	4197	14590	24983	35376	45777	56220	77534	100115
O	−6728	0	4343	14862	25296	35715	46133	56220	77534	99224
H	−6197	0	4197	14590	24983	35376	45769	56162	76948	97734
CO	−8669	0	5929	21686	38848	56739	75023	93542	131026	168929
CO_2	−9364	0	8314	33405	61717	91450	121926	152862	215635	279295
H_2O	−9904	0	6920	25978	48095	72689	98964	126361	183280	241957
OH	−9171	0	5991	20933	36840	53760	71417	89584	126934	165222

* Data from *Introduction to Thermodynamics*, Sonntag and Van Wylen (1991). More detailed tabulations are available in that or other texts such as that by Wark, K., 1989, *Thermodynamics*, McGraw-Hill, New York, and Keenan, J. H. and Kaye, J., *Gas Tables*, Wiley, New York.

Note that N_R excludes liquids and solids whose volume is negligible and N_P excludes any water which has condensed to liquid.

6.7 DISSOCIATION AND CHEMICAL EQUILIBRIUM

At any temperature, a small number of molecules have broken up (i.e. dissociated) into subparticles due to the vibration of the molecule and the proportion increases with temperature increase. The dissociation of gases into various subcomponents absorbs energy and limits flame temperatures and useful energy available to the combustor. Various species exist under any given pressure and temperature conditions both of which independently vary the amount of dissociation. The presence of other gases is also a factor influencing the final dissociated composition. The calculations required appear from the equations to be reasonably simple but are, in fact, quite complex as these equations do not readily form sets amenable to simple iterative or similar types of solutions. However, they are a thermodynamic calculation and do not require experimental data or constants in order that a solution be obtained. This is because, while molecules combine and dissociate continually, the proportion which is dissociated for any given set of restraints remains a constant. Therefore, an equilibrium exists which is termed **chemical or thermodynamic equilibrium** and this implies that rate processes are unimportant in the final analysis.

In an internal combustion engine, the pressure and temperature are already high at the start of the reaction due to the compression process and some minor increase in dissociation levels occurs before combustion takes place. A major increase in temperature occurs during the reaction and this causes a marked increase in dissociation levels. The corresponding increase in pressure usually offsets this slightly although temperatures at every stage are less than those that would be achieved had no dissociation occurred. Thus the effect of dissociation on internal combustion engines is quite severe and more pronounced than in many other reactors. Some of the energy is regained as the molecules recombine during expansion but this does not offset the original loss because there is now an insufficient further expansion ratio to fully extract this energy.

To understand chemical equilibrium, a first law analysis is insufficient and the second law is required. The entropy (or more conveniently, the Gibb's function) is examined which leads to equilibrium constants from which calculations can proceed. The concept of chemical equilibrium will be first discussed in relation to the entropy with which the reader will be more familiar but later equations will be developed from the Gibb's function which is a more convenient although directly related property.

6.7.1 Entropy changes during combustion

The change in entropy during a reaction can be calculated from the individual entropies of the reactants and products as follows.

$$\Delta S = \sum (N_P s_P) - \sum (N_R s_R) \tag{6.18}$$

The second law of thermodynamics, however, relates ΔS to the heat transfer

$$\Delta S \geqslant \int dQ/T \tag{6.19}$$

which can be evaluated. If the process is adiabatic ΔS must be positive in any real system. However, for any process with heat transfer from the system as in most combustion systems, the positive value of ΔS is reduced and may conceivably become negative if the heat transferred is sufficient. In either case the procedure for examining the reaction limitation is the same but the adiabatic case, being simpler, will be used here for illustrative purposes.

Consider a chemical reaction of substances A and B forming C and D. Assume, for simplicity, that only one kmole of each substance satisfies the chemical conditions although it must be recognized that, in general, this is not the case. The reaction relating the start to the end of the process is

$$A + B \longrightarrow C + D \tag{6.20}$$

At the start of the reaction, only A and B exist but as the reaction proceeds, increasing quantities of C and D will form until the reaction is complete when only these last two will exist as shown above. At any intermediate stage, all four substances will coexist with A and B decreasing and C and D increasing in proportion. This can be expressed as

$$A + B \longrightarrow xC + xD + (1 - x)A + (1 - x)B \tag{6.21}$$

where x lies between 0 and 1 and indicates that the reaction is that proportion completed. For an exothermic (combustion) reaction, the enthalpy (or internal energy) of the reactants and products can be evaluated as functions of x and hence the energy liberated can be determined giving, from the procedure adopted in Section 6.5.2, the flame temperature, mixture pressure and partial pressure of the constituents for each stage of the reaction. In addition, the entropy change can be evaluated as

$$\Delta S(x) = x S_C + x S_D + (1 - x) S_A + (1 - x) S_B - S_A - S_B = x(S_C + S_D - S_A - S_B) \tag{6.22}$$

and

$$S(x) = S_1 + \Delta S(x) \tag{6.23}$$

where S_1 is the entropy at the start of the reaction.

As x increases, for the adiabatic condition, ΔS is positive and $S(x)$ must also increase with time as the reaction process proceeds to completion. It is not possible at any stage for $S(x)$ to decrease as this will mean ΔS is negative and, if the calculations show it doing that, this must be the point at which the reaction can proceed no further even if it is not complete at the

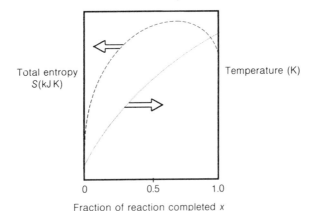

Figure 6.7 Progression of a reaction towards competition in an adiabatic vessel. Typical entropy and temperature plots.

time. That is, the combustion will stop and no further increase in temperature will occur. This is illustrated on Figure 6.7 which shows the $S(x)$ curve peaking (i.e. $\Delta S = 0$) and consequently the reaction ceasing at x somewhere between 0.5 and 1.0. It is typical of all combustion reactions although the exact value of x at which cessation occurs will vary with the reaction, the temperature, pressure and the presence of inert gases.

Obviously, if the process is not adiabatic, the calculation is a little different. Here, the significant plot is $S(x) - \int dQ/T$ versus x, Q being negative for heat transferred from the system. In all cases the maximum value of the curve is the limit of the combustion process.

6.7.2 Chemical equilibrium equations

From the above, it can be seen that reactions stop if a situation is reached when the entropy of the isolated system is decreasing. An alternative approach is to consider, artificially, a fully completed combustion reaction with the products then breaking up (dissociation). Using the Gibbs function

$$G = H - TS \qquad (6.24)$$

At its turning point, when G increases, the process must cease. This criteria is equivalent to the statement in terms of the entropy function but is usually more convenient in developing dissociation equations. From Chapter 3, it will be remembered that, for a single species

$$dG = V dp - S dT \qquad (6.25)$$

If there are a number of species, a property is not only a function of two other properties but of the quantity of each species in the mixture. That is, the Gibbs function for a mixture of gases may be expressed as

$$G = f(T, p, N_1, N_2, \ldots, N_i) \tag{6.26}$$

where $N_1, N_2, \ldots, N_i$ represent the number of moles of each chemical species. The total differential equation is therefore

$$dG = \left(\frac{\partial G}{\partial P}\right)_{T,N_i} dp + \left(\frac{\partial G}{\partial T}\right)_{P,N_i} dT + \sum_i \left(\frac{\partial G}{\partial N_i}\right)_{P,T}$$

$$= V dp - S dT + \mu_i d N_i \tag{6.27}$$

The quantity, μ_i is called the **chemical potential** as it represents the effect on the Gibbs function of the change in composition during a reaction.

When dissociation occurs, there are essentially two governing equations. Imagine a reaction A and B to form C and D with one kmole of each of the substances being involved. Now a kmole is made up of millions of molecules which react on an individual basis, i.e. if one molecule of A collides with one molecule of B, one molecule of each of C and D will form and the original two will disappear. In practice, most reactions will not have unit coefficients but the same principle is involved at a molecular level, the dissociation reaction is always stoichiometric. This can be represented in general terms by

$$v_A A + v_B B = v_C C + v_D D \tag{6.28}$$

This equation merely represents the way molecules of A and B dissociate to form molecules C and D. The actual chemical equation representing the total number of kmoles may be different because the constituents may not be in stoichiometric proportions and because the reaction may not be complete. That is, A and B formed from the combustion only partly dissociate to substances C and D. The equation representing this is

$$N_{A1} A + N_{B1} B \longrightarrow N_C C + N_D D + N_A A + N_B B \tag{6.29}$$

Here N_A and N_B are less than N_{A1} and N_{B1} respectively. Now at constant temperature and pressure, (6.29) gives

$$dG_{T,P} = \sum_i \left(\frac{\partial G}{\partial N_i}\right)_{P,T} dN_i = \sum_i \mu_i dN_i$$

$$= \mu_A d N_A + \mu_B d N_B + \mu_C d N_C + \mu_D d N_D \tag{6.30}$$

Now assume the reaction proceeds by a very small amount indicated by a proportionality constant, k, to the right. This will take place according to the molecular equation and will represent a limited number of molecular collisions. Therefore

$$dN_A = -k v_A \quad dN_B = -k v_B \quad dN_C = +k v_C \quad dN_D = +k v_D \tag{6.31}$$

from which

$$dG_{T,P} = k(-\mu_A v_A - \mu_B v_B + \mu_C v_C + \mu_D v_D) \tag{6.32}$$

At equilibrium, as with but opposite to the entropy, G, is changing from a decreasing to an increasing value. That is

$$dG_{T,P} = 0$$

giving

$$-\mu_A v_A - \mu_B v_B + \mu_C v_C + \mu_D v_D = 0 \tag{6.33}$$

The chemical potential of gas in a mixture at the mixture pressure is the same as the chemical potential of the pure substance measured at its partial pressure. Therefore, to obtain values for the mixture, pure values at the partial pressures are used. If, for a pure substance at constant temperature, T, and partial pressure, p_i, the number of moles is changed by a given multiple, G, is increased in the same proportion. Therefore

$$\mu_{i,T} = \left(\frac{\partial G_i}{\partial N_i}\right)_{T,P_i} = \frac{\Delta G_{i,T}}{\Delta N_i} = \frac{G_{i,T}}{N_i} = g_i \tag{6.34}$$

At a given temperature T

$$g_i = h_{i,T} - T s_{i,T} \tag{6.35}$$

Values are usually tabulated at a standard state of one atmosphere for a range of temperatures and this is indicated by a superscript 0 and a subscript T. The enthalpy of an ideal gas does not vary with pressure (i.e. $h = f(T)$ only) but the entropy does. Substituting from the conventional formulas for these gives

$$\mu_{i,T} = h_{i,T}^0 - T_{i,T}^0 + R_0 T \ln(p_i/p^0) \tag{6.36}$$

where p^0 is the pressure of one atmosphere. If p_i is measured in atmospheres, this becomes

$$\mu_{i,T} = h_{i,T}^0 - T_{i,T}^0 + R_0 T \ln p_i = g_{i,T}^0 + R_0 T \ln p_i \tag{6.37}$$

Substituting now in the molecular equation (6.33) gives

$$-v_A(g_{A,T}^0 + R_0 T \ln p_A) - v_B(g_{B,T}^0 + R_0 T \ln p_B)$$
$$+ v_C(g_{C,T}^0 + R_0 T \ln p_C) + v_D(g_{D,T}^0 + R_0 T \ln p_D) = 0 \tag{6.38}$$

That is

$$\Delta(v_i g_{i,T}^0) + \Delta(R_0 T \ln p_i) = 0 \tag{6.39}$$

But $v_A g_{A,T}^0$ is the number of moles of A undergoing the reaction times the Gibbs function/mole at temperature T. Therefore

$$\Delta(v_i g_{i,T}^0) = \Delta G_T^0 \tag{6.40}$$

This is the change in total standard state Gibbs function for that reaction. Substituting in (6.40) now gives

$$-\Delta G_T^0 = R_0 T(-v_A \ln p_A - v_B \ln p_B + v_C \ln p_C + v_D \ln p_D)$$

$$= R_0 T \ln \left[\frac{(p_C)^{v_C}(p_D)^{v_D}}{(p_A)^{v_A}(p_B)^{v_B}} \right] \tag{6.41}$$

Thus the change in the standard state Gibbs function can be expressed in terms of the temperature and partial pressures of the mixture components.

6.7.3 Equilibrium constants

It is usually more convenient to rewrite the above as

$$\left[\frac{(p_C)^{v_C}(p_D)^{v_D}}{(p_A)^{v_A}(p_B)^{v_B}} \right] = \exp(-\Delta G_T^0 / R_0 T) \tag{6.42}$$

where $\exp(-\Delta G_T^0 / RT)$ is called the equilibrium constant, K_P.

Note that the logarithms may be taken either to the base 10 or base e for practical purposes of tabulation in different texts. Be careful to check the form of any tables used.

Note also the following points.

1. K_P is a function of two properties and is a property itself. It may be obtained either from tabulations or it can be calculated from other properties, i.e.

$$\ln K_P = -\frac{\Delta G_T^0}{R_0 T} = -\frac{\Delta H_T^0}{R_0 T} + \frac{\Delta S_T^0}{R_0} \tag{6.43}$$

2. K_P is independent of pressure during the reaction as the pressure terms were moved to the other side of the equation leaving only the standard state enthalpy and entropy for its definition. However, K_P varies with temperature.
3. This definition uses partial pressures of products in the numerator and reactants in the denominator. This is the most usual. However, some authors define it the other way around. Their value, say, K_P' is then the inverse of K_P.
4. The value of K_P depends on the way the equations are written, e.g.

$$CO + \tfrac{1}{2}O_2 \longrightarrow CO_2 \qquad (a)$$
$$2CO + O_2 \longrightarrow 2CO_2 \qquad (b)$$
$$CO_2 \longrightarrow CO + \tfrac{1}{2}O_2 \qquad (c)$$

These are all essentially the same equation, but Gibbs function for (b) is twice that for (a). That is, $K_{P(b)}$ is $K_{P(a)}^2$. Reversal of the reaction changes the sign of ΔG_T. Therefore, $K_{P(c)}$ is $K_{P(a)}^{-1}$.

from which

$$dG_{T,P} = k(-\mu_A \nu_A - \mu_B \nu_B + \mu_C \nu_C + \mu_D \nu_D) \qquad (6.32)$$

At equilibrium, as with but opposite to the entropy, G, is changing from a decreasing to an increasing value. That is

$$dG_{T,P} = 0$$

giving

$$-\mu_A \nu_A - \mu_B \nu_B + \mu_C \nu_C + \mu_D \nu_D = 0 \qquad (6.33)$$

The chemical potential of gas in a mixture at the mixture pressure is the same as the chemical potential of the pure substance measured at its partial pressure. Therefore, to obtain values for the mixture, pure values at the partial pressures are used. If, for a pure substance at constant temperature, T, and partial pressure, p_i, the number of moles is changed by a given multiple, G, is increased in the same proportion. Therefore

$$\mu_{i,T} = \left(\frac{\partial G_i}{\partial N_i}\right)_{T,P_i} = \frac{\Delta G_{i,T}}{\Delta N_i} = \frac{G_{i,T}}{N_i} = g_i \qquad (6.34)$$

At a given temperature T

$$g_i = h_{i,T} - T s_{i,T} \qquad (6.35)$$

Values are usually tabulated at a standard state of one atmosphere for a range of temperatures and this is indicated by a superscript 0 and a subscript T. The enthalpy of an ideal gas does not vary with pressure (i.e. $h = f(T)$ only) but the entropy does. Substituting from the conventional formulas for these gives

$$\mu_{i,T} = h^0_{i,T} - T^0_{i,T} + R_0 T \ln(p_i/p^0) \qquad (6.36)$$

where p^0 is the pressure of one atmosphere. If p_i is measured in atmospheres, this becomes

$$\mu_{i,T} = h^0_{i,T} - T^0_{i,T} + R_0 T \ln p_i = g^0_{i,T} + R_0 T \ln p_i \qquad (6.37)$$

Substituting now in the molecular equation (6.33) gives

$$-\nu_A(g^0_{A,T} + R_0 T \ln p_A) - \nu_B(g^0_{B,T} + R_0 T \ln p_B)$$
$$+\nu_C(g^0_{C,T} + R_0 T \ln p_C) + \nu_D(g^0_{D,T} + R_0 T \ln p_D) = 0 \qquad (6.38)$$

That is

$$\Delta(\nu_i g^0_{i,T}) + \Delta(R_0 T \ln p_i) = 0 \qquad (6.39)$$

But $\nu_A g^0_{A,T}$ is the number of moles of A undergoing the reaction times the Gibbs function/mole at temperature T. Therefore

$$\Delta(\nu_i g^0_{i,T}) = \Delta G^0_T \qquad (6.40)$$

This is the change in total standard state Gibbs function for that reaction. Substituting in (6.40) now gives

$$- \Delta G_T^0 = R_0 T (-v_A \ln p_A - v_B \ln p_B + v_C \ln p_C + v_D \ln p_D)$$

$$= R_0 T \ln \left[\frac{(p_C)^{v_C} (p_D)^{v_D}}{(p_A)^{v_A} (p_B)^{v_B}} \right] \tag{6.41}$$

Thus the change in the standard state Gibbs function can be expressed in terms of the temperature and partial pressures of the mixture components.

6.7.3 Equilibrium constants

It is usually more convenient to rewrite the above as

$$\left[\frac{(p_C)^{v_C} (p_D)^{v_D}}{(p_A)^{v_A} (p_B)^{v_B}} \right] = \exp(-\Delta G_T^0 / R_0 T) \tag{6.42}$$

where $\exp(-\Delta G_T^0 / RT)$ is called the equilibrium constant, K_P.

Note that the logarithms may be taken either to the base 10 or base e for practical purposes of tabulation in different texts. Be careful to check the form of any tables used.

Note also the following points.

1. K_P is a function of two properties and is a property itself. It may be obtained either from tabulations or it can be calculated from other properties, i.e.

$$\ln K_P = -\frac{\Delta G_T^0}{R_0 T} = -\frac{\Delta H_T^0}{R_0 T} + \frac{\Delta S_T^0}{R_0} \tag{6.43}$$

2. K_P is independent of pressure during the reaction as the pressure terms were moved to the other side of the equation leaving only the standard state enthalpy and entropy for its definition. However, K_P varies with temperature.

3. This definition uses partial pressures of products in the numerator and reactants in the denominator. This is the most usual. However, some authors define it the other way around. Their value, say, K'_P is then the inverse of K_P.

4. The value of K_P depends on the way the equations are written, e.g.

$$CO + \tfrac{1}{2} O_2 \longrightarrow CO_2 \qquad \text{(a)}$$

$$2CO + O_2 \longrightarrow 2CO_2 \qquad \text{(b)}$$

$$CO_2 \longrightarrow CO + \tfrac{1}{2} O_2 \qquad \text{(c)}$$

These are all essentially the same equation, but Gibbs function for (b) is twice that for (a). That is, $K_{P(b)}$ is $K_{P(a)}^2$. Reversal of the reaction changes the sign of ΔG_T. Therefore, $K_{P(c)}$ is $K_{P(a)}^{-1}$.

6.7.4 Equilibrium constants in terms of number of moles

Partial pressures are usually calculated from the mole fraction. For use in calculations it is therefore usually more convenient to bypass this and work directly from the number of moles. Equation (6.43) is then further modified for this purpose:

$$K_P = \left[\frac{(p_C)^{v_C}(p_D)^{v_D} \cdots}{(p_A)^{v_A}(p_B)^{v_B} \cdots} \right] \quad (6.44)$$

The $\cdots$ indicate that there may be more than two substances on either side of the reaction. Now the partial pressures are related to the total pressure by

$$p_i = \frac{N_i}{N} p$$

Substituting gives

$$K_P = \left[\frac{(N_C)^{v_C}(N_D)^{v_D}}{(N_A)^{v_A}(N_B)^{v_B}} \right] \left[\frac{p}{N_m} \right]^{[(v_C + v_D) - (v_A + v_B)]}$$

$$= \left[\frac{(N_C)^{v_C}(N_D)^{v_D}}{(N_A)^{v_A}(N_B)^{v_B}} \right] \left[\frac{p}{N_m} \right]^{\Delta v} \quad (6.45)$$

Here N is the total number of moles in the mixture and includes any inert gases. This equation is now in the most convenient form for calculations.

6.7.5 Numerical values for equilibrium constants

Most thermodynamics texts tabulate K_P for most of the common reactions over a substantial temperature range. However, for the less common reactions, the values need to be calculated from data from other sources. The best source for thermodynamic data is the JANAF Tables. This data has been reduced to a best fit equation by several authors and is then in a suitable form for computer programs. A useful source of these equations is Gordon and McBride (1971), where properties for enthalpy, entropy and Gibbs function are given by standard power series equations as functions of temperature in a dimensionless form. Each has up to seven constants. Two ranges are given and hence fourteen constants are tabulated. The equation for specific Gibbs function, for example, is

$$\frac{g}{RT} = A_1(1 - \ln T) - \frac{A_2 T}{2} - \frac{A_2 T^2}{6} - \frac{A_4 T^3}{12} - \frac{A_5 T^4}{20} - \frac{A_6}{6} - A_7 \quad (6.46)$$

Hence

$$\frac{\Delta G}{RT} = \sum_{iRT} N_i g_i - \sum_{jRT} N_j g_j \quad (6.47)$$

where i, j represent the right and left-hand side of the reaction respectively. This allows the equilibrium constant to be calculated.

6.7.6 Calculation of equilibrium compositions

The equilibrium constants can be determined either from tables or Gibbs function as above. To obtain the equilibrium composition, there are a number of unknowns, these being equal to the number of substances on the right-hand side of the chemical equation. In the example used for the equations above it is four, substances A, B, C, and D respectively with number of kmoles N_{A2}, N_{B2}, N_C, N_D with N_m being the sum of these. The equilibrium constants for the assumed equilibrium temperature are known. Also values of v_A, v_B, v_C and v_D can be determined from the dissociation equations. Thus, three other equations (in this case) are required to complete the set. These are obtained by balancing the number of atoms of each element on either side of the combustion equation and, in general, this is sufficient to obtain a solution.

6.7.7 Effect of pressure changes at constant temperature

Increase in temperature increases the dissociation. That is, it decreases the kmoles of products as calculated originally for the undissociated combustion reaction. It can be seen that the mixture pressure is also a significant variable in the equilibrium equation and it is important to know how pressure affects dissociation. If the exponent of the pressure term is positive as is the case in most reactions, an increase in pressure will reduce the dissociation, that is increase the number of kmoles of the original combustion products and push the overall combustion reaction towards completion. Effectively a positive exponent means that pressure tends to increase as the dissociation proceeds and high pressure helps limit further dissociation. If the exponent is negative the opposite is true. A zero exponent indicates that the reaction is independent of pressure. The addition of an inert gas will decrease the partial pressure of the products and has the same effect as a pressure reduction on the system.

6.7.8 Simultaneous reactions

In many systems, several substances can be involved in equilibrium reactions. These may be due to completely different substances or two alternative dissociations from the one substance. In both cases the treatment is the same, the equilibrium constant being evaluated for each dissociation equation separately. This gives sufficient equations for solution and the calculation is then similar to that when only one substance dissociates.

6.7.9 Significant dissociation equations

In combustion work, a number of dissociation reactions are required. These essentially are the dissociation of the major products of hydrocarbon combustion, water and carbon dioxide. In addition, oxygen and nitrogen combine to form NO which is a major emission product. Also, at very high temperatures, oxygen, nitrogen and hydrogen ionize to atomic forms. The equations are

1. $CO_2 = CO + \frac{1}{2}O_2$
2. $H_2O = H_2 + \frac{1}{2}O_2$
3. $H_2O = OH + \frac{1}{2}H_2$
4. $\frac{1}{2}O_2 + \frac{1}{2}N_2 = NO$
5. $CO_2 + H_2 = CO + H_2O$
6. $N_2 = 2N$
7. $O_2 = 2O$
8. $H_2 = 2H$

Some values of equilibrium constants are given in Table 6.6 using the same numbering system for the reactions. The equals sign designates the possibility of the reaction occurring in both directions.

Dissociation is significant above a value of K_P of about 0.0002 which occurs around 2000 K for the CO_2 and H_2O reactions. The dissociation of H, O and N into ions is not significant until 2500 K, 2600 K and 4500 K respectively. Note that equation 5 can be formed from a combination of equations (1) and (2) and hence it is usually neglected. For $K_P > 1000$, (i.e. $\ln K_P > 6.9$), a reaction is near complete. For $K_P > 0.001$, (i.e. $\ln K_P > -6.9$), the reaction is insignificant. As calculations are complex, it is often convenient to neglect some reactions and the above are observations worth noting for this purpose. Also, in selecting appropriate equations to reduce the total calculations, if there is insufficient oxygen, the burning preference is to give CO, H_2O and CO_2 in that order of priority.

6.8 CHEMICAL KINETICS

The principle noxious emissions from combustion systems are carbon monoxide (CO), unburned hydrocarbons (HC), the nitrous oxide group generally labelled NO_x, sulphur oxides, principally sulphur dioxide and particles of various compounds and sizes. The formation of carbon monoxide and NO_x in particular need careful consideration from a fundamental chemical viewpoint and this requires an examination not only of the equilibrium values but of the rate of formation of the products.

The amount of carbon monoxide has been found to be fundamentally related to the fuel/air ratios, a rich mixture giving high CO emissions and a

Table 6.6 Logarithms* to the base 10 of the equilibrium constant $K_P = ((p_E)^{\nu_E}(p_F)^{\nu_F})/((p_A)^{\nu_A}(p_B)^{\nu_B})$ for the reaction $\nu_A A + \nu_B B \leftrightarrow \nu_E E + \nu_F F$

Temperature (K)	1	2	3	4	5	6	7	8
298	−71.224	−81.208	−159.600	−15.171	−40.048	−46.054	−45.066	−5.018
500	−40.316	−45.880	−92.672	−8.783	−22.886	−26.130	−25.025	−2.139
1000	−17.292	−19.614	−43.056	−4.062	−10.062	−11.280	−10.221	−0.159
1200	−13.414	−15.208	−34.754	−3.275	−7.899	−8.789	−7.764	+0.135
1400	−10.630	−12.054	−28.812	−2.712	−6.347	−7.003	−6.014	+0.333
1600	−8.532	−9.684	−24.350	−2.290	−5.180	−5.662	−4.706	+0.474
1800	−6.896	−7.836	−20.874	−1.962	−4.270	−4.617	−3.693	+0.577
2000	−5.580	−6.356	−18.092	−1.699	−3.540	−3.780	−2.884	+0.656
2200	−4.502	−5.142	−15.810	−1.484	−2.942	−3.095	−2.226	+0.716
2400	−3.600	−4.130	−13.908	−1.305	−2.443	−2.525	−1.679	+0.764
2600	−2.836	−3.272	−12.298	−1.154	−2.021	−2.042	−1.219	+0.802
2800	−2.178	−2.536	−10.914	−1.025	−1.658	−1.628	−0.825	+0.833
3000	−1.606	−1.898	−9.716	−0.913	−1.343	−1.269	−0.485	+0.858
3200	−1.106	−1.340	−8.664	−0.815	−1.067	−0.955	−0.189	+0.878
3400	−0.664	−0.846	−7.736	−0.729	−0.824	−0.679	+0.071	+0.895

* Based on data from the JANAF Tables, NSRDS-NBS-37, 1971, Dow Chemical Co., Midland, Michigan Additional data may be found in Thermodynamics texts such as those by Wark, K., 1989, *Thermodynamics*, McGraw-Hill, New York and Sonntag, R. E. and Van Wylen, G. J., 1991, *Introduction to Thermodynamics*, Wiley New York.

lean mixture giving very little. This is because the CO, CO_2 formation mechanisms are fast; they are substantially within the time scale of the individual combustion processes in most systems and hence the levels obtainable at any particular stage are close to equilibrium values. At low temperatures, CO is predominantly undissociated and hence the level of CO is largely a result of insufficient oxygen to complete the combustion. This is particularly true of lean mixtures, rich mixtures having the effect of introducing a slower rate of recombination and increasing CO above equilibrium values when the exhaust cools to atmospheric temperatures. The NO_x formation is, however, quite different. At low temperatures, very little NO_x-type compounds exist in equilibrium as is evident from the fact that both N_2 and O_2 occur together in abundance in nature without more than minute quantities of NO_x forming. At high temperatures, of the order of those in combustion processes, a comparatively major reaction occurs between N_2 and O_2. If NO_x compounds remain in the exhaust, it is obvious that some non-equilibrium event is occurring to retain them at the low temperatures where they should have substantially reformed to N_2 and O_2. Essentially, the high levels in the exhaust is due to the slow rate at which N_2 and O_2 are reformed when temperatures drop, this dissociation of the NO being considerably slower than its formation. Thus the study of reaction rates (i.e. chemical kinetics) is of particular importance for this exhaust emission product.

6.8.1 Reaction rates

At a molecular level, substances are continually forming and breaking up. Equilibrium constants apply to chemical reactions when the rate of combination of atoms or radicals to form species exactly balances the breaking up of the same species (i.e. destruction) to their original components. These processes, on a microscopic scale, never occur instantaneously after a change in conditions from a previous equilibrium state. The change may be rapid, of the order of nanoseconds, or may take many hours depending on the constituents of the reaction and the conditions. Rapid cooling of a gas in, say, an engine after combustion (or a shock tube for experimental duplication) may 'freeze' the products of the original reaction and they may take considerable time to reach equilibrium at the new, cooler conditions. Consider, for example, air (approximately 3.76 kmol of N_2, 1 kmole O_2) at 1 atmosphere. At a temperature of, say, 298 K the gas will, if given time, come to an equilibrium mixture of 2.44×10^{-16} kmol of NO for each kmole of initial air mixture. Thus there will finally be just a little less N_2 and O_2 at equilibrium than the 3.76 and 1 kmol respectively but the difference will be undetectable. If the temperature is raised to, say, 3000 K and there is sufficient time for equilibrium to be reached, it will now be found that 0.046 kmol of NO exist for each original kmole of mixture. Thus, in comparison with 298 K, there is now a considerable quantity of NO. When cooled again

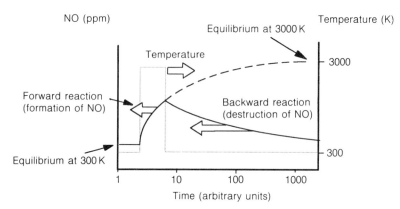

Figure 6.8 Schematic representation of the formation and destruction of NO from N_2 and O_2 with change in temperature.

to 298 K the original equilibrium values will again be reached. However, this takes a very long time and during that period the extra NO may react with other substances (e.g. unburned hydrocarbons, HC) to form irritating substances. In other words, much of it does not have the opportunity to revert to N_2 and O_2. These further reactions form compounds which are the basis for photochemical smog.

The effect of a large step increase in temperature followed at a short time interval by an equivalent decrease is shown diagrammatically on Figure 6.8. Note that this is schematic only and does not represent any actual formation process. As the temperature increases, a substance forms at a rapid rate and approaches but does not reach equilibrium at the new conditions in the time available. On decreasing the temperature, the substance starts to reform to the original components but this is a slow process. Hence, a very long period may occur before the original equilibrium is again reached.

To determine the amount of NO (or any other substance given the relevant reaction) formed in a given time after a given change of temperature occurs, the rate, called the **reaction rate**, at which the substances approach equilibrium must be known. This is usually expressed in terms of reaction rate constants, k_f, for the forward reaction and k_b for the backward one, these being determined experimentally. They are normally expressed in the form

$$k = AT^n\exp(-E/R_0T) \tag{6.48}$$

The term $A\exp(-E/R_0T)$ is called the Arrhenius factor after Swedish physical chemist Svante Arrhenius who first derived that form. Here E is the activation energy which is the minimum energy of a molecular collision which is sufficient for the reaction to occur. Below that value, the molecules will simply rebound. Constants are A and R_0, the universal gas constant.

The addition of the term T^n where n is also a constant has been shown to improve the agreement in many cases.

For a reaction, the following equation may be assumed as previously to govern the formation of substances at a molecular level as they interact (collide). That is, molecules A and B combine to form C and D. Assume C is the substance whose formation is of interest. The values of coefficients, v_i are the numbers of each molecule engaged in the reaction

$$v_A A + v_B B \longrightarrow v_C C + v_D D \qquad (6.49)$$

Assume a reaction rate has been measured experimentally for this reaction and is expressed in terms of the rate of increase of the concentration of C with respect to the concentrations of the forming substances. A bracket [] is used to denote volume concentration in the total mixture.

$$\frac{d[C]}{dt} = k_f [A][B] \qquad (6.50)$$

The constant k_f is the rate for the forward reaction, i.e. that which **forms** substance C. Now the actual quantities available to react are N_{A1} and N_{B1} giving a reaction equation

$$N_{A1} A + N_{B1} B \longrightarrow N_C C + N_D D + N_A A + N_B B \qquad (6.51)$$

This applies at any instant between time $t = 0$ and equilibrium with, of course, the numerical values of the coefficients N_C and N_D increasing with time while N_A and N_B decrease. The concentration of A, for example, in the mixture on the right-hand side is

$$[A] = N_A A / (N_C + N_D + N_A + N_B) \qquad (6.52)$$

This is distinct from A which is often used to signify simply the number of moles of A. Using $[]_0$ to indicate the concentration at time $t = 0$, the mixture equation may be represented then by

$$[A]_0 A + [B]_0 B + [C]_0 C + [D]_0 D \qquad (6.53)$$

For a small time interval Δt, the rate equation gives for the forward part of the reaction only

$$\Delta[C] = k_f [A][B]\Delta t \qquad (6.54)$$

But C is simultaneously being destroyed and a similar equation will work in reverse. This is called the backward reaction, and is governed by

$$\frac{d[C]}{dt} = -k_b [C][D] \qquad (6.55)$$

where k_b is the backward reaction rate constant. The net rate of formation of E at any time is then

$$\Delta[E] = -k_f [A]_0 [B]_0 - [C]_0 [D]_0 \qquad (6.56)$$

from which the new mixture concentration at the end of the first time increment can be found. That is

$$[C] = [C]_0 + \Delta[C] \tag{6.57}$$

Of course, at the beginning of the reaction it is likely that neither $[C]$ nor $[D]$ exists. However, for the second and later time steps, both $[C]$ and $[D]$ will have finite values.

Now $[C]$ is the concentration of the substance of interest which has been formed. The changes in the other constituents may be found by proportion from the molecular equation

$$\Delta[D] = + C\nu_D \Delta[C] \tag{6.58}$$

and so on for the other substances.

The new mixture at time Δt may now be evaluated and the process repeated over a large number of small time steps. The above is very time consuming and simplified schemes have been determined for many practical cases.

6.8.2 Backward reaction rate constants

The missing information is k_b as usually it is convenient to measure a reaction rate in one direction only. However, k_b can be determined from k_f by considering the equilibrium condition.

Generally when substances are reacting, the formation and destruction rates of a substance are not equal and only become so at equilibrium. Essentially, this is what chemical equilibrium means, not that no process is occurring but that the rate of formation and rate of destruction are the same. Expressing this mathematically

$$k_f[A]_e[B]_e = k_b[C]_e[D]_e \tag{6.59}$$

while $[\ ]_e$ represents the equilibrium concentrations. Hence

$$\frac{k_f}{k_b} = \frac{[C]_e[D]_e}{[A]_e[B]_e} \tag{6.60}$$

$$= \frac{N_C N_D}{N_A N_B} \tag{6.61}$$

Note that this is the expression for the equilibrium constant K_p for a simple bimolecular reaction, that is when $\Delta\nu = 0$.

A reaction is not fully described by the overall 'global' equation which is the one normally given. It takes place as a series of molecular collisions and these are generally two species types with unit coefficients. Thus, finding the forward reaction rate for these subreactions is necessary experimentally while the backward reaction rate follows from

$$k_b = \frac{k_f}{K_P} \qquad (6.62)$$

6.8.3 Reaction rate schemes

Simple schemes containing only the required substances can be used, for example oxygen and nitrogen giving NO, but these provide only a small part of a combustion reaction. That is, the results can be inaccurate because the formation of other substances which participate in a total scheme are controlling factors. For example, a second reaction may provide a necessary intermediate species which in turn reacts to produce the substance of interest. If this intermediate species is produced rapidly, the controlling reaction will be the final reaction. However, if it is produced slowly, the final process will be delayed while it is formed and it therefore becomes dominant in determining the overall rate. In combustion systems, there are many intermediate species formed and these can have a substantial impact on the final formation. For the production of NO, schemes range from a simple two-equation scheme, the Zeldovitch scheme, to a quite extensive reaction series. Some typical schemes are as follows.

Zeldovitch

This is a simple scheme and combustion is not necessary

$$O + N_2 = NO + N$$
$$N + O_2 = NO + O$$

Each reaction provides atoms which feed the other reaction.

Extended Zeldovitch

This is more commonly used in combustion as an OH radical is involved, this coming from the HC oxidation

$$N_2 + O = NO + N$$
$$N + O_2 = NO + O$$
$$N + OH = NO + H$$

The rate of formation of the OH radical is an important variable and to a lesser extent formation rates of additional N and O which can also come from the combustion process.

Various combustion schemes which use the extended Zeldovitch mechanism have been proposed with reasonable success and involve many HC reactions. Examples may be found in the work of Bowman (1972) and Smooke, Puri and Seshadri (1986) respectively. Significant reactions for determining the final global reaction rate range from about 17 to about 35. Large computations are therefore involved. As each equation may contribute radicals to feed other processes, the total system should be used at each time step. However, by selectively eliminating very slow reactions, the degree of complexity may be reduced. Note that non-reacting substances can also be involved and effectively play a catalytic role.

6.9 REFERENCES

Bowman, C. T. (1972) Kinetics of nitric oxide formation in combustion processes, *Proc. 14th Symp. (Int.) on Combustion*; The combustion Inst., Pittsburgh, Penn., 729–38.

Gordon, S. and McBride, B. J. (1971) Computer Program for Calculation of Complex Equilibrium Compositions, *NASA SP–273*, USA.

JANAF Thermochemical Tables (1971) US National Board of Standards Tables NSRDS–NBS–37, Dow Chemical, Midland, Mich., USA.

Smooke, M. D., Puri, J. K. and Seshadri, K. (1986) Comparison between numerical calculations and experimental measurements of the structure of a counterflow diffusion flame burning diluted methane in diluted air, *Proc. 21st Symp. (Int.) on Combustion*, The Combustion Inst., Pittsburgh, Penn., 1783–92.

Sonntag, R. E., and Van Wylen, G. J. (1991) *Introduction to Thermodynamics*, Wiley, New York.

Wark, K. (1989) *Thermodynamics*, McGraw-Hill, New York.

PROBLEMS

6.1 A series of hydrocarbon fuels (in the vapour state) are burnt in air with an air/fuel ratio that is (a) stoichiometric, (b) 16:1. The fuels are (i) methane, CH_4, (ii) propane, C_3H_8, (iii) methyl alcohol, CH_3OH, (iv) iso-octane C_8H_{18}, (v) petrol (gasoline) assumed to be C_7H_{13}.

For each combination, determine the air/fuel volume ratio, the percent excess or insufficient air and the fuel: air equivalence ratio. For the stoichiometric case, also determine the air/fuel mass ratio. Molecular masses are $C = 12.01$, $H_2 = 2.016$, $O_2 = 32$, $N_2 = 28.008$.

6.2 For methane, the flammability range in air is approximately 5% (lean) to 14% (rich) where the values are expressed as fuel volume divided by fuel plus air volume. Determine the percent theoretical air for these limits, the air/fuel mass ratio and the appropriate equivalence ratios.

6.3 A burnt gas mixture is at a pressure of 5 MPa, 500 °C and contains CO_2, H_2O, O_2 and N_2 in proportions of 1.954:1:1.154:11.398 by mass. Given that the combustion was in air and that the molecular mass of the hydrocarbon fuel was 58.12, determine (a) the relative proportions by volume of the gases in the burnt mixture (per unit volume of CO_2), (b) the formula for the fuel.

6.4 Calculate the higher and lower enthalpy of combustion of liquid ethanol (ethyl alcohol) on a mass basis. If 1 kg of this ethanol fuel originally at 298 K burns in air at an equivalence ratio of 1.15, determine the chemical equation. For one kmol of fuel, how many kmol of CO_2 and CO are theoretically produced. Calculate the heat available if the products are returned to the original temperature after combustion, assuming that all the water vapour is condensed to liquid at the end of the process.

6.5 Butane gas at 25 °C is mixed with 300% excess air which is at 227 °C. The reaction occurs at a constant pressure of 1 atmosphere in a steady flow process. Determine the approximate adiabatic flame temperature assuming no dissociation occurs.

If the same combustion process occurred in oxygen, rather than air determine the flame temperature.

If the same combustion process in air now occurs at constant volume in a closed system, what is now the temperature and the approximate final pressure if it is assumed that the reactants are, when mixed at 500 K?

6.6 Determine the higher and lower enthalpy of combustion values in $MJ\,kg^{-1}$ of propane gas.

Propane is burnt in a closed piston-cylinder device with 20% excess air. The reactants are at 500 K, 100 kPa prior to combustion. After combustion work is done by the piston and heat is transferred out until the products reach 298 K, 100 kPa. Determine the number of kmoles of water which remain as vapour and the number which have condensed to liquid at this final condition. You may assume that the saturated vapour pressure of water at 300 K is 3.6 kPa.

Determine the total energy transfer (work plus heat) per unit mass of fuel from the system that has occurred in order to reach the final state.

For propane, molecular mass $= 44.08$, $c_v = 1.48\,kJ\,kg^{-1}\,K^{-1}$ approximately in the range 298 to 500 K.

6.7 A liquefied petroleum gas mixture (LPG) consists of 70% propane (C_3H_8) and 30% butane (C_4H_{10}) by volume. The gas is burnt with 50% excess in a steady flow burner at a pressure of 1 atmosphere. It enters the burner at a temperature of 298 K. (a) Determine the appropriate combustion equation assuming no dissociation occurs. How many kmol of CO_2, H_2O, O_2 and N_2 exist in the products of the combustion for each kmole of fuel mixture? (b) Find the air/fuel ratio by mass and the equivalence ratio, (c) Calculate the adiabatic flame temperature for the above reaction.

6.8 A mixture of 2 kmol of nitrogen and 1 kmol of oxygen at 298 K, 101.3 kPa is heated at a constant pressure to 3200 K. It is maintained at that temperature for sufficient time for it to reach equilibrium. (a) List all molecules and radicals that you would expect to exist in any reasonable quantity both before and after heating. (b) Write the overall chemical reaction for the final state. List the relevant dissociation reactions together with their equilibrium constants at the final state as given on Table 6.6. Comment on which is likely to be the most dominant of these. Assuming that only the most dominant applies and all others can be neglected, determine the molar proportions at the initial and final state. (d) If at the final state the pressure is doubled at the same temperature, what change would you expect in the molar proportions calculated in (c).

6.9 A mixture of hydrogen and oxygen is burned in a steady flow combustor at a pressure of 10 atmospheres (1.013 MPa). Sampling of the products of combustion gives the following mole fractions (%):

Substance	Mole Fraction (%)
H_2O	74.00
H_2	6.57
O_2	9.32
OH	10.12

(a) Determine the oxygen/fuel ratio of the reaction on both a volumetric and mass basis. Also determine the equivalence ratio. (b) Calculate the temperature at which the reaction occurred, (c) Determine the heat transferred from the combustion chamber per kmol of hydrogen assuming the reactants enter at 500 K.

6.10 A fuel consists of a stoichiometric mixture in air of two fuels, n-octane C_8H_{18} and benzene C_6H_6. After combustion at a constant pressure of 1 atmosphere, measurements give the following mole fractions (%) in the products:

	CO_2	CO	O_2	H_2O	H_2	OH	N_2
mole fraction (%)	11.807	1.180	0.636	12.619	0.231	0.276	73.250
c_p (kJ kmol^{-1} K^{-1})	54.60	33.72	35.19	43.86	31.55	32.02	33.39

(a) Determine the relative proportions of C_8H_{18} and C_6H_6 in the original fuel. Write the full stoichiometric equation for 1 kmol of the fuel. (b) Write the dissociation equations for all the relevant dissociations that have occurred in the combustion process. Relate each group to their appropriate equilibrium constant. Hence determine the temperature of the mixture at the time of sampling. (c) Calculate the heat transferred from the combustor per kmol of the fuel assuming that the reactants were originally at 25 °C, 1 atmosphere at the start of the reaction and that n-octane and benzene were in the gaseous phase.

Assume that the c_p values given in the table above are accurate for the range of temperatures considered.

6.11 A highly preheated stoichiometric mixture of propane (C_3H_8) and air burns during steady flow in a combustion chamber. Heat transfer is negligible and the maximum temperature is 3000 K when the pressure is 1 atmosphere (101.3 kPa). During the combustion, there has been sufficient mixing and time to ensure all appropriate molecular reactions occur, including the dissociation of some of the CO_2 to CO and O_2. Assuming that no significant dissociation occurs for other products, determine the number of kmol of all products of the combustion per kmol of propane in the original mixture.

6.12 A gas turbine operates using natural gas as a fuel with an air/fuel ratio of 20.6:1 in the primary combustion zone. The natural gas may be assumed to be 100% methane for calculation purposes. The gas turbine has a pressure ratio of 20:1 and inlet air conditions of 297 K, 80 kPa before the compressor. The value of the isentropic index for the compressor may be taken as 1.4. The gas, injected into the combustion chamber is at the same temperature as the air before combustion. It may be assumed that the processes are that of a normal Brayton cycle. (a) Sketch the cycle on p–V and T–s diagrams, (b) Determine the equivalence ratio for the combustion, (c) From the gas turbine compression process, calculate the temperature and pressure entering the primary zone of the combustion chamber, (d) Assuming that no dissociation takes place and that the combustion is adiabatic, write the combustion equation and calculate the temperature of the combustion products at the end of the primary zone. (e) List the likely dissociation reactions that could be of significance at this temperature, (f) Assuming for an initial calculation that only the CO_2 dissociates and that the final temperature under these conditions is 2200 K, determine the proportion of CO in the products at the end of the primary zone.

The value of c_p for methane may be taken as 36.15 kJ kmol^{-1} K^{-1}.

6.13 In the combustion products at start of exhaust in a petrol engine operating at stoichiometric conditions, both of the dissociation reactions for water need to be taken into account for calculation purposes as well as that for the CO_2. Assuming that the exhaust is at 3000 K and 10 atmospheres pressure, calculate the number of kmol of all substances in the exhaust. Assume that the fuel is C_7H_{13}.

7 | Combustion and pollution formation in engines

7.1 ENGINE COMBUSTION SYSTEMS

There is a great diversity in the fuels and fuel preparation systems used, the equivalence ratios which are desirable or achievable and the types of combustion chambers possible for internal combustion engines. To give an indication of the diversity, one needs only to consider the range of each of the above that is associated with the different engine types. Fuels include those currently in common usage which are the familiar petrol (gasoline) and diesel (distillate) for reciprocating engines and the range of kerosene formulations for gas turbines. Then there are the common alternatives such as alcohols or gas which are often proposed for use in the near future in all types of conventional engines and the more exotic substitutes such as the liquid hydrogen/liquid oxygen or solid fuel propellants of rockets. Equivalence ratios vary from lower than 0.2 to about 1.2 in conventional engines and may be very hydrogen rich in rockets. Combustion chambers can be either the continuous flow gas turbine or rocket types of varying sizes and geometries or the unsteady piston/cylinder types for reciprocating engines which have intermittent flow, proportions which change during combustion and single or divided chambers and which range in practical sizes from piston diameters of 40 or 50 mm to about 1 m. Each type, configuration and size has its own desired gas flow patterns, mixing, combustion and heat transfer characteristics which may also vary with the fuel and it is not possible to deal with them all. However, a selective examination of a few of the most common types should elucidate the general principles involved in their design. In line with the emphasis of this text, those which will be discussed here are gas turbines, reciprocating SI engines, reciprocating CI engines.

7.2 GAS TURBINES

7.2.1 Description of gas turbine combustion

Gas turbine combustion is essentially steady flow although transients do occur during both warm-up and load and speed changes. The reader may be familiar with high-output steady-flow combustors which are widely used in industrial applications for fuels which include gas, fuel oils and pulverized coal and the same general design principles which apply to them are also relevant to gas turbines. That is, as with all burners, the combustion must be stabilized by swirl and recirculation in a primary zone with sufficient residence time to allow the fuel to ignite and partially oxidize and a secondary zone must exist where additional air is added to complete the combustion and maintain combustor surface temperatures at reasonable levels. However, the demands are much greater on gas turbines because typically a gas turbine combustor might have a specific power output (i.e. power per unit volume) of about ten times that of an industrial type.

An important restriction on gas turbines exists because they are primarily used for air transport, and this is that the size (cross-sectional area and length) and weight of their combustion chambers must be minimized. This means that they must produce a large amount of power per unit size, referred to as specific ouput and the required, value of around $500 \, \mathrm{MW \, m^{-3}}$ of combustor volume is obtained through high intake pressures and high velocities through the chamber. Gas velocities for flow through the combustion chamber are around $30 \, \mathrm{m \, s^{-1}}$ which is well in excess of the flame velocities of the fuels normally used. A typical laminar flame velocity of a hydrocarbon fuel at normal ambient conditions is about $500 \, \mathrm{mm \, s^{-1}}$ but this will perhaps double at the prevailing unburned gas conditions in the combustion chamber. These are typically pressures in the range up to 20 to 30 atmospheres and temperatures up to 700 K to 800 K. A turbulent flame velocity is an order of magnitude greater than the laminar value but, even at around $10 \, \mathrm{m \, s^{-1}}$, it is still too low relative to the intake air flow velocity. Thus substantial flame stabilization is required with flame blowout being likely. It will be remembered from the discussion in Chapter 6 that blowout is the phenomenon that occurs when the air velocity carrying the flame downstream is greater than the speed with which it can burn back upstream. The opposite phenomenon is flashback and because of the velocities, there is little likelihood of this happening. In fact, because the fuel is injected into the air stream rather than being fully mixed from well upstream, flashback (upstream movement) cannot occur past the injection position. Nevertheless, poor combustion could result if insufficient air is supplied with too low a velocity but this is a less important aspect than that of obtaining high power output.

The simplest concept which could be applied to the design of a gas turbine burner is that of a flame burning in a parallel tube. At an air velocity of, say $30 \, \mathrm{m \, s^{-1}}$, this is not viable. In order to achieve a combustion chamber meeting the necessary air throughput requirements, substantial development of flame stabilization techniques from this simple concept has been necessary. Figure 7.1 shows schematically the evolution of ideas for enhancement of burner ouput and flame stabilization. These include the addition of a shaped chamber which slows the air velocity at the

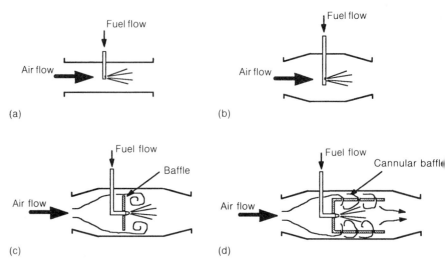

Figure 7.1 Development of a stabilized high performance combustion chamber. (a) straight burner, (b) shaped burner (diffuser slows air velocity at injector), (c) simple baffle (creates recirculation zone at injector), (d) cannular combustion chamber (recirculation zone at injector, with stoichiometric air, additional air added downstream to complete combustion and provide cooling).

appropriate cross-section, a reversed flow burner with air flow through a perforated, external 'can' to provide turbulence, a baffle which creates a vortex pattern to give greater residence time for burning and which provides a vortex driven stabilization regime and, finally, a combination of the two with the addition of further air holes downstream which allow secondary burning to take place. This last shows the elements of a modern gas turbine combustor.

The ability of any fuel to burn is determined by its flammability limits as discussed in Chapter 6. While some fuels such as hydrogen have wide limits, hydrocarbon fuels do not. Typical limits on the lean side of stoichiometric are at equivalence ratios of between 0.6 and 0.7 depending on the fuel composition and this does not vary much with pressure. The best conditions for ignition are about 5% rich at which the fuel will ignite even at very low pressures, down to about 5 kPa absolute. At atmospheric pressure, on the rich side, the flammability limits are around 2.5 to 3.0 and continue to rise with pressure. This is depicted on Figure 7.2. It can be inferred that, for good ignition, the fuel/air ratio needs to be close to stoichiometric. Now in a gas turbine combustor, the temperatures at the end of combustion (i.e. at the inlet to the turbine blades) must be limited by the ability of the metal container to withstand the gas temperature conditions as discussed in Chapter 2. Typical maximum allowable values are currently in the range

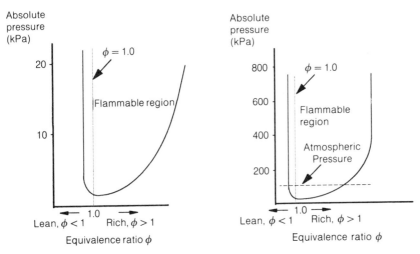

Figure 7.2 Representation of typical flammability limits for hydrocarbon fuels. (a) lower than atmospheric pressure, (b) higher than atmospheric pressure.

1600 °C to 1700 °C. A stoichiometric hydrocarbon flame with reactants at 300 K will produce a flame temperature of over 2000 °C even if dissociation is accounted for and this temperature will be noticeably higher in a gas turbine because of the elevated precombustion conditions. Gas turbines must therefore operate lean to keep the turbine entry temperature to the desired level. Simple calculations show that the overall maximum equivalence ratio that can be tolerated is about 0.4 to 0.45. If the constraints of ignition and turbine entry conditions are considered simultaneously, it can be seen that there needs to be an initial zone in which the mixture is close to stoichiometric where ignition of the fuel and primary burning can take place followed by a further zone or zones where additional air is added giving a lean mixture to complete the oxidation and reduce the final temperatures to the desired values.

A typical modern combustion chamber based on these principles is shown on Figure 7.3. Air enters the chamber through several apertures. The first is through a swirler placed around the fuel injector which creates a vortex about the axis of the chamber, the purpose of the swirl being to prevent combustion fluctuations caused by the irregular shedding of transverse vortices from the slightly later introduction of primary air and from the burning zone itself. Primary air holes around the initial combustion region provide sufficient air to bring the mixture to about stoichiometric conditions. From about 25% to 40% of the total air flow is added through the combined swirler and primary air holes. The primary air creates a toroidal recirculation zone vortex similar to, but more stable than, that of the simple baffle insert depicted in Figure 7.1. The fuel is introduced into

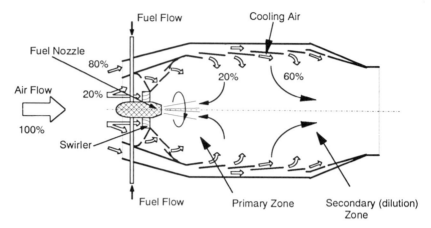

Figure 7.3 Typical cross-section of a gas turbine combustor.

this region so that the primary zone air/fuel ratio is approximately stoichiometric and ignition and combustion are then readily supported here. When secondary air is added, the overall air/fuel ratio is greater than about 37:1 ($\varphi = 0.4$) and more usually about 60:1 ($\varphi = 0.25$) for continuous operation. This gives very complete combustion and keeps the temperatures at the end of combustion at reasonable levels for entry to the turbine.

7.2.2 Fuel introduction and mixing

The fuel, usually a liquid (although it should be noted that gaseous fuels are also commonly used in stationary land-based gas turbines), enters the combustion chamber as a jet and requires atomization and distribution throughout the combustion space. The droplets formed must vaporize and mix in the recirculation region where burning commences. Penetration of the spray through the air zone is important so that the fuel droplets are not clustered in a region close to the nozzle exit thereby partially excluding the air from neighbouring drops. While droplet sizes vary from one injection system to another, the mean droplet sizes are likely to be about 20 μm to 30 μm. In order to obtain a spread of droplets, it is desirable to have a range of drops with diameters distributed about the mean value which varies the individual droplet kinetic energy thereby providing a spray that is more fully distributed throughout the circulating air. Average fuel flow velocities of around 100 m s^{-1} are desirable but these are not easy to achieve because of the great variation in fuel flow rates that any individual gas turbine will use over its operational range. Liquid fuel injectors which can accommodate the wide mass flow rate range from the minimum which is idle at low ambient air density (high altitude) through to the maximum of full power at high density, (take-off conditions) have been developed. These consist of several types, going under the names of simplex, variable port (Lubbock), 'duplex' and airblast injectors.

Before examining these, it is important to first look at the spray process itself. In a spray, liquid flows from a fuel nozzle initially as a liquid stream which starts to break up around its edges some small distance downstream from the nozzle exit. The higher the ratio of the injector pressure to the gas pressure into which it is being sprayed, the quicker the atomization process takes place and the more complete it is. Thus, for good injection, a high injector pressure is required. Early types of fuel injector were the simplex type which consist of a simple nozzle with the fuel upstream being given a swirling motion in an appropriately designed chamber just prior to exit. Their fuel flow rate, from basic fluid dynamics, is proportional to the square root of the pressure difference across the nozzle. Thus a very large variation in the upstream pressure is required to provide the necessary flow rate range. Depending on the nozzle diameter, this is either inadequate at the maximum requirements or results in overfueling at the minimum. As in the variable port and duplex types to be next discussed, this nozzle uses an air flow around the outer edge of the spray to prevent carbon buildup near the nozzle exit but this plays no role in the atomization process. It is obvious that the simplex injector limited gas turbine versatility and, to provide a wider range of flow rates without resorting to an excessive pressure ratio range, variable port nozzles were developed. These incorporate a metering area or port just upstream of the nozzle which is controlled by an axially sliding piston. However, because of the moving component, durability and maintenance problems are inherent in these types. An alternative solution to give an adequate atomization over the whole range is the concentric (duplex) nozzle. These allow a high velocity jet at low fuel flow rates through one nozzle, usually the central one, with the second nozzle inactive. As the flow rates increase, more fuel is allowed to flow to the larger outer nozzle and both then operate simultaneously. These types allow good control over a wide range of conditions and have been extensively used. A modern alternative is the airblast type now used. This supplies fuel at low pressure onto a central cone which is inside an annulus type nozzle. Introduction of compressor air directly around this cone at high velocity provides the atomization due to its shearing action on the liquid. Air velocites of about $100 \, \mathrm{m\,s^{-1}}$ are necessary but these are easily provided from the compressor flow. In fact, the pressure drops in this airstream are similar to those across the combustor liner at about $25 \, \mathrm{kPa}$. The advantage of the airblast type is that the air initially mixed with the spray prevents rich pockets of fuel from forming in the combustion region and hence smoke is reduced. In addition, the low-pressure pump required for the fuel deposition is considerably lighter than that for the high-pressures sprays. A disadvantage is that additional starting sprays are required. Diagrams of different nozzle types are shown on Figure 7.4.

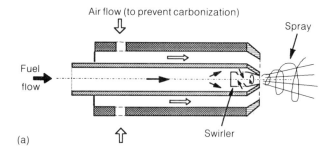

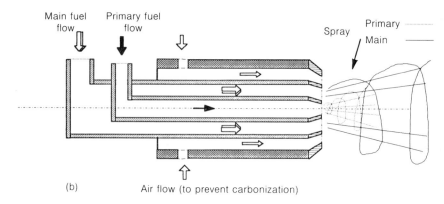

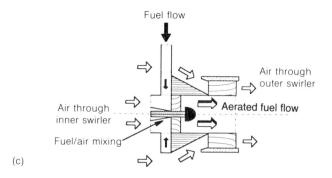

Figure 7.4 Cross-section of the major different fuel nozzle types. (a) simplex nozzle, (b) duplex nozzle, (a) airblast atomizer nozzle.

While the direct liquid injector has been by far the most commonly used type, it has sometimes been replaced by a vaporizing type (as shown on Figure 7.5) for special purposes. Here the fuel pipe passes through the burning region. Some of the heat from the combustion is transferred to the fuel and vaporizes it before it enters the combustion chamber. The advantage of this system is that the gas phase fuel mixes more readily with the air and

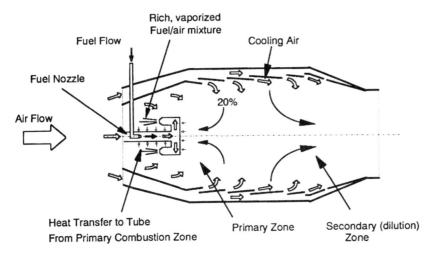

Figure 7.5 A vaporizer type gas turbine combustor.

smoke is then less. Problems of poor response and narrow stability limits make this type of injection less desirable. The poor response is due to the time delay for the new quantities of vaporized fuel required under the changed conditions to reach the combustion chamber from the vaporizing tube and the different heat transfer rate then required for fuel vaporization. That is, as load is increased, the fuel may not be fully vaporized until additional heat supply is available from the enhanced combustion thereby ensuring a period of non-optimal burning while adjustment to the new condition takes place. In regard to the ignition stability, it must be remembered that a rich and lean limit on air/fuel ratios exists. When droplets evaporate directly in a combustion environment, the diffusion of the vaporized fuel from the surface of the droplet into the air always provides a suitable ratio at some local point. That is, there are many ignition points throughout the combustion space. A fully vaporized fuel jet entering the chamber mixes at the periphery of the jet and is more subject to disturbance from the aerodynamic flow patterns within the chamber. Some of the fuel vapour may then exist as a mixture in proportions which are outside the flammability limits. Thus some narrowing of the overall chamber stability limits is likely. However, as mentioned above, a major advantage of the vaporizing system is that smoke from the exhaust tends to be lower. This is because the already vaporized fuel is not subject to pyrolysis by local over-heating of the liquid fuel as it enters the combustion chamber.

7.2.3 Gas turbine combustion chambers

A number of configurations exist for gas turbine combustion chambers. Early gas turbine combustion systems in full scale engines consisted of a series of chamber cans, these being essentially a number of small cylindrical burners placed around the periphery of the engine between the compressor and turbine. Some readers may be familiar with low-output-type gas turbines which have only a single can as these have often been used for demonstration purposes in teaching laboratories but high-output machines in practice use about eight such circumferential chambers. These operate to some extent independently of each other and are termed either **cannular** or **tubular**. The earliest types were reverse flow cannular. Here the inlet air entered at one end in an air tube which surrounded the flame tube. When it reached the far end of the tube, the air was turned through 180° and passed into the flame tube, exiting after combustion at the original entry end for the air. This is compact thereby keeping shaft lengths between compressor and turbine short which reduces vibrational problems and structural mass. However, the pressure loss is high and the engine cross-sectional area is large which is a disadvantage for aircraft units and the arrangement is now rarely used except on very small machines. In general, combustion chambers are now straight-through types with the burnt products exiting from the opposite end to that where the air intake and fuel injection occur. Cannular types consist of the inner flame tube surrounded by an air tube, the latter being interconnected in multican engines so that the supply pressure to the flame tubes equalizes. This type of design is strong and withstands the high temperatures well. It is also easy to develop as only one of the combustion cans, i.e. probably only one-eighth of the engine combustion, needs testing. It is also cheap to maintain.

While relatively simple to develop and build, this multichamber type of arrangement has a large ratio of the wall area to flow cross-sectional area and therefore still leads to a moderately high pressure loss and lower efficiency. The complete engine also has a large overall cross-sectional area for a given combustion chamber air flow. Its mass is also high. In addition, because the flame tubes are separated, each combustion can requires a separate ignition system. Modern design has therefore tended to link all the flame tubes into a single 'annular' chamber to eliminate the part of the wall structure that exists between the cans. That is, the central combustor is fully annular and is surrounded by a single concentric annular air tube. Annular chambers are more difficult to manufacture and can be more subject to thermal distortion because of the lower curvature and greater unsupported surface area but they give lower pressure losses and result in smaller diameter engines, the latter being of great importance in minimizing aerodynamic drag in high speed flight. Mass is lower and ignition can be accomplished by, in principle, one igniter although in practice more are used to provide redundancy. There are many possible problems with annular chambers, one of which is the development of a secondary circumferential flow which results in a non-uniform temperature distribution at the turbine. Nevertheless, these problems have largely been solved and most large engines now use annular combustors.

An intermediate type of combustor with a configuration lying somewhere between the annular and cannular arrangements filled the evolutionary gap between the two types and it provides some of the annular chamber advantages. This is termed *tubo-annular* or *can-annular*. It uses an annular outer shell within which the air is distributed to the individual cannular type flame tubes and therefore has a pressure loss intermediate between the other two types. The different types of chambers are illustrated in Figure 7.6 and their advantages and disadvantages classified on Table 7.1. In this respect, the tubo-annular types lie between the cannular and annular types mentioned on the table.

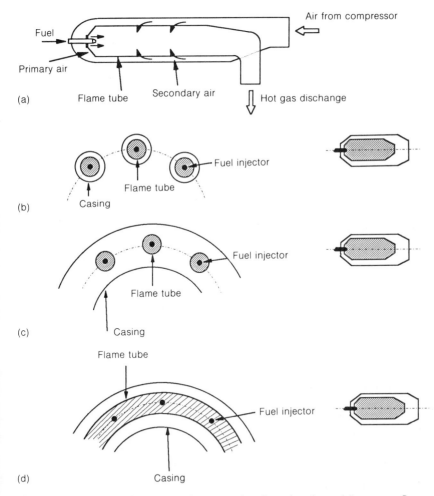

Figure 7.6 Different arrangement of gas combustion chambers. (a) reverse flow cannular (Whittle) type, (b) cannular arrangement, (c) tubo-annular (can-annular) arrangement, (d) annular (i.e. fully annular) arrangement.

Table 7.1

Chamber	Advantages	Disadvantages
Cannular (tubular)	Structurally robust	Bulky and heavy
	Fuel and air flow patterns easily matched	High pressure loss
	Simpler chamber development and testing	Ignition required for each can
Annular	Minimum length and weight	Buckling possible
	Small frontal area	Full engine air flow required for testing
	Minimum pressure loss	Complex air flow patterns must be controlled
	Ignition required at fewer points	Circumferential temperature gradient possible

7.2.4 Flows in gas turbine combustors

The gas flow in the combustion chamber of a gas turbine is complex because of the swirl and recirculation zones in the combustion or primary zone and the addition of secondary air through the chamber casing in the downstream region. Liquid fuel is sprayed into the primary zone where it must evaporate, mix, and burn. This spray must be initially well distributed throughout the primary zone. In developing gas turbine combustors, modelling of the flow is therefore important. This can allow an assessment of the performance of combustors for design purposes and an understanding of the formation of emission products. Early attempts to understand the flow relied on modelling the recirculation region as a free vortex and the fuel spray could then be examined as it moved through this region. However, because of the three-dimensional, compressible and viscous nature of the flow, these methods have been largely superseded by CFD techniques. That is, the flows are fully described by solution of the three-dimensional continuity, Navier–Stokes, energy and species equations (for combustion) as described in Chapter 4. Because the flows are steady, the application of these equations is not as difficult as with reciprocating engines. The solutions allow the performance of the combustor to be examined in relation to the effect of different mass flow rates and pressures. The pressure loss and heat transfer from the combustion chamber are the important results as well as the level of emissions, principally, in the gas turbine, smoke and NO_x. It has also been important to isolate the conditions under which secondary, circumferential flows can exist in annular types of chambers.

Some simpler techniques are possible and these are based on experimental data, are thermodynamic in principle and ignore the flow details.

Empirical models were used for many years and are described in, for example, Oates (1989). These rely on experimental data and are based on a series of dimensionless groups. Perhaps the most widely used relationship is that by Lefebvre (1966) which correlates the combustion efficiency with a reaction rate parameter, Θ given as

$$\Theta = (p^{\alpha} A_R h \exp^{\tau})/\dot{m}_a \qquad (7.1)$$

where p is the combustion pressure, A_R is the combustor cross-sectional area, h is the combustor height, τ is a function of the primary zone combustion temperature which depends on the fuel/air ratio and $\dot{m}_a$ is the mass flow rate through the combustion chamber. Data correlated shows that combustion efficiency, η_c falls as Θ becomes lower as shown on Figure 7.7. That is, low pressures, small combustion chambers, low primary zone temperatures and high mass flow rates reduce combustion efficiency.

An important factor in gas turbine combustion which is quantified parametrically is the blow-off limits. These are related to the equivalence ratio and what is termed the *loading parameter*, usually given as $\dot{m}/(Vp^{1.9})$ although the index of 1.9 is often replaced by 2. Here, V is the chamber volume and p its pressure. A typical diagram showing the relationship is given on Figure 7.8. The term $Vp^{1.9}$ is a function of a 'mass' which can be related to the chamber as well as its pressure. It can be seen that, when the mass flow rate is high, this loading term is also high and so, for a given chamber under a fixed set of conditions, the blow-off limits become narrower with increase in mass flow rate until they reach a condition under which burning can no longer be supported.

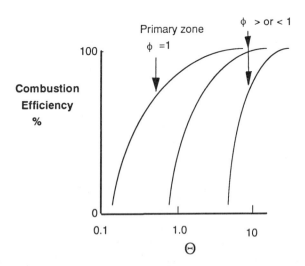

Figure 7.7 Effect of reaction rate parameter θ on combustion chamber efficiency η_c.

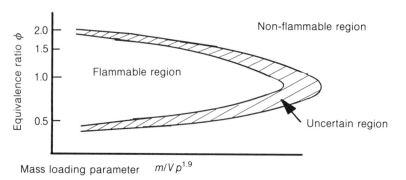

Figure 7.8 Diagrammatic illustration of the effect of the mass loading parameter on the mixture strength able to support combustion.

7.3 SPARK IGNITION ENGINES

Reciprocating engines, whether spark or compression ignition, provide some of the most difficult combustion environments that can be envisaged. This is because the whole process is unsteady, limited time is available for the fuel and air mixing processes and the release of the chemical energy. Ignition must be reliably and regularly accomplished and a wide range of operating conditions must be achieved with good efficiency. In addition, modern engines must produce low exhaust emissions. Thus, research into reciprocating engine combustion is still very active.

7.3.1 Fuel introduction and mixing

Essentially, the usual type of spark ignition engine draws into the cylinder what is, ideally, a homogeneous mixture. In fact, true homogeneity is difficult to achieve unless a gaseous fuel is used. This is because the fuel, introduced as a liquid by either a carburettor, throttle body or port fuel injector, can enter the cylinder as a vapour, as airborne fuel droplets of various sizes or even, under cold transient conditions, as a liquid wall film. The fuel droplets are transported by the aerodynamic drag at approximately the gas stream (i.e. air and vapour) velocity although, due to their greater inertia, they respond more slowly to velocity variations caused by changes in flow area, transients due to throttle opening and closing and pulsations and backflow from the piston motion in the region near the inlet valves. Nevertheless, their behaviour is not such as to directly cause major timewise maldistribution although some spatial variations in mixture homogeneity has been observed in some experiments. Any fuel deposition into films either on a manifold wall, a port or an inlet valve is, however, a greater problem. Films may form either directly by impingement of the spray, or by the deposition of droplets from the gas stream onto the walls in turbulent regions such as those which exist behind a butterfly valve, or at bends and changes in cross-sectional area of the passageways at ports and inlet valves due to the inertia of the droplets. Any surface deposition or film flows several orders of magnitude more slowly than the gas/airborne droplet streams and this can cause problems during acceleration and deceleration transients where a change in the proportion of the overall fuel supply contained in it occurs. That is, the changed quantity of fuel in the film is transported to the engine slowly relative to the airborne fuel and can cause severe excursions in in-cylinder air/fuel ratios which result in engine lag and stumble, reduced efficiency and high emission levels. Port injection (as opposed to carburettors and throttle body injectors) minimizes, but does not entirely eliminate, these response problems. It has the disadvantage that it is more expensive and harder to control but is now widely used especially on the more expensive vehicles.

With all these systems, further mixing and evaporation will occur in the manifold, port and cylinder during intake and compression but complete evaporation and uniform distribution is not entirely ensured under all conditions by the time of ignition. Currently (1993), significant work is underway on direct in-cylinder injection systems where the fuel is introduced into the cylinder as a very finely atomized spray during the compression stroke. Air blast type injection systems are being considered in order to achieve this. This arrangement has many advantages although there is little time for evaporation and mixing to occur prior to ignition and it is even more difficult to ensure that the mixture is homogeneous over the required range of operating conditions. To do this, a very finely atomized spray is essential with a fairly even distribution over the available cylinder volume but without so much penetration that cylinder wall-wetting occurs. A similar spray distribution must be obtainable over all speed/load conditions and during transients where air flow patterns can vary considerably. Thus the problems related to direct in-cylinder injection are severe.

It should be noted that it is a system which is often associated with a non-homogeneous mixture possibility called the **stratified charge** concept which has several possible advantages. Stratified charge is something which is impossible to achieve in any controlled manner with manifold or port injection systems. A stratified charge engine requires the mixture to be about stoichiometric near the spark plug so that ignition is guaranteed but increasingly lean away from it to give high fuel economy as well as clean burning. However, no regions where the air/fuel ratio becomes lean to such an extent that flame quenching occurs should exist. The appropriate fuel distribution is obviously very hard to achieve. In either case, homogeneous or stratified charge, direct in-cylinder injection is a demanding process.

7.3.2 Combustion and flame front movement

With modern engines and fuel introduction systems, the fuel is substantially vaporized and the fuel and air are reasonably well mixed except under cold start or transient conditions. Thus, as a first approximation of the combustion process, it can be assumed that ignition takes place in a homogeneous air/fuel mixture. The flame grows from an initial kernel around the spark and propagates as a deflagration front across the combustion chamber. It needs to be noted that the flame propagation period takes a noticeable proportion (perhaps 10%) of the time for a single revolution of the engine, commencing at spark discharge well before top dead centre is reached and finishing well after this point. That is, the combustion chamber volume and shape vary during the combustion period with modifications to the flow caused by the piston motion and the interaction of the flow produced by it with the wall and the head occurring simultaneously.

The flame propagation controls the rate of pressure rise during combustion which, in a SI engine, is extremely important in determining the efficiency and smooth running of the engine, in the phenomenon of engine knock and in its relation to the local temperatures and hence the rate of NO_x production. Combustion chambers are therefore designed to give the appropriate rate of pressure rise. In order to approach the design rationally, a number of questions need consideration.

- What controls the flame propagation and hence the mass burning and pressure rise rates in the engine. What is their relationship to combustion temperature?
- What happens in the combustion initiating process? Where in the cycle should the spark occur?
- What unusual combustion problems such as knock and abnormal ignition exist? What, if any relationship exists between them?
- What is the effect of the rate of pressure rise on efficiency and emission levels?

These will be dealt with in turn.

7.3.3 A mathematical description of flame propagation

If the pressure in the combustion chamber of an SI engine is monitored at any two widely spaced points using a time base applicable to measurements of events over one or two engine revolutions, no discernible difference will occur at any one instant between the two pressure traces. That is, any pressure rise due to a local 'parcel' of mixture being burnt at the flame front is transmitted by extremely fast compression waves throughout the chamber. The pressure does, however, rise everywhere in the chamber as the combustion proceeds. Temperature, on the other hand, if measured at the same two points, will generally show major differences. In other words, the wave motion does not transmit temperature and there are a number of separate regions at the same pressure in the combustion chamber having different gas conditions and, indeed, different gaseous compositions. The principal division between regions is a moving combustion front.

Figure 7.9 provides a simple representation of the combustion. The flame front, a fairly ragged boundary separating burned and unburned gas regions, propagates from the ignition source by a process of heat transfer and mass diffusion to the gas ahead of it. This brings the unburned gas to a temperature where local ignition of the fuel/air mixture will occur. The change from unburned to burned gas occurs most rapidly in a short distance just behind the front but it is, in fact, not accurately represented by step changes, being to some extent a continuous variation. However, for most purposes, it is simpler to view the various regions as discrete local gaseous 'packets'. These are basically the flame front region itself and zones ahead and behind it which can be termed **pre-** and **post-flame regions**.

The flame front can be loosely defined as the leading edge where the rate of temperature rise becomes high. The flame front region may be subdivided into a preheat zone at the front edge where a noticeable rise in the unburned gas temperature caused by heat transfer from the flame front occurs to bring it to the ignition temperature. Immediately behind the preheat zone, there is a region termed the **reaction zone** where most of the release of the chemical energy occurs. These two zones will generally be dealt with here combined into a single unit. It may range from very thin for laminar flames to quite thick for turbulent ones. The other predominant zones are the ones which contain burned (i.e. post-flame) gas between the back of the reaction zone and the ignition source and unburned (pre-flame) gas ahead of the front. As the pressure rises throughout the combustion chamber, the

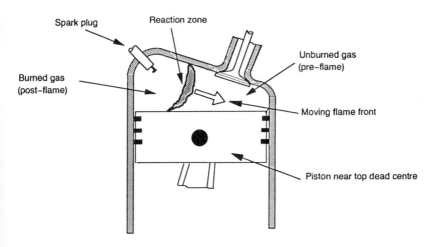

Figure 7.9 Representation of combustion in a spark ignition engine.

temperature of all sections, pre-flame, reaction zone and post-flame, rises further due to the compression created by the energy release. Thus the unburned gas pressure and temperature increases with time even in parts well away from the flame front where direct heat transfer and diffusion from the flame have a relatively small effect. The post-flame gas also is not necessarily at its hottest immediately behind the reaction zone due to the same compression effect. Some of this gas, burned early in the combustion process, will have a considerable temperature rise due to the compression which more than offsets any heat transfer from it to the walls or to the other gaseous regions.

Although a quiescent chamber giving a laminar flame propagation is possible, in general in engines, a highly turbulent combustion takes place. Turbulence in engines arises from a number of causes, these stemming either from the intake flow or from the subsequent piston motion. Turbulence consists of eddies ranging from those with a large scale (called the **integral scale**) with a characteristic dimension approaching that of the engine bore down to decayed vortices of molecular dimensions. The latter are a result of the breakdown of the larger scale turbulence and are responsible for the viscous energy dissipation. At the molecular level, the scale (referred to as **Kolmogorov**) is of the order of microns while intermediate vortices (termed **Taylor microscale**) exist. These scales do not represent discrete elements, there being a continuum from large to small dimensions during the breakdown but the various categories are useful for modelling purposes. In between, a range of turbulence of different dimensions exists. Time scales which give a measure of the time over which an eddy exists are also used to refer to the turbulence. All these scales are related to the magnitude of the velocity fluctuations, u, from the average flow velocity, or, more precisely, to the time averaged rms value of u which is called the turbulence intensity, u'. Typical scales for engines are given on Table 7.2 (from Heywood, 1988).

Much of the turbulence is initially generated directly from the inflow process. For example, the inlet valve airflow jet will produce turbulence of about the scale of the jet width which is about comparable in dimension to the valve lift. More organized motion, such as swirl of dimension up to values approaching the cylinder bore size, may be created by the shaping of inlet port passageways, by shrouding inlet valves to direct the flow tangentially or by adding a swirl control butterfly valve in the inlet port.

Table 7.2 Approximate turbulence scales for engines

Turbulence group	Length of scale	Approximate time scale
Integral	2 to greater than 10 mm	1 ms
Taylor	approximately 1 mm	100 μs
Kolmogorov	approximately 10 μm	10 μs

Alternatively, divided chambers may be used to generate swirl from the in-cylinder gas motion. That is, different swirl levels may be designed into the engine. Another design factor which creates turbulence is 'squish', which is a forcing of flow from the narrowing regions between piston and cylinder head during the upward motion of the piston. This creates a jet-like flow into the main combustion space. Again, some divided chambers are turbulence rather than swirl generators, the details of the fluid motion depending on the shape of the chamber and, more particularly, on the passageway connecting it to the main chamber. It should be noted that divided chambers are little used in SI engines due to their high losses. Inlet generated tumble may also exist. This, like swirl, is a more organized motion but the rotation is normal to the cylinder axis rather than parallel to it. In all these cases, the eddies created are modified and broken down by the compression process but may still exist, even at a large scale during combustion and expansion. The turbulence level markedly affects the flame propagation speed but because of its different generation by many factors, it should be examined in relation to specific engine designs. As a rough indication, the turbulence intensity is about half the mean piston speed in engines without swirl (Heywood, 1988) and is increased by the swirl. Swirl numbers (i.e. the ratio of what is called the 'solid body rotation' of the flow to the angular velocity of the crankshaft) are typically about five in an engine designed for swirl but can be increased well above that to perhaps 15 by piston crown (bowl in piston) and head design. For a thorough examination of turbulence and turbulent combustion in engines, the reader is referred to Heywood (1988).

If the burning velocity is designated by the general term, s_b, the rate of production of burned gas may be expressed as

$$dm_b/dt = A_f \rho_u s_b \qquad (7.2)$$

Here, s_b may be used to represent either a laminar (s_ℓ) or turbulent (s_t) burning velocity as appropriate.

In equation (7.2), the burning velocity is the rate at which the flame moves relative to the unburned gas ahead of it, the result being obtained from a Gallilean transformation rendering the front stationary. It is a result of burning new gas thereby transforming it into a combustion front itself. It is important to distinguish between the above burning velocity which is based on the mass consumed by the flame and the expansion speed which is based on the rate of volume increase of the combustion 'fireball'. The latter includes both the consumption of new unburned gas as described above and the effect of the velocity imparted to the unburned gas by the expansion of the burned gas behind the flame. The last is at a lower density because it is at the combustion chamber pressure but has a higher temperature and lower in average molecular mass following combustion. The difference between the two velocities can best be envisaged by imagining a front where the sole contribution is by heat transfer to the gas behind it. Here, the front would move forward with a velocity due to the volume expansion imparted to both the heated gas behind and the unheated gas ahead without any mass crossing the front. Correction of measured visual or similar volume-based speeds must therefore be carried out to convert them to mass-based burning velocities. The reverse correction must be applied to mass-based burning velocities in order to establish position (i.e. volume-based) coordinates. It should be noted that these corrections will be different for plane, cylindrical or spherical flames. In engines, the flame, although curtailed by the walls, is generally of a roughly spherical form as it develops from a point source ignition.

Laminar flames can be easily handled as the burning velocity is a function only of the fuel, the air/fuel ratio and the unburned gas temperature and pressure. Whilst there is no general theoretical derivation covering all fuels and conditions in engines, there are many experimental correlations. The form developed by (Metghalchi and Keck, 1982) is suggested here although others may also be valid. Here, the laminar burning velocity for a given fuel is expressed as a function of the pressure and the unburned gas temperature as

$$s_l = s_{l,0}(T_u/T_0)^\alpha (p/p_0)^\beta \tag{7.3}$$

Subscripts l, u and 0 represent laminar, unburned and standard state (298 K and 1 atmosphere) values respectively. Note that the pressure p is uniform across both the burned and unburned regions and is therefore unsubscripted. The indices, α and β, are functions of the equivalence ratio, φ, for the particular fuel. Typical values of α, β and $s_{l,0}$ for some commonly used fuels are given in Figure 7.10.

The turbulent flames most commonly found in practice are much more difficult to model. These are described in Heywood (1988) and a condensed description now follows. A macroscopic view of a turbulent flame front is that it is a highly wrinkled region of perhaps 5 mm to 10 mm thick distorted by the turbulence to such an extent that many islands of unburned gas lie within the reaction zone. In fact, experimental photographic evidence indicates that a reaction zone is a complex region interconnecting burned gas surrounding a substantial number of indentations and pockets of unburned gas which are in the process of being consumed. The flame front is therefore difficult to model at this level. However, from a more macroscopic view, a reasonable description can be postulated and this is that the flame front burns locally through these pockets at the laminar flame speed increasing the overall burning rate substantially above the laminar value because of the much larger effective flame front area. The higher turbulent flame velocity is therefore due to the entrainment of unburned gas into the front. The turbulent burning may be viewed as equivalent to a laminar process for a flame of effective area A_l which is much greater than that of the actual flame that would be seen physically. This latter area is called A_f. The

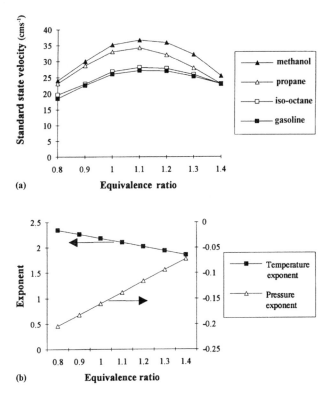

Laminar burning velocity for *methanol, propane, iso-octane and gasoline* as functions of ϕ (from Metghalchi, M and Keck, J.C. (1982))

For *methane* (data for $\phi = 1$ only), $s_{l,0} = 33$ cm s^{-1},

$$\alpha = 0.26 + 1.90\,p - 0.34\,p^2$$
$$\beta = 0.01 - 0.27\,p \qquad \text{where p is in atmospheres}$$

(from Milton, B.E. and Keck, J.C. (1984))

Figure 7.10 Functions for the determination of the laminar burning velocity of propane, methanol, iso-octane and gasoline. (a) standard state laminar burning velocity as a function of equivalence ratio, (b) temperature and pressure exponents for laminar burning velocity as a function of equivalence ratio (see also Heywood, J. B., 1988, *Internal Combustion Engines Fundamentals*, McGraw-Hill, New York).

difference between these two areas, $A_t - A_f$, represents the contribution due to the turbulent entrainment of unburned gas into the front.

Assume now that the combustion front is turbulent and contains a reaction zone, say 5 mm to 10 mm thick. The entrained gas in pockets is being burned locally by laminar flames with a thickness of perhaps 100 μm. The front progresses by a combination of the laminar burning at the outer edge directly into the unburned gas zone in the same manner as a laminar flame

and by unburned gas entrainment into the reaction zone where it is consumed by these local fronts. The relative significance of these two mechanisms will vary with the fuel and conditions. For example, a fuel such as
hydrogen with a high laminar flame speed will have a thinner reaction front
and therefore less entrainment than a fuel with a slower laminar burning
speed under the same conditions. In (7.2), the burning velocity s_b will be
made up in the turbulent case of the sum of these two factors which are the
sum of the laminar burning velocity, s_ℓ, and a turbulence entrainment related term. If an entrainment velocity of unburned gas across the combustion front is defined as u_e, the mass entrained as m_e and the entrainment
rate as dm_e/dt, these can be related by

$$dm_e/dt = A_f \rho_u u_e \qquad (7.4)$$

It is possible to envisage a simple laminar front of area A_l which produces
the same total burned gas mass as the combined mechanism described
above. The rate of production of burned gas, $m_{e,b}$, from the entrained
pockets alone can therefore be described by the difference between A_ℓ and
A_f as

$$dm_{e,b}/dt = \rho_u(A_\ell - A_f)s_\ell \qquad (7.5)$$

If the leading and trailing edges of the front are designated by subscripts f
and b respectively, the reaction zone volume is $V_f - V_b$ of which a proportion is occupied by unburned gas. Multiplying the top and bottom of the
right-hand side of (7.5) by this volume gives

$$dm_{e,b}/dt = [\rho_u(V_f - V_b)(A_\ell - A_f)s_\ell]/(V_f - V_b)$$
$$= [\mu(A_\ell - A_f)s_\ell]/(V_f - V_b) \qquad (7.6)$$

where

$$\mu = \rho_u(V_f - V_b)$$

Here, μ can be seen to be the mass within the reaction zone if it was to be
entirely filled with unburned gas and $(A_\ell - A_f)s_\ell$ is the volume rate at which
the mixture entrained within this zone is burned. If not filled, this mass and
the time reduce proportionally. The reaction zone volume divided by this
volume rate will give the characteristic time which would be taken to con-
sume the unburned gas within the reaction zone, τ. That is, the parameters
μ and τ are characteristic mass and time values for the combustion of the
entrained gas. From the above, the rate of production of burned mass from
the turbulent entrainment is

$$dm_{e,b}/dt = \mu/\tau \qquad (7.7)$$

Hence, the total rate of production of burned mass may be written as

$$dm_b/dt = A_f \rho_u s_\ell + \mu/\tau \qquad (7.8)$$

Essentially, μ is proportional to the mass of unburned gas existing at any instant in the reaction zone and it therefore increases because of any mass entrained and decreases due to the mass consumed. As the flame develops with time, the front changes from laminar without entrained gas to turbulent with substantial entrainment. That is, entrainment of new gas proceeds at a faster rate than the burning of that already entrained. Once developed, the flame may be regarded as quasi-steady as it progresses across a combustion space, both entraining and entrained burning rates being approximately equal. When the combustion is near completion, the entrainment will reduce and the gas will burn out. This would normally occur when the flame reaches the wall. The rate of change of μ with time is equal to the rate of entrainment minus the burning rate and is

$$\mathrm{d}\mu/\mathrm{d}t = \mathrm{d}m_\mathrm{e}/\mathrm{d}t - \mathrm{d}m_\mathrm{e.b}/\mathrm{d}t = A_\mathrm{f}\rho_\mathrm{u}u_\mathrm{e} - \mu/\tau \qquad (7.9)$$

In this equation, a reasonable initial condition is that $\mu = 0$ at $t = 0$. That, is when the flame kernel is very small relative to the turbulent scale, it exists as a laminar flame only. It has been suggested (Keck, 1982) that the entrainment will initially increase at a very high rate, gradually levelling out as the quasi-steady developed flame condition is approached. It can therefore be modelled as an exponential increase. The relationship suggested is

$$u_\mathrm{e} = u'[1 - \exp(-t/\tau)] \qquad (7.10)$$

where u' is a characteristic entrainment velocity at the quasi-steady condition directly related to the turbulence intensity. This equation fulfils a further condition that, as the turbulence level decreases to zero, the flame will be laminar. From experimental observations, Keck has suggested the following empirical relationships for u' and τ

$$u' = 0.08\, u_\mathrm{i}(\rho_\mathrm{u}/\rho_\mathrm{i})^{0.5}$$
$$\tau = 0.8\, L_\mathrm{iv}/s_\ell(\rho_\mathrm{u}/\rho_\mathrm{i})^{0.75} \qquad (7.11)$$

Here u_i is the mean gas velocity through the inlet valve which may be obtained from an approximation of the average volume inflow rate in terms of the mean piston speed. That is

$$A_\mathrm{iv}\, u_\mathrm{i} = \eta_\mathrm{v}A_\mathrm{p}s_\mathrm{p} \qquad (7.12)$$

giving

$$u_\mathrm{i} = \eta_\mathrm{v}(A_\mathrm{p}/A_\mathrm{iv})s_\mathrm{p} \qquad (7.13)$$

The mean piston speed is s_p, A_p and A_iv are the piston and inlet valve opening areas, L is the valve lift and η_v the volumetric efficiency.

These equations describe the initial development of the flame front and its quasi-steady propagation across the combustion chamber as a developed flame. All that remains is to examine the burnout period when the flame reaches the furthest wall and no additional unburned gas entrainment is possible. Here, only the unburned gas trapped in pockets within the front is responsible for further heat release.

It can be seen from (7.9) and (7.10) that, if A_f is zero

$$dm_b/dt = \mu/\tau \tag{7.14}$$

and

$$d\mu/dt = -\mu/\tau \tag{7.15}$$

Separating the variables in (7.15), integrating gives a solution. Evaluation of the integration constant is obtained by taking the mass parameter μ as μ_w and the burning rate as $dm_{b,w}/dt$ at the instant the front completes its movement to the wall (t_w) at the start of the burnout period. Quantities throughout the burnout are then designated generally by μ, dm_b/dt, t, etc. This gives

$$\ln(\mu/\mu_w) = -(t - t_w)/\tau \tag{7.16}$$

from which

$$dm_b/dt = \exp[-(t - t_w)/\tau]\, dm_{b,w}/dt \tag{7.17}$$

These equations allow the burned mass fraction to be calculated as a function of time or crank angle position. A typical curve is shown on Figure 7.11. This follows the type of curve known to exist in practice. It can be seen that the initial period of flame development is slow and has traditionally been referred to as the ignition delay period. It is, however, not distinct from later flame development but is caused by the slow burning in the initial laminar phase. It is best referred to by the now commonly used term, **flame development period.** Its extent is somewhat arbitrary but typically it might be assumed to occur while the initial 10% of the mass is undergoing combustion. At the top end of the curve, the burnout period exists and this again gives a progressively slower mass burned increase. It again is defined arbitrarily and may occupy about the last 10% of the mass burning process. It between, the major combustion process occurs. This can be referred to as the rapid burning phase.

It should be noted that the above approach is a representative model only. It is based on a mixture of application of the known fundamentals and some empirical formulae. That it works well is evident from the agreement with experiment. However, it does not exclude other approaches from being used to describe the combustion process.

7.3.4 Flame propagation effects on engine design and operation

Cylinder pressure versus crank angle

The combustion in an SI engine occurs as a turbulent flame propagating across the chamber. The flame front is initiated by the spark and grows first

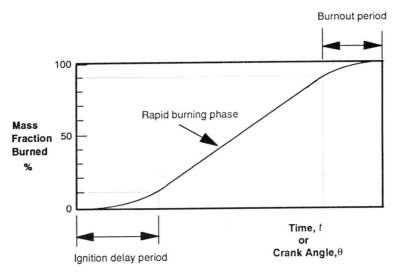

Figure 7.11 Variation in mass burning rate in SI engine combustion.

at a slow rate in a roughly spherical manner. Here the front is small compared to the larger turbulent structures in the combustion chamber and so small-scale turbulence can affect its development while larger turbulent structures are likely to juggle it around to different locations in the field. It therefore has the potential to vary considerably in shape, volume and location in this early stage, affecting the otherwise more stable later combustion which has far greater energy release. The overall variations to the combustion are quite visible when recorded on pressure traces and are known as **cycle-by-cycle variability**.

Examination of engine pressure traces is useful. For some period, perhaps about 10° or more of crankshaft movement, after the spark initiation, there is little departure from the pressure trace for the unfired engine (i.e. the 'motoring' trace). This is the classical 'ignition delay' period discussed above. As the flame front area and the temperature and pressure of the unburned gas all become larger, there is a continual increase in the rate at which the unburned mixture is consumed and the flame front then progresses, constrained by the chamber shape, providing significant pressure rise until the mixture is almost burned out. The flame front during this stage is no longer spherical but is modified by the walls of the combustion chamber. Obviously, a long thin chamber limits the growth in flame front area compared to a more open shape as is illustrated on Figure 7.12 with a slower combustion rate in the former case. As the end of combustion in any configuration is approached, the rate at which unburned mixture is consumed decreases because the flame front area is again limited. Therefore, the mass fraction of the initial mixture burned when plotted against

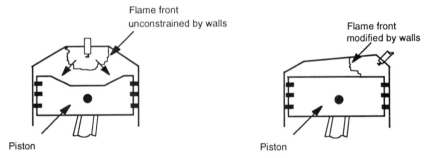

Figure 7.12 Flame fronts either minimally or highly modified by the walls of the combustion chamber.

time (or crank angle degrees) follows the smoothed ramp form as shown on Figure 7.11.

The first two regions can be distinguished as shown on Figure 7.13 where a 'fired' pressure trace is compared with a 'motored' one. The length of the flame development period is basically dependent on the type of fuel, mixture strength, and the cylinder conditions of temperature and pressure, a typical set of curves being shown on Figure 7.14 (after Ricardo, 1933). It can be seen that the flame development time is extended significantly for weak mixtures with a minimum delay occurring for rich mixtures with an equivalence ratio of about $\phi = 1.2$. Above this, the delay again increases although at a slower rate than with weak mixtures. For very long flame development periods, the rapid burning period will not be reached within the time scale available and no net work will be available from the combustion. For a given set of pressure and temperature conditions, the time taken for the

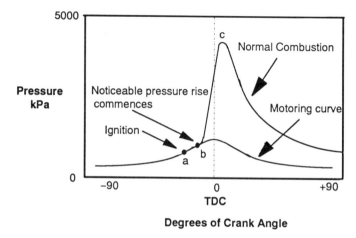

Figure 7.13 Engine pressure traces showing typical ignition and combustion curves.

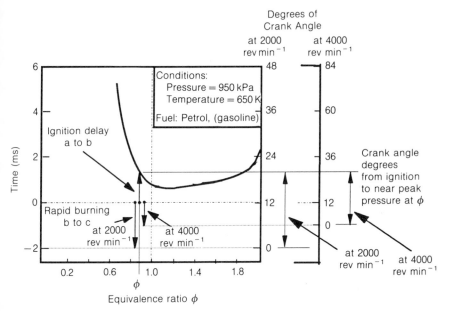

Figure 7.14 Effect of ignition delay and rapid burning on period from ignition to peak pressure (adapted from Ricardo, H., 1933, *The High Speed Internal Combustion Engine*, Blackie, London). Note similarity to flammability limit curve of Figure 7.2.

flame development period is substantially independent of engine speed. This is because the flameball is small and it is substantially unaffected by the large scale turbulence generated by the inflow and piston motion which are themselves functions of engine speed. Consequently an increase in engine speed directly increases the crank angle required for the period.

The rapid burning period is the region designated B to C in Figure 7.14. Most of the unburned mixture is consumed here and the pressure departs noticeably above the motoring curve and reaches a high value. The time taken again depends on fuel, mixture strength, pressure and temperature and it should be noted that the last two vary considerably during the burning period. However, other factors are now important, one being the shape of the combustion chamber and another the engine speed, the former dictating the size of the flame front area in relation to the remaining unburned mixture volume and the latter modifying the turbulence levels within the cylinder. In the rapid-burning region, the burning velocity is highly dependent on engine speed. For a given engine and fuel, the number of degrees of crank angle over which this phase occurs show little dependence on engine speed. That is, the time taken for flame propagation decreases at roughly the same rate as the engine speed increases. This is because the turbulence generated by the inflow process and modified by the compression stroke are related more or less directly to the engine speed. Obviously, this is only a rough approximation and there is some variation from engine to engine.

If the position in the cycle at which peak pressure is desired can be fixed, it is then possible to determine the spark timing for any set of conditions. As a first approximation, the total burning time of the mixture, A to C, can be found by adding the appropriate roughly constant crank angle for the rapid burning period B to C (for example, say $12°$) to the crank angle at the appropriate engine speed for the flame development period A to B, which is a variable.

Fuel, diluent and oxygenation effects

If the preceding curves are typical of a conventional spark ignition engine fuel, i.e. a petroleum-based fuel, it may not be representative of other substances, which have quite different burning characteristics. For example, hydrogen burns very rapidly and has wide flammability limits and this can be seen on the diagram of Figure 7.15. That is, it has a reduced flame development period when compared with petrol particularly away from stoichiometric air/fuel ratios. The flame propagation period may also be substantially reduced, the amount depending on the turbulence levels as the laminar flame speed of hydrogen is extremely high.

The rate of flame propagation increases with the flame temperature because of the more rapid heating of the unburned gas ahead of the flame front to its ignition temperature. It also increases when there is less inert gas

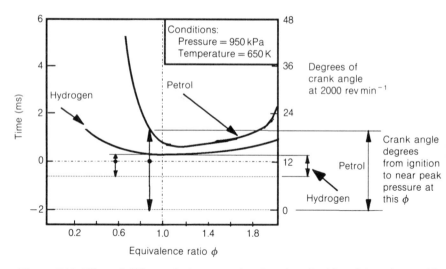

Figure 7.15 Effect of different fuels on combustion time (ignition delay plus rapid burning) (adapted from Ricardo, H., 1933, *The High Speed Internal Combustion Engine*, Blackie, London).

to heat in the unburned mixture ahead of the flame. Flame temperatures are highest in pure oxygen and become progressively lower as increasing quantities of inert gas are added to the unburned mixture. Consequently, the addition of oxygen can be shown to reduce the flame propagation period. Alternatively, the addition of an inert gas increases this period. Oxygen enriched combustion is uncommon in all engines except those for rockets although the possibility of easily and cheaply increasing the oxygen level of air through the use of membrane separation techniques now exists. While some experiments have been carried out, principally on diesel engines, in order to reduce smoke levels, it is not clear that there is a major advantage in this technology. Inert gas levels above those of normal ambient air are far more likely and these are referred to as diluents. The most important diluent of this type is the exhaust gas which is always present to some extent in a normally aspirated SI engine as residual gas from the exhaust. Alternatively, additional exhaust gas may be deliberately added for emission control purposes (EGR). Thus, a slow burning rate may then occur which can have some effect on reducing the cycle efficiency. The effect of oxygen addition and diluents is also shown on Figure 7.16.

Pressure rise during combustion

The rate of pressure rise in an engine during combustion can be modified to some extent by the design details. For example, the size (i.e. the bore),

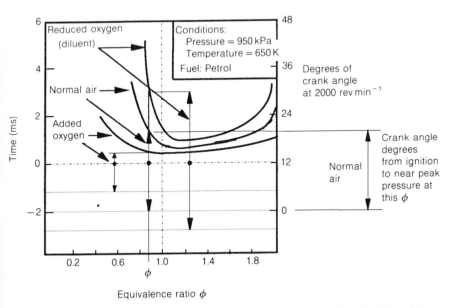

Figure 7.16 Effect of changes in oxygen levels on combustion time (ignition delay plus rapid burning) (adapted from Ricardo, H., 1933, *The High Speed Internal Combustion Engine*, Blackie, London).

operating speed, air/fuel ratios, type of fuel, compression ratio and operating temperature, all play some role in determining the number of crank angle degrees from the start of a discernible pressure increase above the motoring trace to the position of peak pressure. However, all the above are constrained by other design parameters. The turbulence built into the combustion chamber is a variable with which there is some freedom and can therefore be used to modify, within limits, this rate of pressure rise.

Typically, the pressure rise occurs over about crank angle of 10° to 20° and the desirable rate may be about 250 kPa to 300 kPa per degree of crank angle movement with the maximum pressure being reached at about 10° ATDC. Obviously, the faster the rate of pressure rise, the closer the engine approaches the constant volume combustion which, from the thermodynamic analysis, provides the most efficient cycle. However, if the rise is faster than that indicated above, excessive turbulence occurs in the combusting gases and the motor will run roughly with the rate of heat transfer to the walls becoming excessive. For low compression ratios, the flame propagation is slow and substantial turbulence needs to be designed into the head. At higher compression ratios, this becomes less important although most engines operating on compression ratios of about 8:1 have some 'squish' area.

The peak pressure has a maximum value when it occurs at TDC. For smooth running it is generally a little later than this, about the 10° mentioned before. This has little adverse effect on efficiency as the combustion chamber volume has increased very little at this point. Putting the maximum pressure before or well after TDC causes excessive heat transfer as large areas of cylinder wall are exposed during the major part of combustion. Also, for the maximum pressure to occur at TDC, the ignition point is early in the cycle and low pressures and temperatures then cause a long flame development period and initially a slow flame propagation.

Ignition timing

The major factors in determining spark timing are now apparent. On the assumption that the maximum pressure rise should, wherever practicable, occur a little after TDC, it is possible to work backwards to find the appropriate ignition point.

It can be seen that, for a given mixture strength, the spark must be advanced as engine speed rises to cope with the longer crank angle for the same delay time (Figure 7.14). On the largely superseded mechanical type distributors, this is the purpose of the centrifugal weights. At part-throttle operation, the proportion of residual gases in the mixture increases. Consequently, both the ignition delay and the flame propagation period become longer even at the same revolution speed. Again, the spark must be advanced and this is accomplished by a vacuum device operated from the intake manifold. Mixture strength usually plays little part in determining ignition timing as most spark ignition engines operate in the range $\simeq 0.95$ to 1.2 where the curve is relatively flat. However, with the modern lean-burn engines, which is a technique used to lower emissions while retaining high efficiency, some retarding factor may need to be introduced on acceleration when the mixture is enriched. This is best accomplished electronically.

It should be noted that with very weak mixtures, a large spark advance is required and this is compounded by the fact that, at this position, the pressure and temperature of the only partially compressed gas is very low, further slowing flame propagation and increasing delay time. Consequently, extremely large advances are not used. If then, for a normal spark setting, a very weak mixture is used, the slow combustion can leave gas burning during the exhaust stroke. Some of this may be trapped in crevices (e.g. spark plug) in the engine and can ignite the incoming mixture on the next stroke causing the engine to backfire. The slower increase of delay time with change in at the rich end prevents this from happening there.

In turbocharged engines, the system is often designed so that little pressure rise occurs until a certain engine speed is reached after which the boost pressure is considerable. The higher pressures and temperatures then decrease the total time from ignition to peak pressure and a retard needs to be induced into the timing system at this point. Careful modification of the distributor is therefore required when fitting a turbocharger to an existing engine.

Finally, it should be noted that, while about 10° ATCD may be ideal from a performance viewpoint, control of NO and other emissions sometimes requires a retarded spark to reduce peak pressures and hence temperatures at some parts of the cycle. The characteristics of distributors on modern cars are therefore a little more complex than the system above envisaged and the necessary retard will depend on how much the unmodified design exceeds the legislation requirements.

From Figure 7.16 it can be seen that for a given mixture strength, the spark must be advanced as engine speed rises or the throttle is closed in order to cope with the longer delay.

7.3.5 Spark ignition engine knock

Spark ignition engines are subject to combustion noise which is generally termed *knock*. While there are a number of types of knock, two are most frequently encountered, these being **pre-ignition** and **end-gas autoignition**. The high frequency pinging noise in an engine is end-gas autoignition, sometimes referred to as detonation. Pure pre-ignition either does not cause a noise or produces only a dull thud.

Pre-ignition

This occurs when a hot spot in the combustion chamber (e.g. carbon, spark plug) ignites the charge before the spark plug fires. There is no essential difference to the flame front and combustion except that the ignition source is different. When it occurs just a little before normal ignition, it is difficult to notice any effect on the engine. The danger lies in that, with early ignition, more heat is transferred to the walls, the hot spot becomes hotter and the pre-ignition point moves further forward in the cycle causing more heating and eventual engine damage. This is particularly serious when only one cylinder is pre-igniting. Generally, for an engine which does not autoignite when the peak pressure is near TDC, pre-ignition will not promote it. If, because of the available fuels being of too low an octane rating, the engine has been deliberately retarded to prevent end-gas autoignition, the pre-ignition may promote it. The hot spot may have many causes mostly related to poor maintenance or design. However, one possible cause of pre-ignition is prolonged end-gas autoignition. The end-gas autoignition waves may disturb the cool layer near the wall and promote greater heat transfer, hence starting a hot spot.

End-gas autoignition

When the flame front moves across the combustion chamber, the unburned gas is compressed rising in temperature as well as pressure. Radiation from the flame front also affects this gas. The critical section is the end gas, the last portion to be burned. If its pressure and temperature becomes too high, it will ignite spontaneously and high-pressure severely-peaked waves will cross the chamber fed by the hot mixture ahead of them. These are similar to severe shock waves and there is some similarity to classical detonation waves although they are not identical. In the engine case, the wave is unconfined and is fed by both the combustion energy behind it and the additional compression and heat transfer from the flame front.

Waves of this type can occur when as little as 5% of the end gas ignites. They provide sound of audible magnitude, usually regarded as a ringing of the surrounding metal. As mentioned previously, if prolonged the waves may promote pre-ignition. While short bursts of this type of knock behaviour appear to be of little consequence, prolonged knocking will cause a slight pitting of the surface of the piston crown and eventually severe erosion. This is probably because of additional local high temperatures and high heat transfer rates.

The cause of end-gas autoignition is either too high a compression ratio, too long a flame front travel or poor cooling in the region of the end gas. The high compression ratio starts the end gas at more severe conditions where it is easily overheated during combustion. A long combustion chamber with its increased time for flame front travel isolates a small portion of the end gas until it becomes severely compressed and can ignite spontaneously. Poor cooling has the obvious effect of providing high end-gas temperatures.

For a given engine, fuel type is the most important factor in promoting or preventing end-gas autoignition. Octane rating is used as a measure of the fuel's ability to resist end-gas autoignition. Pure iso-octane has a rating of 100 while the addition of lead compounds (or thallium or some other substances) can raise the rating above this or increase that of inferior fuels without otherwise becoming involved in the combustion process. Some fuels have an octane rating well above iso-octane. Examples of these are benzene, methane and the alcohols. Note that the use of a weak mixture in the end gas allows a higher compression ratio to be used. Thus stratified charge engines, designed to operate with a weak mixture to reduce CO, have this as an added advantage.

In general, for hydrocarbon fuels, the following apply.

- The knock resistance of paraffins increases with decreasing carbon atom.
- Olefins show similar behaviour to paraffins. Below about four carbon atoms/molecule, paraffins are better. Above this olefins are usually better.
- The addition of tetra-ethyl-lead has the largest effect with paraffins. Aromatics are less affected while olefins and acetylenes are only very slightly improved. The effect of the lead seems to be that the reaction rate is reduced.
- Knocking is promoted by adding ozone, nitrites, peroxides which increase the reaction rate.

For a chosen compression ratio, in order to reduce the tendency to detonate, the following design features should be considered.

- The length of flame travel should be kept small so that no significant end gas has a high post-ignition compression. A centrally located spark plug or two plugs can be used.
- Surfaces should be kept cool in the end gas region.
- The exhaust valve, which is at a high temperature because of the high velocity exhaust gas flow over it and the difficulty in providing adequate cooling for it, is best located away from the end gas.
- Sufficient turbulence is required to give the fastest flame front propagation consistent with the other design features of the engine. That is, the correct rate of pressure rise is desirable.
- Diluents may be used to reduce flame temperature. Note that this reduces power and restricts the available equivalence ratios for good ignition.
- Alcohols, alcohol–water or acetone–water may be used as an additive to the fuel. This may be emulsified with the fuel or used separately through injectors. Generally, water–alcohol injectors have been found to be most beneficial with supercharged engines and were originally developed for the high-output aero engines that existed prior to the gas turbine. Note that they may cause corrosion and they require additional maintenance. Thus water or water–alcohol addition is not widely used.
- The spark may be retarded. This is basically a method of adjustment for an engine which otherwise does not match the fuel available. It reduces the maximum engine pressure and temperature but may result in reduced

power and efficiency. Some additional heat transfer from the exhaust gas may cause the exhaust valves to overheat. It is not recommended for any but a minor adjustment.

7.3.6 Combustion chamber shapes

A great many different combustion shapes are possible and it is not the intention to discuss all possibilities here. However, to provide some insight into the importance of shape, some of the designs that have been commonly used over the years are discussed here. Figure 7.17 shows a number of such combustion chamber designs. Their various advantages and disadvantages are listed below.

L head (side valve)

This is now rarely used. It has the advantage of being a simple design with the valves connecting directly to a camshaft in the block. It can have good turbulence characteristics but a long flame path travel. Consequently compression ratios are limited. Valve sizes are also severely restricted by being located in the overlapping area at the side. A 'squish' area may be added to increase turbulence if necessary.

F head (overhead, inlet, side exhaust valve)

A development of the L head, this retains most of its general characteristics but generally has a more compact combustion chamber. However valve size, particularly that of the inlet valve, can now be increased but the overhead location requires rockers or a separate camshaft. A slight reduction in end gas allows a slightly higher compression ratio.

Overhead valve

Nearly all engines are now overhead valve types. The basic type allows large valves and has a short flame path length; hence it has good compression ratio characteristics. However, it may have insufficient turbulence and is usually modified with a 'squish' region as shown. Mechanically it is more complex than the L head as it requires rockers or an overhead camshaft. One or two camshafts may be used.

Inclined overhead valve

This type with variations goes under names such as 'hemispherical' head, 'pent roof' etc. It allows the largest valves, has a short flame path length. If insufficient turbulence is available, this may be increased with squish areas. It is mechanically the most complex, generally requiring two overhead camshafts, but is becoming more commonly used.

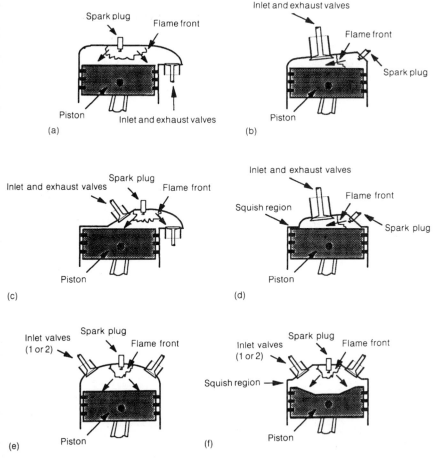

Figure 7.17 Examples of different types of spark ignition combustion chambers. (a) side valve (L head) single camshaft, (b) basic overhead valve single camshaft, (c) overhead inlet, side exhaust valve (F head) single or double camshaft, (d) overhead valve with squish single camshaft, (e) inclined overhead valve (hemispherical head) double overhead camshaft, (f) inclined overhead valve (pent roof head) double overhead camshaft.

7.3.7 Basic thermodynamic analysis from pressure trace data

Combustion chamber pressures being uniform at any one instant are relatively easy to measure but temperatures which vary considerably both spatially and timewise are not. Since the realization that NO plays a crucial role in the formation of photochemical smog, renewed attempts have been made to examine the combustion space mathematically during the flame front movement. The reason for this is that NO formation is very sensitive to the local temperature within the combustion chamber. Two approaches are used. These are:

- With a given engine and a measured (or assumed) pressure trace, the mass fraction burnt can be determined as a function of time and the temperatures of the various portions follows. This is basically a thermodynamic model used for analysis.
- A complete fluid mechanical model is possible by setting up the full equations of motion of the fluid in three dimensions. These include the continuity, momentum and energy and mass diffusivity equations and, in addition, a mathematical formulation of flame front propagation velocity, combustion chemistry and heat transfer. This method does not have the disadvantage of using a measured pressure trace but the accurate assessment of flame propagation speeds and a knowledge of the fluid mechanics of the gas throughout the combustion chamber is required. It is therefore a very complex procedure and results have not to date been entirely satisfactory. It is, however, a rapidly developing area of CFD and some proprietary codes are available. The reader is referred to more specialized texts and research papers for this purpose.

Only the thermodynamic analysis will be dealt with here.

Thermodynamic approach

This method relies on the input of a measured (or otherwise known) pressure trace to the fundamental thermodynamic calculations and thereby bypasses knowledge of both the position of the flame front and the burning velocity. A typical example of this method is that described by Lavoie, Heywood and Keck (1970). The thin reaction region is ignored and the combustion space is broken up into two zones, these containing the mass fractions of burnt gas, designated as x_b, and that of unburned gas x_u, respectively. That is

$$x_b + x_u = 1$$

Using subscripts b for the burnt and u for the unburnt sections, the following equations can be obtained.

Average specific volume, $v =$ chamber volume/mass in chamber

$$= V/m$$

$$= \int_0^{x_b} v_b dx + \int_{x_b}^1 v_u dx \qquad (7.20)$$

The average temperature in each region is given by

$$T = 1/x \int T \, dx \qquad (7.21)$$

Now multiplying through by the instantaneously uniform combustion pressure $p(t)$, (i.e. remembering that p is a function of time)

$$p(t)V/m = R_b T_b x_b + R_u T_u (1 - x_b) \qquad (7.22)$$

Using the First Law of Thermodynamics, the heat and work transfer can be related to the total internal energy in the combustion chamber

$$Q - W = U_2 - U_1$$
$$= m(u_2 - u_1) \qquad (7.23)$$

where

$$u_2 = \int_0^{x_b} u_b \, dx + \int_{x_b}^1 u_u \, dx \qquad (7.24)$$

Note that Q would be negative as heat is transferred from the system. Now the internal energy can be approximated by assuming specific heat is constant over the temperature range considered, i.e.

$$u_b = c_{v,b} T_b + h_{f,b} \qquad (7.25)$$
$$u_u = c_{v,u} T_u + h_{f,u} \qquad (7.26)$$

Here $h_{f,b}$, $h_{f,u}$ are constants and T_b, T_u are mean temperatures. Obviously, variable specific heats could be used instead of the constant values specified but with an increase in complexity. Combining the above gives

$$(U_1 + Q - W)/m = x_b(c_{v,b}T_b + h_{f,b}) + (1 - x_b)(c_{v,u}T_u + h_{f,u}) \qquad (7.27)$$

Equations (7.21) and (7.25) have three unknowns, x_b, T_b, T_u, if it is assumed that the work term can be evaluated from

$$W = m \int p \, dv \qquad (7.28)$$

A third equation is therefore required. This is obtained by assuming that the average unburned gas temperature T_u can be evaluated by isentropic compression from the pressure at the beginning of combustion, i.e.

$$T_u = T_0(p/p_0)^{(\gamma_u - 1)/\gamma_u} \qquad (7.29)$$

Hence, at any known pressure, $p(t)$, x_b and T_b may be obtained as functions of time. Typical results are shown on Figures 7.18 and 7.19.

This method provides a very useful but simple method for estimating the mean gas temperatures in the two zones. It essentially assumes that all the burnt gas from any instant in the progressive combustion is instantaneously fully mixed with all previously burnt gas. This is a limiting case only. The other limit is that each element of burnt gas retains its individual temperature which is raised by the further compression. An additional equation similar to (7.29) is therefore required for each of the individual elements of burnt gas.

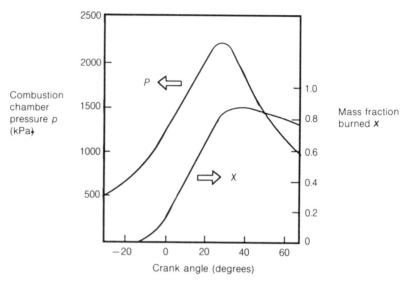

Figure 7.18 Calculation of the mass fraction burned from a known pressure trace (adapted from Lavoie, G. A., Heywood, J. B. and Keck, J. C., 1970, Experimental and theoretical study of NO formation in IC engines, *Combustion Sci. Technol.* **1**, 313.

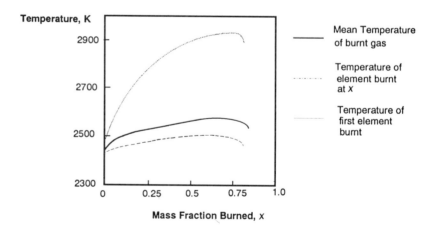

Figure 7.19 Relationship between the temperature of the first and current elements of fluid burned and the mean gas temperature (adapted from Lavoie, G. A., Heywood, J. B. and Keck, J. C., 1970, Experimental and theoretical study of NO formation in IC engines, *Combustion Sci. Technol.* **1**, 313.

A full description of this procedure can be found in Heywood (1988).

7.4 COMPRESSION IGNITION ENGINES

With compression ignition engines, air alone is admitted to the cylinder where it is compressed until it reaches a temperature sufficient to ignite the fuel which is sprayed directly into the combustion chamber near to that point. Temperatures for ignition are approximately 800 K upwards and hence compression ratios of more than about 14:1 are required. Starting devices such as glow plugs which are an electrical heating device within the combustion space are often necessary for cold starting. While simple theory indicates that a continuous increase in efficiency takes place as compression ratio is raised, in practice the maximum efficiency of a reciprocating engine occurs with a compression ratio somewhere around 15 to 17:1 depending on the details of the particular engine. This is because, at higher values, the increased heat transfer and friction losses offset the gains from the improved thermodynamic performance. Higher compression ratios up to about 23:1 are used in some smaller engines which have divided combustion chambers, these being called indirect injection (IDI) types. These high values are required to offset heat transfer losses to keep air temperatures high enough for ignition. The now more common direct injection (DI) types with a single combustion chamber use the more optimal lower values.

The fundamental combustion process in a compression ignition engine takes place as a diffusion flame. The fuel enters the combustion chamber from a high pressure injector nozzle where it rapidly breaks up into a fine spray of droplets. Before combustion can occur, the droplets must partially evaporate and the vapour must diffuse into the air until a suitable air/fuel ratio is reached locally to support combustion. Once combustion commences, the increased temperature promotes evaporation in other droplets and the burning rate increases. There are many aspects which control the rate of combustion. These are the rate of injection, the size (degree of atomization) of the droplets, the engine temperature, the fuel type, the forced mixing patterns and level of turbulence in the engine. Combustion is therefore generally slower than in spark ignition engines. In very large diesel engines, this is an advantage as it keeps peak pressures low. However, efficiency is better with rapid combustion and small high-speed diesels are designed to make the burning process as rapid as possible. In some small engines, the rate of pressure rise is not all that much lower than that in spark ignition engines.

7.4.1 Fuel injection in CI engines

A fine, highly atomized spray is obviously an advantage from the point of view that it gives a very large surface area of droplets for a given mass of fuel entering the combustion chamber. This then allows rapid vaporization, mixing and diffusion and so combustion can occur rapidly. However, a very fine spray or mist has drops with low inertia and these then have little penetration into the combustion space. Hence a compromise is necessary so that the spray can be spread through the chamber. A range of droplet sizes is preferred as some (the smaller) will then reside near the injector, evaporate rapidly and commence combustion while the larger will penetrate to the far side of the combustion space allowing good air utilization.

7.4.2 CI engine combustion modes

The heat release rate is important in diesel engine analysis. It is obtained from the First Law of Thermodynamics for a closed system which may be written in time dependent form as

$$\frac{dQ}{dt} = \frac{dw}{dt} + \frac{dU}{dt} = P\frac{dV}{dt} + mc_v\frac{dT}{dt}$$

Substituting from the ideal gas equation for dT gives

$$\frac{dQ}{dt} = p\frac{dV}{dt} + \frac{mc_v}{mR}\left(\frac{vdp}{dt} + \frac{pdV}{dt}\right)$$

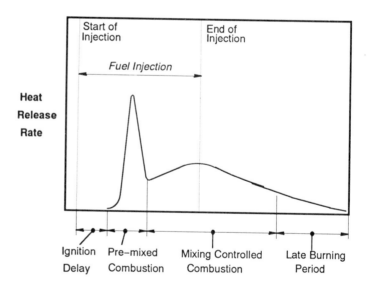

Crank Angle Degrees

Figure 7.20 Typical heat release diagram for a direct ignition diesel engine.

$$= p\left(1 + \frac{c_v}{R}\right)\frac{dV}{dt} + \frac{c_v}{R}\frac{V dp}{dt}$$

$$= \left(\frac{\gamma}{\gamma - 1}\right)\frac{p \, dV}{dt} + \left(\frac{1}{\gamma - 1}\right)\frac{V dp}{dt} \qquad (7.30)$$

The right-hand side of (7.30) is now in terms of pressure p, volume v and time, t only. The pattern of combustion related to the injection rate can then be seen from the heat release diagram developed from a pressure trace on Figure 7.20. It can be seen that the combustion process in diesel engines has three fairly distinct phases. These are the ignition delay, premixed burning and diffusion burning periods. While not so clearly distinguishable on the heat release diagram, a further period called the burnout (or late burning) phase is sometimes distinguished as a separate mode. The important aspects related to CI engine combustion that need to be studied can then be broken into four groups: fuel injection; ignition delay period; premixed combustion phase; diffusion (or mixing) controlled period. These are discussed below.

Fuel injection

Fuel injector systems consist of a basic means of delivering and distributing the fuel to each cylinder, a method of raising its pressure to that necessary for injection and an appropriate nozzle so that the required spray pattern exists within the cylinder. In addition, a governing system must be incorporated so that maximum rotational speeds are not exceeded, so that the minimum idle speed is maintained and so that the fuel supply does not exceed the smoke limits as the engine changes speed and load. It is not the purpose of this text to discuss these aspects and the reader is referred to any diesel engineers handbook for detailed information. All that needs comment here is that the pressure rise and distribution may be achieved in a single high-pressure pump with individual lines to each cylinder or a low-pressure supply pump can be used with the pressure raised in the individual injector. The quantity of fuel is most often regulated within the injector itself but systems exist where this control is located within the high pressure, distributor pump. A typical control varies the end cut-off point of the injector flow by rotation of the central plunger which has a helically cut recess in it. Once the cut-off point is reached, the additional flow spills and is drained back to the fuel tank. Typical injection pressures have been 60 MPa to 75 MPa although, in order to provide more finely atomized sprays which lower smoke emissions, some modern systems inject at 100 MPa to 150 MPa.

Although the flow in an injector is variable, having to start and stop in the short time period available, an approximation of the mass flow rate can be expressed by the normal incompressible nozzle flow equation

$$\dot{m} = C_D A_n \sqrt{(2\rho_f \Delta p)} \qquad (7.31)$$

More accurate models also consider the effect of the pressure pulse in the line between the distributor pump and injector.

The nozzles may be of the multihole type with a needle which acts as a valve on their upstream side. This needle is lifted at the start of injection and lowered at the end. An alternative type, usually used in IDI diesels, is the pintle nozzle which has a single hole and central needle with a shaped external pin which closes on the outer side of the nozzle.

The important characteristics of the spray are the spray angle, the droplet distribution and the penetration of the spray through the central core region of the air in the chamber. The spray angle θ depends on the L/D ratio of the nozzle and the ratio of gas to liquid densities, ρ_g and ρ_ℓ, the penetration S on the nozzle pressure ratio Δp, liquid and gas densities and time, t. A number of essentially empirical formula are available and are given in Heywood (1988). These can be summarized for the case without swirl in the following equations. The spray angle, θ, is given by

$$\tan(\theta/2) = 1.182(1/A)\sqrt{(\rho_g/\rho_\ell)} \tag{7.32}$$

where

$$A = 3.0 + 0.28(L/D)_{\text{nozzle}}$$

The penetration distance can be expressed in various forms but is best divided into that before the spray breaks up into droplets and that after breakup. These equations are

$$S = 0.39\,t\,\sqrt{(2\Delta p/\rho_\ell)} \tag{7.33}$$

$$S = 2.95(\Delta p/\rho_g)^{0.25}\sqrt{(d_n t)} \tag{7.34}$$

The time at which breakup occurs is small and is given by

$$t_{\text{breakup}} = \frac{29\,\rho_\ell d_n}{\sqrt{(\rho_g\Delta p)}} \tag{7.35}$$

It should be noted that the airflow patterns, particularly swirl will affect the penetration distance.

Droplet velocities are related to the Sauter mean diameter of the spray, this being defined as the diameter of a droplet that has the same volume to surface ratio as the spray as a whole. It is therefore given by

$$D_{\text{SM}} = \sum n d_d^3 \Big/ \sum n d_d^2 \tag{7.36}$$

and has been evaluated by the empirical equation of Hiroyasu and Kadota as

$$D_{\text{SM}} = C(\Delta p)^{-0.135}\rho^{0.121}V_f^{0.131} \tag{7.37}$$

where $C = 25.1$ and 23.9 for pintle and hole nozzles respectively. The droplet spray velocity is then given by

$$\frac{dV}{V} = 13.5 D^3 \exp[-3D]\, dD \qquad (7.38)$$

where $D = d_d/D_{SM}$.

It should be noted that these equations are at least partially empirical and that other equations have been given by various investigators to characterize sprays.

Ignition delay

As the fuel is injected, a considerable proportion accumulates before any noticeable pressure rise due to combustion occurs. The delay period is usually defined as being from the start of injection to the first noticeable pressure increase above the motoring trace although the appearance of a luminous flame has also been used to determine the latter point. Delays defined by the pressure rise are fractionally shorter and appear to give more consistent results. The delay occurs because of both the time related to the physical processes such as the evaporation and the mixing which are necessary prior to burning and to some chemical processes such as the rate of the pre-combustion reactions. These are therefore sometimes classified as a physical delay and a chemical delay, the former depending on the environment to which the fuel is subjected while the latter is related to the reactive nature of the fuel. Ignition delay is a function of the cylinder pressure and temperature at the time of injection, the fluid motion (swirl and turbulence) in the chamber and the properties of the fuel. Various correlations exist to evaluate this period, some of the best known being the Wolfer (1938) correlation in pressure p and temperature T only which does not consider either the fluid motion or fuel and that of Hardenberg and Hase (1979) which does to some extent by the inclusion of the mean piston speed S_p and the fuel activation energy E_A. These equations are

$$\tau_{id} = \frac{382 \exp(2100.8/T)}{p^{1.0218}} \quad \text{(Wolfer type equation)} \qquad (7.39)$$

$$\tau_{id} = (0.36 + 0.22 S_p) \exp\left[E_A\left(\frac{1}{RT} - \frac{1}{17\,190}\right)\left(\frac{2120}{p - 1240}\right)^{0.63}\right] \qquad (7.40)$$

(Hardenberg and Hase equation)

In these equations, τ_{id} is in ms (7.39) and in degrees of crank angle (7.40), T is in K and p in kPa. The activation energy E_A (called an apparent activation energy) for the latter has been approximated in terms of the cetane number (CN) as

$$E_A = \frac{618\,840}{CN + 25} \qquad (7.41)$$

Premixed combustion

During the ignition delay period, much of the fuel injected by that time has been able to evaporate and mix and the ensuing combustion which follows is extremely rapid, something characteristic of premixed fuel and air in general. It is difficult to model precisely because the initial ignition point or points are not known. As a first approximation, it could be assumed that all of the premixed fuel will reach readiness for ignition simultaneously and burn instantly but this would then overestimate the effect. Under these conditions, the total amount of fuel injected during the delay period could be used to approximate that undergoing premixed combustion. In reality, the time period for premixed combustion is small but finite indicating that not all of it is ready to autoignite simultaneously. Also, some of the fuel that has been injected will not be fully prepared for burning at the end of the delay period. A more accurate assessment of the process follows that given by Lyn (1962) who divided the injection into a series of discrete mass elements. Each incremental mass of fuel reaches readiness for combustion at a time dependent on its initial point of injection and the prevailing in-cylinder conditions. At the time the ignition delay period is completed, combustion of the prepared fuel commences. As it is consumed, more fuel is added during this additional short period. It should be noted that, as would be expected, the total mass of fuel prepared is identical to that in the total injection process.

Mixing controlled combustion

The initial rate of combustion is quite rapid. Once this phase is over, the remainder of the fuel must diffuse and mix with the air in order to support the already existing flame. This then controls the rate of further reaction which is much slower than the initial phase. Most of the fuel is burned in this mode, typical estimates from heat-release calculations being around 75%. The fuel is now entering the combustion chamber as a liquid spray surrounded by hot burning gases and so the rate of mixing is dependent on the spray characteristics and the air flow patterns within the combustion space. The processes are variable with different injector and engine design and are therefore complex making it difficult to apply a simple approach for its quantification. What is clear is that increased swirl and turbulence, particularly the former increase the mixing rate. Also, a higher spray velocity, finer atomization and good spray distribution without wall impingement are important. Each situation needs consideration on its own merits and a full CFD analysis provides the best approach. Note that, under some circumstances, the combustion in this mode is likely to be fuel rich and a small quantity of the liquid fuel may be affected by the heat, pyrolysing to other compounds which may be less combustible.

7.4.3 Combustion chambers for compression ignition engines

There is considerable variation in the combustion chambers used in compression ignition engines, particularly at the smaller engine capacity end of the range. While modern methods of computer simulation of the flow and the combustion process are being increasingly used, in many ways the design of CI engine combustion chambers is still somewhat of an art. The larger engines present fewer problems. Here, it is best to start by considering these and then move progressively to the smaller types.

Very large diesel engines which are used for marine or stationary purposes may have cylinders of up to a metre in diameter. Under these circumstances, a considerable amount of fuel is injected in each cycle, even at low load. Thus the injector nozzles are reasonably sized with multiple orifices which can distribute the spray quite widely. Multiple injectors are also quite feasible and are sometimes used. In addition, the engine speed is low, in some cases being perhaps around 100 rpm, and so there is considerable time available for mixing of fuel and air after injection. With such large cylinders, too rapid a combustion process would generate high peak pressures and hence cause immense forces to be developed and a slightly slower combustion is desirable, in spite of the thermodynamic advantages of fast burn. The slow speed means that the ignition delay occupies relatively few crank angle degrees. Given that, very large diesel engines require single chambers with little turbulence, typical designs are depicted in Figure 7.21. These limiting large chambers are sometimes termed **quiescent**.

As the engine becomes smaller and faster, the combustion chamber needs modification so as to increase the mixing rate in order to compensate for the poorer spray patterns and to provide an increasingly rapid combustion. Various patterns for achieving this, still using a single chamber, are shown. These usually have some method of increasing swirl rates at the time of injection. Inlet flow direction can generate the initial swirl which can be increased near the end of compression by forcing the flow into, say a deep toroidal bowl in the piston crown. A certain amount of squish area associated with the bowl will also increase the turbulence. A number of variations of the bowl-in-piston designs with different amounts of swirl enhancement and turbulence generation are possible. A further possibility is a design which uses a pintle injector with the spray directed to a hot section of the wall, usually the base of the piston bowl. Here, the fuel evaporates off at a rapid but to some extent controlled rate and the mixing is in the gas phase.

All the above types of single chamber engines are called direct injection (DI) diesels. The minimum cylinder size at which these are appropriate is becoming progressively smaller as the technology of injector design and the understanding of in-cylinder, flows improves and DI chambers are now possible in cylinders of less than 0.75 litre capacity.

For the very small engines, single orifice injectors spraying from the nozzle past central, axially located dividers (pins) are used. These are called *pintle injectors.* Below what is regarded as a minimum optimal size for a DI chamber which will vary depending on the design specification of the engine, the demands of the mixing process become very severe and can only be met by splitting the combustion chamber into two sections, a main chamber directly above the piston and an auxiliary chamber connected to it. Note that the main chamber can be and often is smaller than the auxiliary chamber.

These types are called indirect injection IDI diesels (Figure 7.22) and have a lower efficiency than direct injection due to the flow and heat transfer

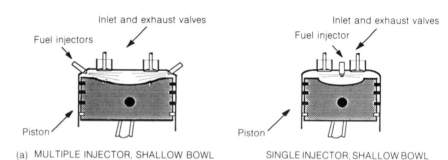

(a) MULTIPLE INJECTOR, SHALLOW BOWL   SINGLE INJECTOR, SHALLOW BOWL

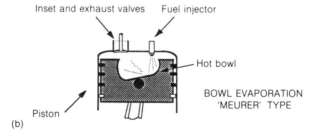

DEEP TOROIDAL BOWL   PLAN VIEW OF HEAD   SQUISH LIP

BOWL EVAPORATION 'MEURER' TYPE

(b)

Figure 7.21 Examples of different types of compression ignition combustion chambers: direct ignition, open chamber types. (a) very large engines with quiescent heads, (b) deep bowl types, medium size engines.

losses between the chambers both during compression, combustion and expansion. For many years, some manufacturers used IDI designs to moderately large sizes, say 1.5 litre or more per cylinder but they are now generally confined to cylinder sizes well under 1 litre capacity. A number of variations in IDI design philosophy are possible and these are often classified into prechamber, air-cell and swirl chamber types. The differences are perhaps subtle. Prechambers use a small throat from the main chamber to the auxiliary chamber which is sometimes further subdivided into several small apertures through which the air passes to initiate turbulence. Injection occurs with a relatively rich spray into the auxiliary chamber giving a mixture which is roughly stoichiometric. The combustion which follows then forces the flame back out through the throat into the main chamber where it is diluted to a lean combustion process. Swirl chambers are very

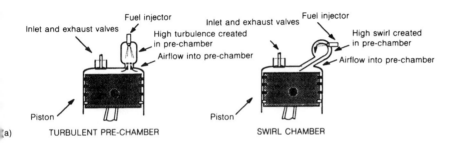

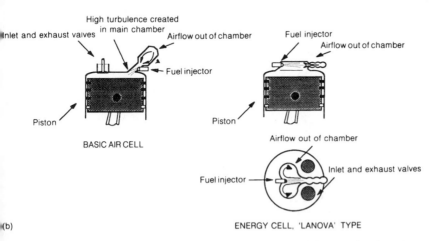

Figure 7.22 Examples of different types of compression ignition combustion chambers: indirect injection, divided chamber types. (a) pre-chamber types, (b) air cell types.

similar except that they are designed to create a much more organized fluid motion for mixing but are otherwise similar to prechambers. In general, swirl provides better mixing and these are now the most common IDI types. The pintle injector is oriented so that its spray is in the direction of the swirling air so that the droplets are carried away from the nozzle. Air-cell types are a little different in that they spray into either the throat between the chambers or the main chamber rather than the auxiliary chamber and rely on air from the cell being forced out from the latter as a jet when the pressure rises. Generally, on modern small diesels, prechambers and swirl chambers dominate. Air-cells, popular some time ago, are now rarely used.

It should be noted that, regardless of the combustion chamber type or shape, the maximum equivalence ratio for a CI engine is about 0.8 (i.e. fuel/air ratio 80% of stoichiometric). Above this, excessive smoke occurs. In some circumstances, for example at high speed where the thermal loading on the engine may become excessive, the fuelling level may be reduced further. For this reason, the specific power (power per unit mass of the engine) for a diesel is generally considerable lower than that of an SI engine.

7.4.4 Knock in diesel engines

A major advantage of a diesel engine is that, because it compresses air alone, there is essentially no hot end gas containing combustible compounds. Thus the type of knock that occurs in SI engines cannot happen in a diesel. Because of the effect of a long flame path in increasing SI knock, there is a limiting cylinder size for an SI engine. This may be up to, say, about 120 mm bore if the compression ratio is kept at moderate levels but it is generally smaller in order to obtain high efficiency. Thus very large SI engines require many cylinders with an enormous increase in the mechanical complexity of the engine. CI engines have no such size limitation and very large types with relatively few cylinders exist.

However, knock can occur in a diesel. Unlike that in SI engines, it occurs because of a very high rate of pressure rise at the beginning, not at the end of the combustion process. This high initial pressure rise rate is evident in the heat release diagrams studied earlier and it is with the premixed combustion phase. The knock in CI engines is a result of a large amount of fuel reacting in this way. The premixed combustion increases in proportion when the ignition delay period is lengthened and hence CI knock is a result of a long ignition delay period. It is a result of the injected fuel being subjected to an excessive delay period during which too much premixing can take place. Shortening the delay by either changes to the fuel or the engine reduce or eliminate this noise. The factors which reduce the ignition delay, as discussed previously, can be classified as either physical or chemical. Higher precombustion pressures and temperatures increase the ability of the fuel to autoignite and hence shorten the delay. It should be noted that this is the opposite to SI engines where these factors increase the knocking tendency. In addition, a good spray distribution into a turbulent air motion reduces the physical factors in the delay.

As far as the fuel is concerned, the ability to autoignite rapidly is measured by a comparative test with standard fuels as described in Chapter 6. The appropriate rating given to a CI engine fuel as a measure of its ability to resist knock is the cetane number, a higher value indicating a better knock resistance. Cetane number values should be in the 40 to 50 range.

7.5 ENGINE EMISSIONS

Since the 1950s it has been recognized that transportation engines in the developed countries are the major source of air pollution. While it is apparent that the proportions to be attributed to various causes vary both in time and from place to place, typical USA figures are available (Ledbetter, 1972) as shown in Table 7.3. It can be seen that transportation is responsible for the biggest share of CO, HC and NO in the atmosphere as well as a large proportion of the particulate matter. The particulates in the table do not include dust from the road, rubber particles from the tyres, photochemical smog particles or asbestos from brake linings. They are merely the particles, mainly carbon, directly attributable to the exhaust system. In addition to the above, it must be noted that automotive engines have been a major contributor to lead in the atmosphere although this is now generally being phased out and that transport in general is instrumental in much noise pollution.

Several philosophical questions, however, should be considered when assessing what is a pollutant. Should, for example, carbon dioxide be considered as such, given the adverse effects of increasing CO_2 proportions in the atmosphere on the global weather patterns? If so, all combustion engines except those burning hydrogen or a totally renewable hydrocarbon must always be major polluters. Is carbon monoxide always a pollutant given that most of it will eventually oxidize to CO_2? That is, is it a pollutant in rural areas or only in cities where the concentrations become large? What of noise? Where and when does it cause problems? These questions need to

Table 7.3

Source	CO (%)	Sulphur oxides (%)	Hydrocarbons (%)	Particulates (%)	NO (%)
Transport	92	4	65	14	42
Industry	4	32	26	44	21
Generation of electricity	0	48	0	21	32
Space heating	3	12	3	14	5
Refuse burning	1	4	6	7	0

be considered in drawing up legislation to maintain acceptable environmental standards. However, in relation to the current studies, only noise and the less nebulous term than pollution, i.e. emissions, will be considered as these can be evaluated directly.

7.5.1 Noise

Although not an important factor in this course, noise should always be considered in the design of engines. Some typical figures are as follows.

- **Rockets**. These are extremely noisy and large rockets such as those used to launch spacecraft can be heard to a distance of 50 km or more from the launch site. They are usually not important in the total noise spectrum as firing sites are remote and firing is an infrequent event.
- **Jet aircraft**. These are a more common although fairly localized problem near major airports. Modern technology, high bypass engines have reduced noise levels dramatically from those which prevailed in the 1970s. Noise levels are now commonly in the range 90 to 140 epndb. This scale is **effective perceived noise decibels** which is a subjective scale based on the noise field but weighted to the parts of the noise spectrum which are perceived to be the most offensive. Military aircraft are at the upper end of this range but again airfields are usually remote. The problems with commercial aircraft where airports must be sited near major population centres has increased with greater frequency of operation but modern planes are much less noisy. Noise levels are proportional to approximately the seventh power of jet velocity which is the reason why the high bypass engines with lower jet velocities are quieter. The whine from the fans ahead of the aircraft is now one of the more objectionable noises. Older commercial aircraft with low bypass engines are being phased out or re-engined for both noise and fuel economy requirements. Legislation requiring noise abatement procedures in takeoff and landing has reduced the major noise footprint affecting densely populated areas.
- **Surface vehicles**. Typical noise levels from individual vehicles are of the following levels at a distance of 15 m: motorcycles 84 dbA; trucks 84 dbA; cars (100 k/hr) 75 dbA.

Generally, the solution is to provide more effective exhaust silencing without increasing the exhaust back pressure and to improve the sound insulation around the engine compartment. With large CI engines, reduction of the ignition delay period can reduce some of the most objectionable noise which is the incipient CI engine knock.

7.5.2 Engine emissions

Uncontrolled engines produce significant exhaust emissions, typical values (1968 figures, USA) for spark ignition and compression ignition engines of the exhaust emissions being given in Table 7.4.

Table 7.4

Pollutant	SI Engines (kg/1000 litres)	CI Engines (kg/1000 litres)
Aldehydes	0.5	1.2
Carbon monoxide	276	7.2
Hydrocarbons	24	16
Nitrogen oxides	14	27
Particulates	1.4	13
Organic acids	0.5	3.7
Sulphur oxides	1.1	5

In addition, evaporation from the fuelling system and petrol tanks of SI engines can substantially increase the unburned hydrocarbons in the atmosphere. The principal concern is with CO, HC and NO which, as can be seen from Table 7.4 make up 99% of the SI and 69% of CI emissions. Their effects are as follows.

- **Carbon monoxide**. This is present in combustion in significant quantities when insufficient air is available. Toxicity occurs because the haemoglobin in the blood has a higher affinity for CO than for oxygen. Figure 7.23 shows some of the effects.
- **Unburned HC**. The amounts are small compared with CO but are objectionable because of the odour, their influence in the formation of photochemical smog and possible carcinogenic effects. The odour is more noticeable with extremely lean mixtures typical of diesel or stratified charge engines. Hydrocarbons may also show up as particulate matter and in older engines may contain lead and engine scavenging compounds.
- **NO_x (N_2O, NO, NO_2)**. These are formed at high combustion temperatures, usually as NO which later oxidizes to the other compounds. All NO_x compounds are destroyed only very slowly at room temperatures and so remain in the atmosphere for considerable periods. They can settle on the haemoglobin in a similar matter to CO. The NO can form dilute nitric acid in the lungs. They are also an important constituent of photochemical smog.
- **Photochemical smog**. This is a mixture of ozone, aldehydes, oxides of nitrogen and hydrocarbon. It results from reaction of these compounds in the atmosphere via a complex chain mechanism which requires photolysis due to the action of sunlight. The amount of photochemical smog depends on the concentration of reactants, their reactivity (unsaturated hydrocarbons are more reactive than saturated), the temperature and light intensity. The characteristic brown colour associated with photochemical smog is due to NO in the atmosphere. This smog causes severe irritation of the eyes, throat and respiratory system, as well as damage to

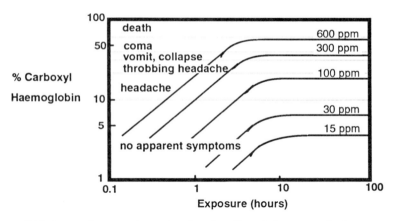

Figure 7.23 Approximate regimes showing the effect of the atmospheric content of carbon dioxide and the time duration of exposure (from Springer, G. and Patterson, D. (1973) *Engine Emissions*, Plenum, New York).

some materials and is thus a major problem especially when it is held down in the local atmosphere of major cities by a temperature inversion.

7.5.3 Emission legislation

Emission control legislation for vehicles began in California due to the severe smog problems in the Los Angeles basin where high automobile usage is combined with difficult geography (a basin surrounded by mountains) and climatic factors (frequent temperature inversions).

The installation of controls on new cars and light trucks began in California with crankcase emission controls in 1961 and exhaust emission controls in 1966. These were revised making them more stringent in 1970 and evaporative controls were added for the fuel tank and the fuelling system. Evaporation is particularly evident during the hot soak period which is the period, usually about one hour, after the engine has been turned off before it fully cools. The controls have been progressively tightened and currently (1993) proposals for ultralow emission vehicles (ULEV) and zero-emission vehicles (ZEV) are to be implemented during the 1990s. The rest of the USA adopted similar legislation two years after California and has generally followed it as standards have been tightened. Many other countries, including the European Economic Community (EEC), Japan and Australia have followed suit. Standards generally have been progressively tightened worldwide although test methods and details of the legislation vary from country to country.

The basic test pattern is that vehicles are to be tested over a standard city driving schedule (the Californian or EPA schedule which is also widely used elsewhere prescribes approximately 16 km and 1372 seconds) which includes starting, idling, acceleration, deceleration and constant speed sections and a shorter highway driving schedule with a range of higher speed and speed change modes. The vehicle is loaded with an approximate inertia value on the dynamometer to simulate the effects of its mass. The above EPA cycle is based on observed driving patterns in California and, while it is used in other parts of the world, it is noted that substantial differences in driving may occur there. Other countries have adopted simpler systems. Testing usually includes evaporative emissions as well as exhaust analysis. The test fuel is closely controlled. Hydrocarbons, carbon monoxide and oxides of nitrogen are detected using flame ionization detectors, infrared analysers and chemiluminescent equipment respectively.

7.5.4 Formation of noxious exhaust emissions

Carbon monoxide

At the high temperatures and pressures during the combustion process, significant quantities of CO form even when there is sufficient oxygen for complete combustion to occur. This is due to dissociation, the equilibrium composition of the products of combustion having significant CO at the high temperatures. The majority of the rate equations involved in CO formation and oxidation to CO_2 are sufficiently fast so that equilibrium occurs within the time scale of the engine processes. Thus, even as the products of combustion cool during expansion within the cylinder, the equilibrium curve is followed and the CO is converted to CO_2 principally as in the following scheme.

$$OH + CO \longrightarrow H + CO_2$$

$$H + OH \longrightarrow H_2O$$

Once the exhaust valve opens, measurements indicate that the rapid fall in pressure and temperature due to exhaust blowdown may cause some departure from equilibrium. However, the reaction rates for the above process are fast enough so that, even here, equilibrium should be maintained. The controlling processes therefore seem to be the ones forming the OH and H radicals. In particular, it seems that the ratio between the concentrations of H and OH is important. Nevertheless, in spite of this departure from equilibrium, reaction continues in the exhaust system and the final CO emission is predominantly controlled by the equivalence ratio. Typical experimental results are shown on Figure 7.24.

It can be seen that, at slightly lean air/fuel ratios with equivalence ratios of about 0.9, very little CO should exist in the exhaust gases. As the mixtures becomes richer, the CO concentration increases at a progressively faster rate and rapidly becomes quite significant. Much of this is simply due to inadequate oxygen to complete the combustion. However, if the calculated curves for the equilibrium concentration at combustion and exhaust conditions are examined, it can be seen that the experimental measurements of Figure 7.24 give values very close to calculated exhaust equilibrium values for lean mixtures but well above them for rich mixtures. That is, the rate controlled non-equilibrium behaviour, which probably occurs during the blowdown process as described above, is more evident as the mixture becomes richer, thus compounding the effect of insufficient oxygen.

Unburned hydrocarbons

While much of the unburned hydrocarbons from engine sources originate as evaporation from the fuel tank and fuelling system, they also exist in substantial quantities in the exhaust gas. The latter are of interest here.

How does some of the fuel remain unburned in the confines of a hot high-pressure environment such as exists in internal combustion engines? There are several reasons which differ between gas turbines and reciprocating engines. In the former, the answer is that too much fuel is being added to the combustion chamber and this cannot be handled in the residence time available. Thus, it progresses, probably in a partially burned state, through to the exhaust. Much of this will appear as black smoke and carbon particles which is likely to be most pronounced under the heaviest loads. These, in aircraft, are the take-off conditions. The control measure is essentially to improve the spray distribution and the gas/liquid mixing processes and to properly control the fuelling process under all speed and

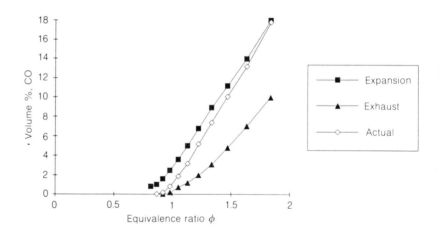

Series 1--Calculations at the conditions at the start of expansion
Series 2--Calculations at the conditions at entry to the exhaust
Series 3--Measurements in the exhaust system

Figure 7.24 Comparison of equilibrium calculations of exhaust CO (%) with values measured in the exhaust system (adapted from Springer, G. and Patterson, D. (1973) *Engine Emissions*, Plenum, New York).

load conditions. Further comments relevent to both gas turbines and CI engines will be made under the topic of particulates.

With reciprocating engines, particularly SI engines, the topic is more complex. Here, even with good control of the fuelling and mixing, substantial quantities of unburned hydrocarbons still appear in the exhaust. What is the reason? Much of it is due to what can be termed **quench zones** within the combustion chamber. These require some attention.

In any confined combustion, such as in an internal combustion engine, a severe temperature gradient always exists near the wall due to the large quantity of heat transferred to the engine's cooling system. This region of rapidly decreasing temperature is called a quench zone and most of the unburned HC initially exist here. Crevice regions such as the space above the upper ring and within the spark plug are important quench areas but some exist simply due to the confining walls of the combustion space. All these can best be understood by consideration of the general problem of quenching.

Quench zones can basically be defined as the regions where a flame cannot be supported. For example, a flame can be allowed to propagate between two cool surfaces. If these are brought progressively closer together, the flame will continue to progress until they reach a critical distance apart. The distance at which the flame can no longer be supported is termed the **quench distance**. This is depicted in Figure 7.25. Essentially, the cooling effect of the walls is now absorbing most or all the heat liberated by the combustion and there is insufficient energy being radiated to the mixture ahead to continue the propagation process. The quenching distance, q_d, is defined as the minimum distance two plates must be apart for flame propagation to continue. Obviously, this distance will depend on the

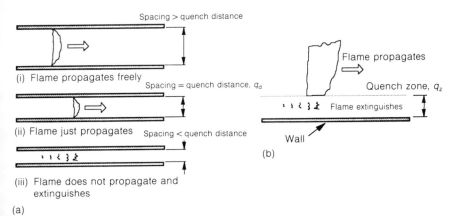

Figure 7.25 Effect of walls on flame quenching. (a) flames propagating between walls, (b) flames propagating near a wall ($q_z = (0.3$ to $1.0) q_d$)

pressure and temperature of the combustion, the fuel and mixture strength and the wall temperature. Detailed quantification is not necessary here. In a similar manner, a quench diameter, q_{dia}, can be defined for a tube and it is slightly larger, perhaps by 30%, than q_d. Also, for a single wall, the quench zone thickness q_z, is again different from q_d, probably being slightly less than half its value. There can be many quench regions of various shape and size in a reciprocating engine combustion chamber.

Typical values of q_d found from experiments for the conditions relevent to engines are in the range 0.125 mm to 1.25 mm. This is a significant proportion of the volume of the combustion space at the end of compression as it occurs all around the periphery of the chamber. It should be noted that the pressure in the quench region is identical to the instantaneous value in the chamber but the temperature is lower. Hence, the density of the gas in the quench region is higher than in the burning zone and the mass proportion within it will be higher than the volume proportion. The quench distance decreases with the fuel/air equivalence ratio from lean to about stoichiometric below which it is thereafter about constant. It also decreases as pressure, p, and temperature, T, increase. However, it should be noted that, as the mixture becomes richer and/or more dense the absolute amount of HC in the quench zone increases even if the zone remains the same thickness.

Unburned hydrocarbons from within quench regions in engines are expelled during the exhaust process. With slightly lean mixtures, although the quench zone is large, the excess oxygen destroys much of the unburned HC when they are mixed later in the exhaust system. With very lean mixtures, combustion generally becomes erratic and the amount of HC (not only from the quench zone) increases. Thus due to both this and the thickening effect, a minimum value of unburned hydrocarbons tends to occur at an equivalence ratio slightly less (leaner than) than one. This is shown in Figure 7.26.

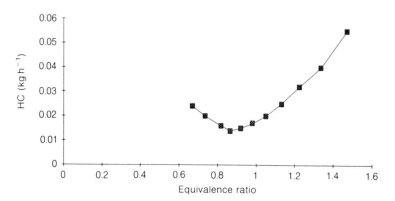

Figure 7.26 Variation of unburned hydrocarbon in a engine with equivalence ratio (typical values).

It has been noted experimentally that, in an SI engine, the concentration of unburned hydrocarbons varies throughout any single exhaust stroke. This can be explained, based on the work of Tabaczynski, Hoult and Kech (1970) as follows. In general, four basic quench zone regions can be identified as on Figure 7.27. The gasdynamic behaviour of each of these could be expected to be different in the exhaust flow. The regions are (1) and (2), the head and the side wall of the cylinder, (3) the piston face, and (4) the quench zone around the piston above the top ring. During exhaust, region (1) and the top part of (2) leave early in the stroke. Vortices form during the stroke which collect the gas in regions (3),(4) and the lower part of (2) and this is expelled right at the end of exhaust. This explains why observed concentrations of unburned HC are high near the end of exhaust and why reducing the crevice above the upper ring substantially reduces HC in the exhaust.

The major engine variables affecting unburned HC in an SI engine are equivalence ratio, compression ratio, engine speed and spark timing. Throttle opening has little effect. The equivalence ratio contribution has already been discussed (Figure 7.26). The trends associated with the other three variables are shown on Figure 7.28. A high compression ratio increases unburned hydrocarbons because, with the smaller combustion space at TDC, the quench zone makes up a larger proportion of the total volume. This is particularly noticeable in the squish region although the high turbulence there tends to mix quench zone and other gases. Unburned hydrocarbons decrease with engine speed because the greater turbulence promotes combustion and decreases the quench zone. Also they decrease as the spark is retarded because more combustion occurs after TDC while the combustion space is enlarging with the quench zone becoming a decreasing proportion of it. Exhaust gases are hotter with a

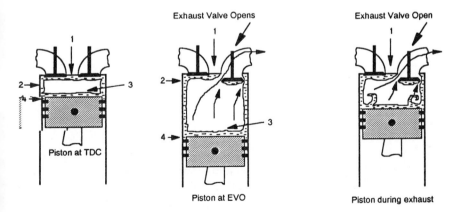

Figure 7.27 Quench zone flow from an engine during exhaust (adapted from Tabaczynski, R., Hoult, D. and Keck, J. C., 1970, High Reynolds Number Flow in a Moving Corner, *J. Fluid Mech.*, **42**, 249–55).

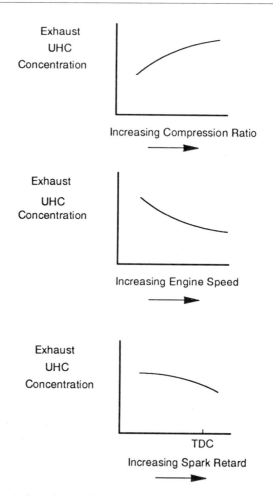

Figure 7.28 Trends in unburned hydrocarbon (UHC) levels with variation in some of the major engine parameters.

retarded spark and more oxidation of the HC can also take place in the exhaust system.

It should be noted that the above explanation is not the complete answer to HC formation and that there are other reasons which help explain HC formation. For example, absorption and desorption by the engine oil during a single engine cycle may play a role. Some tests also show HC increasing by up to 20% when deposits form on the walls and piston crown. The exact reason for this is not clear but it appears that a clean engine helps to prevent high HC emissions.

Oxides of nitrogen

The formation of oxides of nitrogen is a highly temperature-dependent phenomenon and occurs because the equilibrium concentrations of the various NO_x compounds formed when oxygen and nitrogen are mixed are high at temperatures of, say, 2000 K to 3000 K but low at ambient temperatures of, say, 300 K. However, the reaction rates are relatively slow and equilibrium is not fully attained in the time available under most engine conditions. This applies to some extent to the forward reaction but is particularly true of the backward reaction and the NO_x is effectively 'frozen' for a long period after it is exhausted from the engine giving it time to react with other substances to form photochemical smog.

The rate of NO_x formation is coupled to the turbulence in the flow. At the peak combustion temperatures, small changes in temperature make large differences to the amount of NO_x formed. This is predominantly due to the fact that turbulent combustion is faster and hotter but also due to the temperature fluctuations. That is, even at the same mean temperature as a laminar combustion, in turbulent combustion the region above the mean value increases the amount more than the region below the mean decreases it. Total levels of NO_x in turbulent combustion are therefore noticeably higher than in nonturbulent and a full investigation requires temperatures in the entire flow field to be investigated. This is sometimes too complex and the detailed turbulent structure is often ignored.

Various mechanisms for rate studies of NO_x have been given in Chapter 6. These are usually adequate for fuel lean calculations. However, extrapolation of measured values to the zero time point indicate that some NO_x appears much more quickly than these schemes show. This is called **prompt** NO_x and is a maximum at about $\phi = 1.4$. It has only been found in hydrocarbon flames. The most widely recognized theory for its formation suggests that additional routes of NO_x production occur via the formulation of such substances as HCN. Also what is termed superequilibrium (i.e. above equilibrium) levels of atomic oxygen may form early in the combustion process due to its fast forward and slow backward reaction. This distorts the production of NO_x early in the time period. This could be significant as most calculations of NO_x assume equilibrium atomic oxygen concentrations exist throughout. A further possible mechanism relates the prompt NO_x to additional fuel bound nitrogen atoms.

While the most significant feature of NO_x formation is the temperature of the post-flame gases, variation with equivalence ratio, as on Figure 7.29, is important. It can be seen that NO_x concentrations peak at just below stoichiometric. This is due to the fact that, for lean operation, the additional air acts as a buffer and reduces combustion temperatures while for fuel rich operation the latent heat requirements of the additional fuel also keep temperatures lower than stoichiometric.

Summary

A summary of the CO, UHC and NO_x emissions from an SI engine is shown on Figure 7.30. The vertical scale represents relative values only but the variation with equivalence ratio is clear. The value of operating the engine slightly lean (say $\phi = 0.9$) is apparent for CO and UHC but this is near the peak NO_x value. Thus, to overcome this by fuel/air ratio control alone, even leaner operation to say, $\phi = 0.8$, is required. This is in the region where combustion is difficult to support and below that value UHC begins to rise again.

7.5.5 Control of emissions

Cycle by cycle variability

From the preceding section it is apparent that good control of emissions can be obtained by careful consideration of equivalence ratios, combustion and exhaust temperatures and design features to reduce the quench zones in an engine. However, one further factor needs considering. This is cycle by cycle variability (CBCV).

In any engine, even a single cylinder engine, the combustion is not uniform from cycle to cycle as can readily be seen by examining the pressure traces. It is worst at lean mixtures and low load. The cause of this cycle by cycle variability is not clear but the following factors are considered important.

- Charge composition, which may vary from cycle to cycle due to changes in the local fuel/air ratios or to different quantities of residuals remaining at the end of exhaust, is significant. Distribution between cylinders is obviously important in multicylinder engines.
- The gas velocity in the cylinder is significant, both the average velocity and that in the region of the ignition.
- The ignition itself may be an important factor as the ignition energy may vary. Small variations in the delay period of combustion are believed to be important and these are due to both charge composition and gas velocity.

Whatever its cause, which may be any combination of the above, cycle by cycle variability is instrumental in increasing unburned hydrocarbons and carbon monoxide due to less complete combustion in some of the cycles.

Crankcase ventilation

All reciprocating engines require crankcase ventilation to prevent a buildup of pressure on the underside of the piston from gases which blow past the piston rings. This becomes more important as the engine becomes worn. Earlier engines used a simple ventilating pipe but positive crankcase

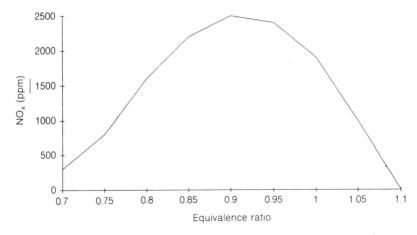

Figure 7.29 Variation of unburned NO_x in an engine with equivalence ratio.

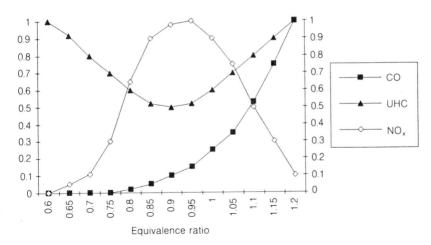

Figure 7.30 Summary of trends in the emissions of CO, UHC and NO_x with equivalence ratio for SI engines. Note that the vertical scale represents values for each particular emission compared with the maximum for it in the equivalence ratio range.

ventilation (PCV) was introduced in the late 1940s on some British cars as a better means of controlling crankcase gases. As these gases are high in carbon monoxide and hydrocarbons, they are an important source of emissions. In California, the use of PCV became mandatory in 1961 and is now universally used.

A PCV system consists simply of a duct from the crankcase, often indirectly via the pushrods and rocker cover, through a PCV valve to the inlet manifold. The lower than atmospheric manifold pressure then draws the crankcase gases into the cylinders where they are oxidized. The purpose of the valve is to control the flow rate so that it is a constant proportion of the intake flow. The PCV valve is therefore closed progressively as the inlet manifold pressure falls.

Mixture control

An examination of the formation of the different emissions shows that NO_x is a maximum and unburned HC is a minimum just lean of stoichiometric. Carbon monoxide increases for rich mixtures. It is therefore apparent that operating at leaner than stoichiometric will reduce both HC and CO as long as CBCV can be controlled.

To avoid misfiring at these lean operating conditions, fuel systems are made with tighter tolerances. Many manufacturers use a proportion of preheated air which enters the air cleaner system via a valve from a 'stove' over the exhaust manifold. Temperatures may be controlled by mixing the preheated and normal air streams to 38 °C within reasonably close limits. This allows better fuel system calibration than does normal operation where ambient air temperatures may vary from less than 0 °C to over 40 °C. Also, the heated air provides better mixing characteristics for the fuel in the intake manifold. Some manufacturers use water-heated intake manifolds for this reason. The disadvantage of preheated air is that it raises maximum engine temperatures slightly, thereby increasing NO_x. It also slightly decreases power output. At wide-open throttle, some systems block the heated air allowing only the cool dense air to enter the manifold.

Because of the adverse effects of different fuel distribution to different cylinders in multicylinder engines, careful consideration is now paid to intake manifold design to keep the flow as uniform as possible in each branch. Obviously, is an advantage multipoint fuel injection over throttle body types.

Spark timing

While, for most operations, the spark timing should be such that the engine works at maximum efficiency, a retarded spark has been used under some conditions by most manufacturers. This has several beneficial effects. First, it keeps peak exhaust pressures and temperature high which provides favourable conditions for oxidation of CO and HC in the exhaust system. One practice is to retard the spark about 10 at idle (i.e. at low manifold pressures) but to bring it back to the normal position under full throttle. This may be achieved by a vacuum retard system, by an electric solenoid which blocks the vacuum advance under some conditions, by solenoid operation of the advance–retard system or by dual contact breakers one of which may be deactivated by a switch. In some systems there may be a thermal override of the system so that spark advance is always normal when the engine is very hot or cold. Others retain the retarded spark in all but top gear.

Exhaust gas recirculation

The addition of an inert gas to the cylinders lowers the peak combustion temperatures and reduces NO_x. An ample supply of buffer gas of this type is the exhaust gas and this is now universally used. The amount of exhaust gas recirculation (EGR) varies from engine to engine, but about 5% would be typical. The gas is drawn from the exhaust manifold by suction from the inlet manifold and a valve system similar to the PCV valve is used to keep the proportion constant.

Temperatures

To decrease the quench zone, higher engine temperatures are used with tighter control than in the past. Obviously a major source of emissions is in the warmup period when the mixture is rich and manifolds and cylinder walls are cool. Many engines are designed for rapid warmup and viscous fans with an additional electric fan to cope with high temperatures are used.

Engine design features

These are generally to reduce the quench zone area and are aimed at reducing unburned hydrocarbons. A minimum surface area to volume is desirable and this is achieved by using fewer cylinders for the same capacity and longer strokes relative to bore. The crevice region above the top piston ring is kept as small as possible and squish areas are generally opened up in depth. Compression ratios may be reduced. This reduces HC and NO_x and gives higher exhaust temperatures.

Idling and deceleration

Very low idle speeds promote CBCV due partially to very high residuals in the cylinder and difficult combustion conditions. High idle seeds (800 to 900 rpm) are now common. Antidieseling devices are often used. Under certain conditions most engines will 'run-on' after the ignition is turned off due to hot spots in one or more cylinders. Emissions during the few cycles where this occurs are extremely high. This dieseling can be stopped by either moving the throttling idle stop to a much lower setting by use of a solenoid or by including a direct mechanical cut-off or similar device to prevent fuel flow to the engine.

When an engine is running with a high manifold pressure, there is usually fuel condensed on the manifold or inlet port walls. On deceleration, the fall in manifold pressure causes this fuel to 'flash' off creating a temporary rich mixture. To prevent this, throttle body systems may be fitted with a bypass system, either valves in the butterfly or a separate bypass duct which allows the vacuum to increase only slowly.

Evaporative emission control

Fuel tanks require venting to operate successfully as do carburettor float bowls. These are now vented to canisters containing activated carbon which, in effect, contain a very large surface area of porous carbon. Hydrocarbons become loosely bonded to the carbon but can be removed by a rapid air flow (or purge) across it. This is achieved during engine operation and the hydrocarbons are burnt off in the engine.

Exhaust reactors and catalysts

Thermal exhaust reactors were commonly used in the USA in the early days of emission control. Exhaust reactors depend on high exhaust temperatures and usually have an insulated chamber into which additional air is blown with an air pump. Temperatures in the reactor are kept at about 750 °C. Engines are often run rich when fitted with a thermal reactor and a good deal of secondary, lean combustion takes place in the reactor.

Catalytic reactors are now the most common means of exhaust emission control. They contain inert metals which alter the various reactions. They help convert hydrocarbons and carbon monoxide to carbon dioxide and water and reduce nitrogen oxides to nitrogen, and oxygen. They are 'poisoned' by lead in the fuel and require lead-free petrol. They are probably the least detrimental control in regard to engine efficiency but the fact that they require either lower compression ratios with the same cut of gasoline or a more specialized cut to retain the octane rating without the addition of lead has given rise to some debate on the overall effect. The general consensus seems to favour them. Lead-free fuel, of course, has the advantage of lowering airborne lead in cities. they may be oxidizing catalysts for CO and HC with NO_x controlled by EGR. More commonly, three-way catalysts are used which also reduce NO_x but these require tight air/fuel ratio control near stoichiometric.

Improved combustion

Emission characteristics can be improved in several ways. The principal one in use is staged combustion, or stratified charge. Here the combustion is initiated in a fuel rich chamber or region near the spark and progresses through a fuel lean region as the flame propagates. Stoichiometric conditions are avoided.

Another alternative is turbocharging where an engine can be kept small and hence can have low emissions during the city driving conditions. When high output is required, the manifold pressure is boosted by the turbocharger. The high pressure also helps to promote good combustion conditions but it should be noted that the temperatures may rise increasing NO_x.

7.6 REFERENCES

Hardenberg, H. O. and Hase F. W. (1979) An Empirical Formula for Computing the Pressure Rise Delay for a Fuel from its Cetane Number and from the Relevant Parameters of Direct Injection Diesel Engines, *SAE Paper 790493*, Society of automative Engineers, USA.

Heywood, J. B. (1988) *Internal Combustion Engine Fundamentals*, McGraw-Hill, New York.

Hiroyasu, H. and Kadota, T. (1974) Fuel Draplet size Distribution in a Diesel Combustion Chamber, *SAE Paper 740715*, Society of automative Engineers, USA.

Keck, J. C. (1982) Turbulent flame structure and speed in SI engines, *Proc. 19th Symp. (Int.) on Combustion*, The Combustion Inst. Pittsburgh, Penn., 1451–66.

Lavoie, G. A., Heywood, J. B. and Keck, J. C. (1970) Experimental and theoretical study of NO formation in IC engines, *Combustion Sci. Technol.*, **1**, 313.

Ledbetter, J. O. (1972) *Air Pollution– Part 1*, Dekker, New York.

Lefebvre, A. H. (1966) *Theoretical Aspects of Gas Turbine Combustion Performance*, Aero Note 163, College of Aeronautics, Cranfield, England.

Lyn, W. T. (1962), Study of burning rate and nature of combustion in diesel engines, *Proc. of 9th Symp. (Int.) on Combustion*, the combustion Institute, Pittsburgh, Penn., USA.

Metghalchi, M. and Keck, J. C. (1982) Burning velocities of air with methanol isooctane and indolene at high temperature and pressure, *Combustion and Flame*, **48**, 191–210.

Milton, B. E. and Keck, J. C. (1984) Laminar burning velocities in Stoichiometric hydrogen and hydrogen-hydrocarbon gas mixtures, *Combustion and Flame*, **58**, 13–22.

Oates, G. C. (1989) *Aircraft Propulsion System Technology and Design*, AIAA Education Series, USA.

Ricardo, H. (1933) *The High Speed Internal Combustion Engine*, Blackie, London.

Tabaczynski, R., Hoult, D. and Keck, J. C. (1970) High Reynolds number flow in a moving corner, *J. Fluid Mech.*, **42**, 249–55.

Wolfer, H. H. (1938), Ignition Lag in Diesel Engines, VDI-Forochungsheft 392 (English Translation, R.A.E. Farnborough, *Lib.* No. 359, UDC 621–436.047, 1959).

PROBLEMS

7.1 Fuel is burnt in a gas turbine combustor under stoichiometric conditions in the primary zone using 25% of the available air. Assuming that the fuel is C_7H_{13}, determine the number of kmol of exhaust products. If the fuel provides $43 \, MJ \, kg^{-1}$ of energy, determine the final temperature of these products at the end of the primary zone assuming that the air enters at $830 \, K$ and that the c_p value for the product mixture is $1.267 \, kJ \, kg^{-1} \, K^{-1}$. If the remainder of the air, still at $830 \, K$ is then added, determine the c_p value of the combined air and products and their final temperature at the combustion chamber exit assuming no heat transfer from the chamber. For air at these temperatures, assume c_p is $1.142\,\text{kJ}\,\text{kg}^{-1}\,\text{K}^{-1}$.

7.2 A stoichiometric propane gas and air mixture is ignited at the centre of a spherical combustion bomb of radius 60 mm. The initial temperature and pressure is 300 K, 100 kPa and the mixture is stoichiometric. If at a particular instant, the pressure, uniform throughout the combustion space is 250 kPa and the temperature behind the burnt gas is 1500 K, calculate the radial position of the spherical flame front at the instant considered and the mass fraction burned. Assume that the gas constants for the mixture of unburnt and burnt gases are respectively $282\,\text{J}\,\text{kg}^{-1}\,\text{K}^{-1}$ and $294\,\text{J}\,\text{kg}^{-1}\,\text{K}^{-1}$. Heat transfer to the gas ahead of the flame may be ignored and the unburned gas ahead of the flame front may be assumed to be compressed isentropically according to the normal relationship with an isentropic index of 1.4.

7.3 If, for the combustion in Problem 7.2, the gas has a burning velocity which is given by (7.3), calculate, for the position designated, the laminar burning velocity relative to the gas ahead of the flame front. Also, by considering the volume of the unburnt gas at an instant when the rate of pressure rise dp/dt is known, develop an equation for the movement of the unburnt gas, dr/dt. Hence, estimate the absolute velocity of the flame front if dp/dt is measured as $24\,\text{kPa}\,\text{ms}^{-1}$.

7.4 In an engine an approximation of the unburnt gas condition may be obtained by assuming an isentropic compression right from the inlet conditions. An engine has a volumetric compression ratio of 8:1 during which an isentropic index of 1.38 applies followed by a flame front compression of pressure ratio 1.8:1, isentropic index 1.35 to the position where 80% of the mass has been burnt. Determine the unburnt gas condition if the initial pressure is 90 kPa, temperature 300 K. Also determine the total cylinder mass, the mass burnt and the mass unburnt if the cylinder has a swept volume of 0.5 l, and that residual gases occupying 50% of the clearance volume at the intake conditions remain in the cylinder. The incoming mixture may be assumed to have an average molecular mass of 29.

If, at the above condition, the end gas reacts instantaneously throughout to form a local peak pressure at first only in the end gas region which may be assumed to be constant volume, determine the corresponding values of temperature and pressure.

Assume that the fuel provides $1.5\,\text{MJ}\,\text{kg}^{-1}$ of energy, that the gas is ideal with $c_v = 0.96\,\text{kJ}\,\text{kg}^{-1}\,\text{K}^{-1}$ and the molecular mass in the burnt state is 29.

If now this high pressure region forms a compression wave moving across the rest of the mixture, determine its pressure and velocity. Assume it to be a shock and that the isentropic index throughout is 1.35 at the time.

7.5 An SI engine running under stoichiometric conditions has its fuel/air ratio altered to: (a) 10% lean, (b) 10% rich. Using the data of

Figure 7.30, calculate the percentage change in each of the emissions depicted on the figure.

If now, during transient (acceleration/deceleration) operation, the engine settings are altered by the above amount (plus or minus 10%) in such a way that 25% of the engine cycles are rich, 25% lean while the remaining 50% are stoichiometric, assess now the change in emission levels.

7.6 Determine the concentration of NO in atmospheric air at a temperature of 298 K and a pressure of 1 atmosphere, (101.3 kPa). The air is then heated to 3500 K at a pressure of 10 atmospheres (1.013 MPa). Determine the concentration of NO that is formed if it is allowed to come to equilibrium at those conditions and express it as a percentage of the atmospheric value. What is the equilibrium value if it expands to 1 atmosphere, 500 K?

<table>
<tr><td>

8

</td><td>

Heat transfer and other energy losses in engines

</td></tr>
</table>

The maximum brake thermal efficiency of a typical spark ignition engine is about 30%, that of a gas turbine over 35% while for a CI engine, efficiency may be over 40%. In all cases, noticeably lower values are possible under some operating conditions. While engine efficiencies are still improving for all types, it can be seen that, of the energy released in the combustion process, some 60 to 70% is not transferred as useful work and can be regarded as an energy loss. This includes direct heat transfer losses and those due to mechanical friction, the latter being themselves reduced via a heat transfer process to an increase in internal energy in either the exhaust gas, engine coolant or wall heat transfer. The above efficiencies mean that, for every unit of useful power, somewhere over 1.5 to nearly 2.5 units, depending on the engine type, are wasted unless, as happens in some fixed installations, a proportion of this heat can be used for other purposes. One typical example of this is the combined cycle system used for power generation where the exhaust heat from the gas turbines supplies the energy for a steam cycle. Another less sophisticated example is the use in a vehicle of the heat flow to the engine coolant for interior space heating.

The energy can be lost in several ways: as energy carried away in the hot exhaust gas, heat transfer loss directly from the engine walls and sump or heat transfer to the coolant. The important questions are, can some of this wastage be avoided by transferring it to the engine work output? If so, how much improvement in efficiency is possible by reducing friction and heat transfer? Also of importance is the effect of heat transfer on the engine modelling process and of friction in reducing the output from the indicated to the brake values.

8.1 MECHANICAL FRICTION

As far as obtaining an understanding of the thermofluid processes in the modelling of an engine, heat transfer is of much more importance than mechanical friction, as distinct from fluid friction which is predominantly associated with the mechanics of the system. Nevertheless, from an overall energy perspective, friction losses can be quite important and some brief comments on friction in engines are pertinent here. With friction, some improvements are possible although these are becoming more limited as many gains have already been made and it will be recognized that it is not possible to avoid friction losses completely. A reasonable estimate may be that, at average speed/load conditions, some 15% of the fuel energy is dissipated directly by friction but this may vary from a low value of less than 3% at high load to perhaps 30% at idle. In the latter case, all the energy is dissipated either to the exhaust gases, through direct heat transfer to the atmosphere, or to friction. In an engine test, it is difficult to distinguish between friction *per se* and the pumping energy used in the gas exchange process and these are often lumped together as a single value. The total of these is the difference between the indicated power (or mean effective pressure) and the brake values and is called the friction power (or friction MEP). Also, the power required for the ancillary equipment such as alternators, fuel pumps, fans and water pumps is normally included in the friction MEP although their magnitude can readily be established and separated with individual tests. Although both pumping and ancillary work are both important, the friction due to the actual engine-components themselves is of interest here.

Within the engine, mechanical frictional dissipation of energy takes place at the rotating and sliding surfaces. In a gas turbine, the shaft bearings are roller types and their friction is already low and it is difficult to see any major improvements. The important problems are therefore to provide adequate lubrication to prevent wear and subsequent failure and to absorb any vibrations due to out-of-balance in the rotors. One approach is the squeeze film bearings where a thin layer of oil is used to prevent direct metal-to-metal contact of the bearing components. These are now widely used on gas turbines. In reciprocating engines, the bearings are normally sliding types which rely on high levels of lubricant flow to lift the journal from the bush in which it rotates. This is called **hydrodynamic lubrication** as the hydraulic lift prevents the metal-to-metal contact and any friction is essentially fluid. Hydrodynamic lubrication can also occur with moving flat surfaces, such as that of a piston moving in a cylinder, a wedge of lubricant supplying the lift. Even though this fluid film exists, it has been established that the principal cause of friction in reciprocating engines is at the piston/cylinder liner interface where up to perhaps 70% of the frictional losses originate. This is because the contact pressures between these components are high and also because it is a difficult region to lubricate. As the piston approaches top dead centre, its velocity reduces to zero and the hydrodynamic lift from the lubricating oil disappears. Thus, this region is subject to high wear rates, called **scuffing**, and to high friction.

As mentioned above, a well-lubricated pair of sliding surfaces, whether their motion be rotational or linear, have an oil film between them. Their relative velocity generates a hydrodynamic force which lifts one surface off the other and there is no direct metal-to-metal contact. This is hydrodynamic lubrication. With the breakdown of the hydrodynamic lift, metal-to-metal contact commences. This can occur because the pressure forcing one surface towards the other is too high, because the relative velocity causing lift becomes too slow or because there is insufficient oil flow to the interface. It is known as the **boundary lubrication regime**. Under these conditions, the frictional forces increase severely. A semipermanent lubricant of greater viscosity such as a grease might minimize this effect but normal lubricants of this type are not useful in an engine as they may be affected by the high temperatures of the combustion and are too viscous. Modern oils contain some additives which are effective here and these are generally termed **friction modifiers**. They must be able to block up the microscopic hollows in the metal surfaces giving an effectively smoother surface and survive the severe conditions to which they are subject. Simultaneously, they must not increase the viscosity of the lubricating oil which is formulated correctly for the hydrodynamic lubrication regime required in other parts of the engine. Developing appropriate lubricating oils to suit modern engines is a major enterprise.

While friction will always occur, its effects can be ameliorated. This can be achieved either by improving the lubricant properties as mentioned above or redesigning engine components. For example, piston rings have received attention in recent years to minimize their frictional effects. This may mean fewer rings as long as adequate compression sealing can still be provided to the combustion space or, on the outer ring surface, different shapes to or materials to lower its local frictional value. Other areas which have received attention are the camshaft, rockers and valve gear. It is not the intention of this text to provide a mathematical treatment of friction and lubrication and for more detail, the reader is referred to a basic text on lubrication such as that by Cameron (1966).

8.2 HEAT TRANSFER IN ENGINES

While the quantity of heat transferred from an engine cylinder is substantial, general experience has shown that reducing it can only provide small improvements in efficiency and hence only minor changes to the losses can be envisaged. Apart from large combined cycle units, the useful contribution of the heat loss is usually limited to the warming of the vehicle interior. The purpose of examining heat transfer in engines is therefore only marginally aimed at improving efficiency. It is rather to gain further insight into the operation of an engine, to help understand the cooling system, to prevent metallurgical problems due to local overheating and to understand where hot spots will occur. In addition, it helps give design information to minimize overheating of the end gas with its consequent detonation problems and allows better and more accurate combustion models to be formulated. Although much of the work to be discussed is specifically directed at reciprocating engines as these produce the most challenging problems, many of the above comments are also relevant to gas turbines. These latter in particular are performance limited by high-temperature metallurgical problems as much as by any other factor.

8.2.1 Energy flows in an engine

There are several categories in which the energy flow in an engine can be placed: energy to useful work, energy to the exhaust gas, energy to the coolant, direct energy loss through the walls, energy to the sump, etc. To some extent, the choice of groupings depends on the problem to be examined. For the purpose of this discussion, the first three categories will be used with the rest classified simply as the remainder. The proportion which is transferred to each obviously depends upon the design of the engine and the conditions under which it is running. A typical distribution at WOT (wide-open throttle) and medium speed for an SI, a CI engine and a gas turbine might be roughly attributable as shown in Table 8.1. The exact values can vary considerably from engine to engine for both types and should be regarded as indicative only. The gas turbine exhaust loss is based on a typical exhaust temperature of 750 K and an air/fuel ratio of 40:1. Note that this table specifies the indicated power (IHP). An additional portion of this, possibly about one-fifth (say 7% of the fuel energy for the SI, 10% for the CI), is used up by friction and also must be attributed as a loss. This will be distributed among the losses to coolant, exhaust, sump and walls.

8.2.2 Heat transfer during reciprocating engine processes

The losses can also be categorized according to the engine process. For example, for an overhead valve SI engine, the heat loss to the walls might be up to 20% of the combustion energy which is made up of perhaps 15% direct heat loss and 5% as a friction contribution. Proportioning the direct heat loss to each process could then be approximately as shown in Table 8.2. Here the positive sign indicates heat flow to the coolant. Again, this is an indicative distribution only. It should be noted that, as the throttle opening is reduced, a greater proportion of the combustion energy is 'lost' although the absolute value of the losses may fall due to the smaller thermal input. Obviously an engine that is idling produces no useful work and all the

Table 8.1 Energy flows in engines

	SI	CI	Gas turbine
Energy to IHP	36%	50%	45%
Energy to coolant (water or air)	24%	20%	—
Energy to exhaust	30%	20%	45%
Remainder (through walls or to sump)	10%	10%	10%

Table 8.2 Heat transfer process by process in an SI engine

Inlet	−0.5%
Compression	+0.5%
Combustion	+2.0%
Expansion	+10.0%
Blowdown	+12.0%
Exhaust	+1.0%
Total	+25.0%

combustion energy therefore goes into pumping, heat transfer losses and exhaust gas losses, the latter two including the frictional effects.

Suction

During suction, wall temperatures are generally higher than gas temperatures and the heat loss is negative. That is, heat is transferred to the gas. This is detrimental to air utilization as it decreases the mass of gas in the cylinder due to its lower density at the same pressure. Note that heating prior to the mixture entering the cylinder raises the acoustic velocity thereby reducing the Mach number for the same flow velocity and hence the flow has less tendency to choke and to experience high Mach number flow losses at the inlet valve at high engine speeds. That is, the volumetric losses are not inversely proportional to the absolute temperature of the inlet charge as would be expected from the ideal gas density relationship but approximately to the square root of it. During inlet, the magnitude of the heat transfer to the gas is small compared to that which exists in other processes.

Compression

During this phase, the gas temperature at some stage rises to greater than the instantaneous value of the wall temperature. Heat flow from the gas now commences. This heat flow during the stroke is small for SI engines although it increases with compression ratio because the gas temperatures then reach higher levels. The higher compression ratios of CI engines makes this effect then more significant. In these, there is a most important effect, not on efficiency, but on the air temperature available for ignition when the fuel is sprayed into the combustion chamber. That is, too much heat transfer from the air in the cylinder cannot be allowed to occur with CI engines during compression, otherwise ignition failure may occur.

Combustion

During combustion the temperature rises enormously although the average may be lower than exhaust (Figure 8.1). Also the gases are now at a high density and have a severe turbulent motion imparted to them which promotes heat transfer. Loss of energy by heat transfer during this phase lowers efficiency but also lowers the gas temperature. This can help to reduce the tendency to knock in SI engines and to reduce the amount of NO_x which is formed. However, low temperature may promote the formation of other emissions, particularly unburnt HC as the quench zones then increase. In an SI engine, end-gas autoignition knock may itself cause increased heat transfer losses as the wave motion it produces is likely to scour off the thermal boundary layer. Local overheating of the engine components, for example the part of the piston crown near the end-gas, may occur and sustained knock can result in severe erosion and pitting of the metal surface. This is regarded as one of the important damage mechanisms of knock.

Expansion

Heat transfer during the early stage of expansion is an energy loss that may have some potential for conversion into useful work. During the latter stage of expansion with little piston motion and hence potential work output left, it is of less importance and so is not then as detrimental to efficiency. At first glance it would appear that, due to the high temperatures at the start of expansion, heat transfer would be greater there. However, as expansion proceeds, more cylinder wall area is uncovered. Also the temperature does not drop as rapidly as would be expected from a simple estimate due to recombination of the dissociated products of combustion. Hence, the combination of moderately high temperatures and large wall area mean that a high rate of heat transfer continues throughout this process.

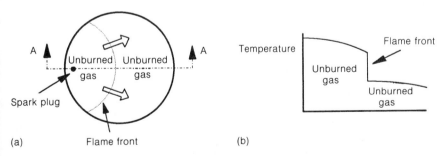

Figure 8.1 Diagrammatic representation of the variation in temperature across the combustion chamber (a) plan of combustion chamber, (b) temperature profile on section A-A.

Of this expansion heat transfer loss, some could perhaps be used as work if the energy could be retained in the working substance. This would keep the gas pressures higher as the expansion stroke proceeds. Furthermore, the exhaust gas would be noticeably hotter and its additional energy could therefore be used in an exhaust turbine. The concept of an engine that is either 'insulated' or 'adiabatic' has been studied, particularly in relation to CI engines. The first has insulating material around a metal cylinder and therefore allows heat exchange between (i.e. to and from, but not through), these walls and the gas within the cylinder. That is, the walls act as a thermal capacitor. The second type has walls of material such as ceramic which do not accept a heat flow at all (Wallace, Way and Vallmert, 1979). The insulated diesel shows little promise while the adiabatic diesel, with turbocompounding, has a theoretical advantage of at least several percent over conventional technology. Nevertheless, the problems and expense of turbocompounding and the difficulty in obtaining suitable ceramic or similar materials with appropriate strength and which can match any metal engine components in coefficient of expansion, has prevented its practical development to date.

Blowdown

Blowdown, like expansion, is also a process where much of the heat transfer takes place, it being caused here by the very high velocities of the gas as it passes over the exhaust valve and ports while still at a reasonably high temperature. The high Reynolds number and high temperature of the flow are ideal for promoting high heat transfer. In general, substantial heat flow takes place to the exhaust valves and either they require cooling or they must be constructed of high temperature resistant alloys for durability. These losses cannot in general be returned to the in-cylinder gases in a useful manner. In fact, the opposite happens and it should be noted that a hot exhaust valve near the end gas in an SI engine will exacerbate any tendency to knock. The use of inlet gas flowing over the hot exhaust valve to provide cooling is effective for that purpose but the consequent heating of the inlet charge then reduces the volumetric efficiency of the engine. In other words, the blowdown heat transfer cannot be used to improve engine performance.

Exhaust

Blowdown is only the initial phase of exhaust. During the exhaust stroke itself, the pressure in the cylinder has fallen to a little above atmospheric pressure and the gases are expelled at about the piston velocity or perhaps a little higher. This is much slower than during blowdown. Also the temperature has dropped due to the expansion of the gas which remained in the cylinder after blowdown. Thus, a heat transfer loss occurs but it is much lower than during blowdown.

In summary, prevention of heat transfer might increase efficiency slightly but by less than about 1% during each of the in-cylinder combustion and expansion processes. This would increase the wall temperature, perhaps dangerously and could cause either pre-ignition or knock in SI engines. Also further dissociation would occur due to the higher combustion temperature thus further limiting any potential improvement. New developments in the use of ceramic materials in engines allow higher wall temperatures and some increase in efficiency may then be possible. Due to knock problems in SI engines, this approach is really only feasible for CI engines. Further gains can then be theoretically achieved by using the higher exhaust gas energy from an adiabatic engine in a turbocompounded system.

8.2.3 Heat transfer variation effects of engine parameters

As with most other engine parameters, engine design and operational changes can be important with heat transfer. Too much heat transfer can dramatically reduce the efficiency and cause such problems as ignition failure while too little can result in component failure and abnormal combustion such as, for example, pre-ignition or knock. This section briefly discusses some engine parameters which affect the heat transfer.

Scale effect

One of the important considerations is how the size of an engine modifies the heat transfer rates. If the cylinder size is increased in a geometrically similar manner, the ratio of the surface area (which is proportional to one of its dimension squared) to volume (proportional to the dimension cubed) decreases. If everything else, such as gas and coolant temperatures, pressure, engine speed and gas flow velocities, are held constant, the reduced surface area to volume will decrease the relative heat transfer. A similar scale effect applies to gas turbines.

In general, smaller reciprocating engines are designed to run faster and so the surfaces are exposed for shorter periods. Turbulence within a given cylinder increases with piston speed. Greater turbulence increases heat flow and the net result when taken with the reduced time available is that the increase in speed produces a relatively small change in heat transfer. Thus, in general, small cylinders tend to lose more heat because of their higher surface to volume ratio. This includes heat during combustion and early expansion which could be useable. 'Over-square' engines (i.e. ones in which the bore is greater than the stroke) have a higher ratio cylinder head plus piston area to cylinder wall area. Thus they lose more heat than longer stroke engines of the same capacity because of the proportionally larger quench zones near TDC where temperatures are highest.

Effect of mixture on heat transfer

The maximum heat flow to the walls occurs for stoichiometric mixtures because combustion temperatures are then at a maximum. This is usually about the level at which modern SI engines operate. In some SI engines, particularly older types and those for light aircraft where rich operation is common and on engines set for maximum power levels which occur at about 5% rich, a slight leaning of the mixture will cause a small rise in engine temperature. This has led to the popular but incorrect idea that lean mixtures cause overheating. Mixtures richer than stoichiometric have lower combustion temperatures due to vaporization of the extra fuel, particularly any of which may temporarily deposit on the walls, while leaner mixtures have unused air which must be heated and acts as a buffer. In other words, the amount of heat transfer is directly related to the temperature of the combustion process. On either side of stoichiometric, the heat flow to the walls drops. On the weak side it varies as a direct function of fuel consumption, on the rich side it is governed by the latent heat of vaporization of the fuel. This is important only with SI engines, the combustion in CI engines and gas turbines being always lean and so, as loading is reduced, they run cooler.

Compression ratio

An increase in compression ratio reduces the heat loss to the walls. This is fundamentally because the higher compression ratio allows more usable expansion and the temperature at the end of expansion is lower. That is, the heat loss to the walls during the blowdown processes in particular is less. It is, however, greater during combustion near top dead centre where the higher temperature and proportionally bigger surface to volume at that position has an effect. Thus heat flow to the cylinder head is greater with higher compression ratios but it is smaller in magnitude than that during blowdown giving a net lower overall heat transfer. Also, although less significant, the direct energy loss in the gas carried out through the exhaust system is less with higher compression ratios because of the greater energy extraction for useful work during expansion. Compression ratio is also important in relationship to SI engine knock and will relate to those effects as discussed above.

Timing

Advancing the timing from the optimum increases the heat flow to the walls because it exposes greater relative areas (i.e. area to volume ratio) to the combustion which starts well before top dead centre. In fact, this is one of the reasons that peak efficiency is reached with a timing that gives maximum pressures some degrees of crank angle after top dead centre. In general, an advance in timing should mean that less heat is lost to the exhaust, more to the coolant. Conversely, retarding the spark increases temperature and the heat loss to the exhaust system and exhaust valves. Also, an over-retarded timing causes greater heat loss in the later part of the cycle (expansion, blowdown) because the higher temperatures at this late time exist simultaneously with large areas of exposed walls.

Engine load

An increase in load at a fixed speed in a given engine requires an increase in air mass flow rate per unit area of piston in an SI type or an increase in equivalence ratio in a CI type or a gas turbine. This always increases the heat transfer as it allows more fuel to be burned and the temperatures to rise. However, the MEP and power also rise and the increase in heat transfer is proportionally less than the increase in power. This is because the power is a stronger function of the increased fuel burning amount than the heat transfer. That is, relative to power, the engine will have lower losses at high load. Figure 8.2 shows typical trends.

Exhaust back pressure

With the exhaust, the basic change that can take place is when the back pressure of the system is altered. This can be either due to poor design of the system including the silencers or the use of an exhaust turbine. An increase in exhaust back pressure ensures that more high temperature residuals from the previous cycle are retained in the cylinder. This increases the average gas temperatures and hence the heat transfer.

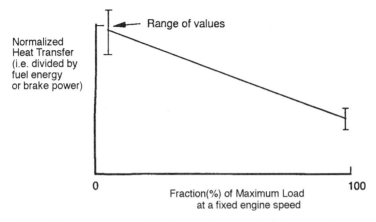

Figure 8.2 Relative proportion of fuel energy lost to direct heat transfer as engine load increases.

Effect of engine speed

Typical variations in the heat transfer per unit of mean effective pressure are depicted against engine speed on Figure 8.3. It can be seen that, at low engine speed, the rate of increase in the heat transfer with decrease in engine speed is noticeable due to the long period of exposure of the walls to the high temperatures. At higher engine speeds, the curve flattens out and, with SI engines, may then increase slightly due to the effect of turbulence. It should be noted that the higher relative heat transfer at low speeds is one of the main reasons why the MEP and hence torque curves fall from their maxima as speed is reduced. For speeds above the maxima, heat transfer is not the cause of the decline this then being due to the poorer volumetric efficiency and hence lower charge density.

8.3 MODELLING OF ENGINE HEAT TRANSFER

As in all other situations, heat transfer in engines can only take place due to conduction, convection and radiation. While the predominant modes are the convection through the thermal boundary layers either on the gas side of the wall or the coolant side together with the conduction through the wall itself, radiation is sometimes significant and can be up to about 20% of the total in CI engines, a few percent in gas turbines and negligible in SI engines. Direct gas radiation is small and the radiation heat transfer is important when there are sufficient carbon particles to provide luminous radiation. Therefore, it is predominantly a factor associated with CI engines, to a small extent with gas turbines and is of little significance with SI engines.

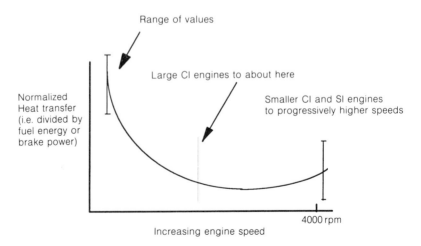

Figure 8.3 Relative proportion of fuel energy lost to direct heat transfer as engine speed increases.

The basic equations for heat transfer are well known for one-dimensional heat transfer but may alternatively be written in multidimensional form. For the latter, the reader is referred to a specialist heat transfer text such as Incropera and de Witt (1990). The one-dimensional forms are sufficient for the present descriptive purposes. These are, for conduction

$$\dot{q} = -k\frac{dT}{dx} \tag{8.1}$$

for convection

$$\dot{q} = -h\Delta T \tag{8.2}$$

for radiation

$$\dot{q} = \varepsilon\sigma(T_1^4 - T_2^4) \tag{8.3}$$

Here $\dot{q}$ is the heat transfer rate per unit area, T the absolute temperature, x a linear distance in the direction of the temperature gradient, k the thermal conductivity, h a film coefficient, ε the emissivity of the radiating substance and σ the Stefan–Boltzmann constant (numerical value $5.67 \times 10^{-11}\,kW\,m^{-2}\,K^{-4}$). For forced convection which is of importance here, the film coefficient is usually found from a correlation of the form

$$Nu = f(Re, Pr) \tag{8.4}$$

where the Nusselt number is $Nu = hD/k$, the Reynolds number is $Re = vD\rho/\mu$ and the Prandtl number is $Pr = \mu c_p/k$. D is a characteristic dimension, v the velocity, ρ the density, μ the absolute viscosity and c_p the specific heat at constant pressure.

8.3.1 Simple models

The simplest model of an engine cylinder would assume it to consist of a series of steady state volumes with the piston in different positions during the cycle. That is, the surface area which allows heat transfer would consist of the fixed head and piston crown areas and the time-varying cylinder wall area. Typical temperature profiles from the gas, through the thermal boundary layer and wall to the coolant are as shown in Figure 8.4. The well-known pipe flow correlation using the relation for forced convection in a heated tube can then be used. That is

$$Nu = 0.023(Re)^{0.8}(Pr)^{0.4} \tag{8.5}$$

To evaluate the dimensionless groups in (8.5), the gas properties can be estimated from average or estimated temperature and pressure conditions during the cycle, the bore of the cylinder for the characteristic dimension is known and a velocity is required. Choice of this velocity is difficult as the

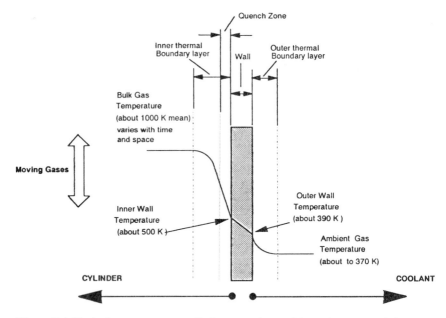

Figure 8.4 Typical temperature profile between the working substance and the coolant in an engine.

gas flows at different speeds in different parts of the cylinder. However, a choice of the mean piston speed is usually made for it with a calibration from engine test data being required for the final heat transfer result. This approach allows calculation of the *'film'* coefficient h. A similar approach but one with more validity because of the steadier conditions there can be applied on the coolant side. These film coefficients together with the equation for conduction through the wall can be used to give the overall heat transfer. The approach does however, ignore radiation.

Differences between this model and the real situation are as follows.

- Fluid temperatures and velocities are unsteady particularly on the gas side. Also the gas properties vary with time during the cycle.
- The real geometry is very complex, particularly the surface area of the cylinder head and piston crown.
- Heat does not necessarily flow at right angles to a surface.
- Local temperature variations exist on the surface. This is particularly noticeable where the gas velocity is high (e.g. at the exhaust valve) or where the heat path length is long (e.g. at the centre of the piston crown where heat flows to the edge, across a lubricating film to the cylinder), or where conductivity is poor (e.g. at spark plugs which are electrically and thermally insulated).

- The time varying heat flux is unlikely to coincide exactly with the temperature profile which drives it. That is, it will lag behind the temperature fluctuation.

8.3.2 More complex models

A comprehensive approach to engine heat transfer is not available although it is now becoming possible, at least in principle, to divide the gas and coolant flow fields and the intervening solid surface into sufficiently small grids for computational analysis. Here, the fundamental, one, two or three-dimensional time dependent equations can be applied giving instantaneous and local treatment of the heat transfer mechanism. This allows a more accurate reflection of the physical processes according to the geometry of the intake or exhaust system and combustion chamber. Obviously, the most appropriate configurations for this approach are simple geometries and operating conditions.

Even then, the calculations are still complex and many simplifications are necessary. For example, the construction of the grid is important and it is usual to orient it so that the conduction through each within the wall is as near to one-dimensional as possible. Two and three-dimensional effects at corners etc. are then ignored and are treated only by the diverging or converging cell structure. The convection and radiation equations are applied to the cells in the gas or liquid near the wall to give the local heat transfer coefficient. The convective heat transfer correlations may use an effective velocity made up of the components of the flow which are parallel to the wall surface. With the radiation, the calculation procedure can be simplified by excluding the wall to wall, gas to gas and gas to wall shape factors by assuming that the transmitting portion of the incoming heat flux in a gas side cell is scattered evenly to the neighbouring cells or wall. Wall heat transfer through it occurs by conduction and the conventional equations appropriately discretized can be used. Once the heat transfer to a cell at any instant has been determined, the rate of change of wall temperature can then be calculated by dividing it by the cell volume and the local material density and specific heat.

An approach such as this is complex and consumes much computer time. It has been applied to simplified situations such as unsteady combustion in constant volume combustion bombs. In engines, it is most feasible for gas turbines because generally the gas conditions are approximately steady and only the multidimensionality adds complexity. This is then a general approach for these engines although experimental calibration is still desirable. The approach is not as yet widely used for reciprocating engines where the unsteadiness of the flow renders the calculations far too complex and other approaches have been developed for general purpose use. These are discussed in the next section.

8.4 IMPROVED HEAT TRANSFER FORMULA FOR RECIPROCATING ENGINES

Comprehensive approaches based on the concepts of the simple model of section 8.3.1 have been attempted and several more sophisticated models have been developed. These all have their advantages and limitations and none can be said to be, at this stage, universal. The original references are listed at the end of the chapter.

8.4.1 Review of some early heat transfer formulas

At least ten formulae exist for engine heat transfer, all of which are basically empirical although they may have been developed with the basis of the fundamental theory in mind. The major references are given at the end of this chapter. The paper by Eichelberg (1939) is of historic interest as it led to the development of many equations and for many years was the dominant equation used, particularly in the USA but it was preceded by a number of other earlier empirical formulas. As is usually the case, there has been some doubt as to the generality of empirical formula, as outlined by Annand (1967). This is not on the grounds of inaccuracy in predicting heat transfer rates because, as he states, large variations exist from engine to engine making it impossible to tell if an error exists or not in the original formulation. It is because the theoretical background is inconsistent. The two major formulas that are now used are those by Annand (1963) and by Woschni (1967) and these will be discussed in more detail a little later in this text.

It is interesting to note that the earliest attempt to find a reasonable formula for engine heat transfer was that by Nusselt. He is regarded as one of the major developers of the convection heat transfer methods now used for general purposes but it should be noted that he carried out his work in order to assess the heat transfer in CI engines. His approach was to determine the heat loss from experiments carried out in cylindrical combustion bombs after the ignition of the air/fuel mixture contained within them. He separated radiation from convection by two tests, the first using an internally blackened bomb, the second a gold plated bomb. The formula he obtained was

$$\dot{q} = C_1(1 + 0.38\,v_p)(p^2 T)^{1/3}(T - T_w) + C_2(T^4 - T_w^4) \qquad (8.6)$$

Note that the constants are not dimensionless.

At a later stage, Eichelberg and Pflaum (1951) developed equations which, as mentioned above, were widely used for many years. Eichelberg developed his equation by choosing a point on the surface of the combustion chamber where it would be expected that heat flow will be normal to the wall. Variation with time of the temperature of the metal surface inside and outside was observed and harmonic analysis used to develop the formula

$$\dot{q} = C v_p^{1/3} (pT)^{1/2} (T - T_w) \tag{8.7}$$

This covers both convection and radiation but is not dimensionlessly correct. It is reported to work reasonably well with engines of low piston speed but not with high speed engines where the factor $v_p^{0.5 \text{ to } 0.8}$ seems to be more reasonable. In (8.6) and (8.7), v_p is the mean piston speed, p and T the gas temperature and T_w the wall temperature. Note that, as these are given here for historical rather than practical purposes, the constants C_1, C_2, C have not been evaluated in the SI system.

Others have developed Eichelberg's approach further. Pflaum (1961) amended (8.7) by applying an intake-manifold pressure correction factor and adding modified constants to account for different flow patterns within the cylinder and different surface finishes at different locations. Overbye, Bennethum, Uyehara and Myers (1961) used similar experimental methods but with improved surface thermocouples to obtain an equation for small four-stroke SI engines. These equations are complex and not widely used and so are not given here. The interested reader is referred to the original sources.

Others, notably Elser (1955) and Oguri (1960) also have used the similar methods to Eichelberg but with dimensional analysis to aid the development of the formula. Elser's equation is

$$\text{Nu} = 6.5 \left(1 + 0.5 \frac{\Delta s}{c_p}\right) (\text{Re Pr})^{1/2} \tag{8.8}$$

where Δs is the increase in entropy from start of compression while gas properties are taken as the average of the wall and bulk gas temperatures. Annand points out that he used the reversible equation for entropy change for the irreversible combustion process. Agreement is poor on four stroke engines but good on two strokes. Oguri (1960) modified this expression to

$$\text{Nu} = 1.75 \left(1 + \frac{\Delta s}{c_p}\right) (\text{Re Pr})^{1/2} [2 + \cos(\theta - 20°)] \tag{8.9}$$

where θ = crank angle in degrees from TDC.

8.4.2 Currently used heat transfer formulas

While the above equations have been used with some degree of success in particular applications, the two most commonly used formulas are those of Annand and Woschni. These are now examined.

Annand's formula

Annand discusses a rational basis for developing heat transfer equations for reciprocating engines. His approach is to ignore the relatively unimportant suction and exhaust strokes and only the compression, combustion, expansion and blowdown processes are considered. This is essentially because his techniques were developed for thermodynamic cycle analysis which normally ignores the gas exchange processes. Also, heat transfer during intake and exhaust is small. While Annand's approach is not as widely used as Woschni's, it is useful to discuss it here because of the logic of its development.

The theoretical considerations are broken into those for convection and radiation, before combining into a single formulation. First, convection will be considered. In reciprocating engines, conditions are clearly unsteady. However, if the processes can be regarded as quasi-steady, then (8.2) can be used at the time considered with an instantaneous film coefficient applied. In fact, Annand points out that the real process is not even a quasi-steady one, there being, as mentioned previously, a phase lag between the driving temperature difference ΔT and the heat transfer flux at the surface. That is, ΔT and $\dot{q}$ are not zero together. A quasi-steady process with this relationship would mean that the film coefficient becomes infinity when the temperature difference is zero.

For a logical development, a dimensional analysis is required and the important parameters related to the heat transfer are given in Table 8.3

In addition, a number of ratios can be used to describe the engine geometry and the intricacies of the flow. These are given in Table 8.4.

Dimensional analysis then gives the following groups.

$$\frac{hD}{k}, \; \frac{\rho v D}{\mu}, \; \frac{\mu c_p}{k}, \; \frac{c_p T}{v^2}, \; \frac{ND}{v}, \; \frac{q}{\rho c_p N T}, \; y_1 \rightarrow y_m, \; u_1 \rightarrow u_m, \; \theta$$

These can be recognized as Nu, Re, Pr numbers as expected, an energy ratio which is inversely proportional to the Mach number squared, rotational velocity or frequency of recurrence ratio and an energy production rate ratio existing only during combustion. In addition, there are a series of descriptive geometric and flow ratios together with the dimensionless angular position in the cycle. While the geometric and flow ratios could be useful

Table 8.3

Parameter	Symbol	Units
Bore	D	m
Velocity	v	$\mathrm{m\,s^{-1}}$
Conductivity	k	$\mathrm{kW\,m^{-2}\,K^{-1}}$
Specific heat	c_p	$\mathrm{J\,kg^{-1}\,K^{-1}}$
Density	ρ	$\mathrm{kg\,m^{-3}}$
Heat release per unit volume	q	$\mathrm{J\,m^{-3}}$
Rotational speed	N	$\mathrm{s^{-1}}$
Position in cycle	θ	—

for examining very specific, local aspects of the heat transfer, engines are, in general, too complex to be able to describe in detail while retaining the concept of a simple model useful for cycle analysis. This applies to either their complex geometries or the flows represented by the various ratios. The major simplification suggested is to revert to a basic geometry, Y/D represented by the major engine dimensions of stroke and bore and a mean piston velocity, $v_p = 2LN/60$.

In the reciprocating internal combustion engine, the problem involves not only convection but also radiation to the walls from a mixture of gases at high and varying temperatures. It should be noted that this is very dependent on the combustion products that exist within the cylinder at any instant. The combustion process changes the composition of the cylinder contents but does not normally do so uniformly throughout the combustion space. Because of this, Annand makes, the following points.

- It is unreasonable to expect to be able to calculate accurately radiation shape factors for use during combustion and expansion. The best that can be done is to use an average factor.
- The importance of radiant transfer will be very different in SI and CI engines due to the deflagration front in one and the diffusion flame in the other. This can be a cause of confusion and it must be recognized that the different types of combustion produce different radiation heat transfer.

A number of properties are required for use in heat transfer relationships. The temperature is important in all forms of heat transfer and particularly so in radiation due to the fourth-power relationship. Thus, small variations in temperature can result in large differences in the heat transferred. Also, the transport properties, viscosity, thermal conductivity and density are very temperature dependent. Instantaneous temperature varies spatially throughout the combustion space and is not easily measured. It can be calculated as an averaged value from the uniform, time-dependent pressure trace which can be easily obtained. During compression and later

Table 8.4

Ratios	Symbol	Comments
Length ratios	y	These describe the geometry. An example is the axial length to the diameter. One basic length and a number of ratios are required
Velocity ratios	v	These describe the flow. An example is the velocity of the intake jet to the mean piston speed. One basic velocity and a number of ratios are required

expansion, the use of a spatially averaged, time-dependent temperature is plausible. Due to the flame front movement in SI engines or its location in CI ones, it is not a reasonable assumption during combustion but is the only simple alternative.

Gas composition is also important, the particular property values required being those for the molecular mass, specific heat, and viscosity. These can be averaged in the normal manner for a mixture. The thermal conductivity, k, is best obtained from

$$k = \frac{\mu c_p}{\text{Pr}}$$

where the Prandtl number, Pr varies little from a value of 0.7.

Using these, Annand gives the total heat transfer as

$$\dot{q} = a \frac{L}{D} (\text{Re})^b (T - T_w) + c (T^4 - T_w^4) \tag{8.10}$$

where a, b and c are constants evaluated from engine data. Note that a includes the Prandtl number effect in the conventional convection formulation because it is approximately constant while the radiation constant c includes the shape factor, emissivity and Stefan–Boltzmann constant. During compression $c = 0$, but it has a constant finite value for combustion and expansion.

Annand evaluates these constants as $a = 0.38$, for two strokes, $a = 0.49$, for four strokes, $c = 1.6 \times 10^{-12}$, for CI engines for combustion and expansion only, $c = 2.1 \times 10^{-13}$, for SI engines for combustion and expansion only. Note that the SI value of c is very small as most radiation occurs from the luminous soot that occurs only in CI engines. The constant a can vary from 0.35 to 0.8 increasing with increasing intensity of the fluid motion.

Woschni's formula

Probably the heat transfer coefficient now most commonly used in engine calculations was developed by Woschni (1967) for CI engines but has application to all types and is referred to by its author as a universally applicable equation. It is based on the fact that experimental data can be taken and analysed to separate the effects of a great many variables. Substantial differences in the heat transfer coefficient can be obtained from different experimental techniques. For example, even simple experiments in a combustion bomb can be assessed in two ways, those which look at the rate of heat transfer from the bomb after burning is completed by evaluating the internal energy from the pressure decay and those using surface wall temperatures measured by thermocouples to give boundary conditions for analysis. These two techniques have been found to give different heat transfer coefficients.

One major difference is that, even for a symmetrical combustion bomb (for example a spherical one with central ignition), higher temperatures exist at the top due to the buoyancy effects in the burnt gas. In internal combustion engines, the differences can be even more substantial. It should be noted that rapid changes in the heat transfer rate can occur when the flame itself, rather than the compressed gas ahead of it, contacts the wall. To correct the differences, a very large number of distributed thermocouples would be required.

The different formulas that have been developed for the heat transfer coefficient, h, therefore can give substantially different values for the total cycle heat transfer even for the same engine under identical conditions. This can be seen from Woschni's calculations which are given in Table 8.5

Woschni proposed a formula which is intended to take into account convection, radiation and the influence of rapid changes in gas temperature with time. It is based on the normal heat transfer correlation for turbulent, forced flow

$$Nu = C\,Re^m \tag{8.11}$$

where C is a constant and the index m is normally 0.8 as in (8.5). The Prandtl number is therefore assumed to be constant and is absorbed into C. This, on rearranging, gives

$$h = C\frac{k}{D}\left(\frac{vD\rho}{\mu}\right)^m \tag{8.12}$$

where D is the principal dimension (the bore) and v is the local average gas velocity rather than the mean piston speed, v_p, as is usually used. By expressing the variables for the properties composing the dimensionless groups as functions of pressure and temperature (i.e. viscosity μ proportional to $T^{0.62}$, density ρ proportional to pT^{-1}, conductivity k proportional to $T^{0.75}$) this equation becomes

$$h = CD^{m-1}v^m p^m T^{0.75-1.62m} \tag{8.13}$$

Table 8.5 Heat transfer in an engine under fixed load/speed conditions

Formula	Heat transfer to walls as percentage of fuel heat
Nusselt	15
Eichelberg	13
Pflaum	23
Annand	22

Woschni relates the value of v to the piston speed v_p by

$$v = C_1 v_p + v_c \qquad (8.14)$$

where v_c is the additional gas velocity due to the combustion imposed on that from the piston motion. Its value increases from zero at the beginning of combustion, reaches a maximum and then decreases during expansion. Woschni suggested an empirical expression for it which is

$$v_c = C_2 \frac{V_s T_1}{V_1 p_1}(p - p_0) \qquad (8.15)$$

where V_1, T_1 and p_1 are the volume, temperature and pressure conditions at the beginning of combustion, p is the pressure at the instant (that is, crank angle position) considered and p_0 is the corresponding pressure in the 'motored' engine.

A typical value for the constant C is 3.26. Other constants as given by Woschni are $C_1 = 6.18, C_2 = 0$ (for scavenging or gas exchange), $C_1 = 2.28, C_2 = 0$ (for compression), $C_1 = 2.28, C_2 = 3.24 \times 10^{-3}$ (for combustion and expansion). These constants were evaluated by careful experimentation where the individual engine parameters were varied one by one. The engine used was a diesel with zero swirl. While they should be applicable to geometrically similar engines, a universally valid heat transfer coefficient can only be obtained by evaluating different constants for each geometrically similar group of engines. For engines with high swirl (velocity v_s)

$$C_1 = 6.18 + 0.417 \frac{v_s}{v_p} \qquad \text{for the scavenging and gas exchange}$$

$$C_1 = 2.28 + 0.308 \frac{v_s}{v_p} \qquad \text{for elsewhere in the cycle}$$

The above formula does not contain any separate expression for radiative heat transfer. This is because it evaluates the coefficients from the total heat transfer experimentally. While radiative heat transfer is proportional to T^4, not to T, it is a relatively small component of the total heat transfer (perhaps a maximum of 20%) and Woschni believes that any error involved in including it in the use of the single correlation is small.

The value of the heat transfer coefficient given by Woschni lies somewhere between those of Eichelberg and Nusselt. It would therefore indicate that a total heat transfer of about 14% is realistic for the CI engine considered by him.

8.5 REFERENCES

Annand, W. (1963) Heat transfer in the cylinders of reciprocating internal combustion engines, *Proc. I. Mech. E.*, **177** 36, 973.
Cameron, A. (1966) *The Principles of Lubrication*, Longman, London.

Eichelberg, G. (1939) Some new investigations on old combustion engine problems, *Engineering*, **148**, 463.

Eichelberg, G. and Pflaum, W. (1951) Untersuchung eines hochaufgeladenen Diesel-motors, *Z. Ver. Dtsch. Ing.*, **93**, 1113, (English Translation *Mot. Ship*, 1952, **33**, 18).

Elser, K., (1955) Instationare Warmuebertragung bei periodisch adiabater Verdich-tung turbulenter gase, *Forsch. Ing. Wes.*, **21**, 65.

Oguri, T. (1960) Theory of heat transfer in the working gases of internal combustion engines, *Bull. Jap. Soc. Mech. Eng.*, **3**, 370.

Overbye, V., Bennethum, J., Uyehara, O. and Myers, P. (1961) Unsteady heat trans-fer in engines, *Trans. Soc. Auto. Eng.*, **69**, 461.

Pflaum, W., (1961) Warmeubergang bei Dieselmaschinen mit und ohne Aufladung *M.T.Z.*, **22**, 70 (English abstract in *Engineer's Digest*, NY, 1961, **22**, 86).

Wallace, F. J., Way, R. J. B. and Vallmert, H. (1979) Effect of partial suppression of heat loss to coolant on high output Diesel cycle, *SAE Paper No. 790823*, Society of Automotive Engineers, USA.

Woschni, G. (1967) A universally applicable equation for the instantaneous heat transfer coefficient in the IC engine, *SAE Paper No. 670931*, Society of Automo-tive Engineers, USA.

PROBLEMS

8.1 Determine the heat transfer rate (in kW) from a cylinder under the following conditions. The correlation for pipe flow

$$\mathrm{Nu} = 0.023 \, \mathrm{Re}^{0.8} \, \mathrm{Pr}^{0.4}$$

may be assumed to apply. In all cases, ignore end effects.

(a) gas turbine combustor, diameter 150 mm, length 350 mm, average gas velocity 30 ms^{-1}, pressure 1.5 MPa, average gas temperature, 1500 K, inner wall temperature, 560 K.

(b) SI piston engine cylinder during expansion, diameter 80 mm, average length for exposed walls during stroke 45 mm, gas speed equal to piston speed 10 ms^{-1}, average pressure and temperature during stroke, 1.3 MPa, 1000 K, inner wall temperature, 500 K.

Assume in both cases that the gas constant is 290 J kg^{-1}K^{-1} with an isentropic index of 1.3, its viscosity is 10^{-4} kg ms^{-1} and its ther-mal conductivity is 0.08 W mK^{-1}.

8.2 A CI engine has dimensions:

Diameter of cylinder (bore) 160 mm;
Length of stroke 200 mm;
Compression Ratio 15:1;
Rotational speed 1500 rev min^{-1};
Mean piston speed 10 ms^{-1};
Power output 15 kW l^{-1}.

Assume that the head can be represented by the flat end of the cylinder plus the clearance volume walls, that the piston crown is flat and that

50 mm. If the mean gas conditions during combustion are temperature 1200 K, density 5.5 kg m^{-3}, estimate the heat transfer rates to the head, the side walls and the piston crown due to (a) gas radiation (b) convection during the combustion (c) both combined. Also estimate the total heat transfer rate during this period as a percentage of the power output. Does this seem reasonable?

The emissivity of the soot particles in the flame may be taken as 0.6 and the Stefan–Boltzmann constant is 5.670×10^{-8} Wm^{-2}K^4. For the convection, the film coefficient obtained from the pipe flow correlation of Problem 8.1 may be assumed to apply to all surfaces with its constant 0.023 replaced by a value of 0.035 which has been found to be more representative of reciprocating engines. The viscosity, thermal conductivity and specific heat (calculated from R and γ) of Problem 8.1 may also be assumed to be valid.

8.3 The walls of a particular engine cylinder consist of a cast iron liner, 8 mm thick. At the time when the mean gas temperature is 727 °C while the coolant is at 95 °C, estimate, for the side walls of the cylinder only, the heat transfer rate from gas to coolant. Also calculate the temperatures at the inner and outer wall surface. Assume that the value of the film coefficient is, on the gas side, 160 W m^{-2}K^{-1}, on the water side is 2000 W m^{-2}K^{-1} and the thermal conductivity of cast iron is 70 W m^{-1}K^{-1}.

8.4 The wall area exposed by the piston movement in an engine with a connecting rod length twice the crank throw (which itself is half the stroke) and the piston velocity are given by the equations for A and v below. In addition, the spatially resolved mean gas temperature and the pressure in an SI engine cylinder for a particular engine was roughly approximated during a cycle for calculation purposes by the expression for T(K) and p(KPa), these peaking at 15°ATDC. In all these equations, the crank angle is in degrees and is measured from zero at top dead centre.

$$T = 1400 + 1100 \sin(\theta + 75)$$

$$p = 1900 + 1200 \sin(\theta + 75) \quad \text{for } 0 < \theta < +180$$

$$A = \pi b \frac{s}{2} [3 - \{\cos\theta + (4 - \sin^2\theta)^{0.5}\}] \quad \text{for } -180 < \theta < +180$$

$$V = s\pi \frac{N}{60} \sin\theta \left[1 + \frac{\cos\theta}{(4 - \sin^2\theta)^{0.5}}\right] \quad \text{for } -180 < \theta < 180$$

The gas velocity may be assumed to 1.5 times the piston velocity due to turbulence. Estimate the instantaneous convective film coefficient from the correlation of Problem 8.1 (with the constant 0.023 replaced by 0.035) and the heat transfer rate (to the cylinder walls only) for crank angles of 15°, 30°, 60°, 90°, 120°, 150° and 180° assuming that the inner wall temperature is constant at 450 K. The following data applies to the engine: bore $b = 80$ mm, stroke $s = 90$ mm, engine speed $N = 3600$ rev min^{-1}.

8.5 Using the Woschni formulae for engines, estimate the heat transfer coefficient and the heat transfer rate through the cylinder walls during combustion and expansion for the same case as Problem 8.4. In order to obtain the conditions at the start of combustion, assume that the heat transfer begins at 10° BTDC. That is, assume isentropic compression from 100 kPa, 300 K to that value with an isentropic index of 1.39. Use the same technique for the motored trace to the piston position at the instant considered. Pressure, temperature and piston speed should be used at each location as in Problem 8.4. The values for the constants in the formula may be taken as $C = 3.26$, $C_1 = 2.28$, $C_2 = 3.24 \times 10^{-3}$.

Further reading

Listed below is a selection of texts related to the basic material of this book. The list is not exclusive and many other excellent texts are available in each section. The list is devised merely to give the reader an appropriate introduction to source material.

Some texts have already been quoted as references in the main sections of this book whilst others are regarded as basic source material. As this text is not designed to stand alone but to provide some relevant fundamentals, to orient the reader towards and summarize some work for application particularly in reciprocating engines and gas turbines, it is recommended that at least one general text from each section be examined by the reader.

THERMODYNAMICS

Black, W. Z. and Hartley, J. G. (1985) *Thermodynamics*, Harper and Row, New York.

Doolittle, J. S. and Hale, F. J. (1984) *Thermodynamics for Engineers*, Wiley, New York.

Fenn, J. B. (1982) *Engines, Energy and Entropy*, W. H. Freeman, New York.

Reynolds, W. C. (1968) *Thermodynamics*, 2nd ed., McGraw-Hill, New York.

Rogers, G. F. C. and Mayhew, Y. R. (1967) *Engineering Thermodynamics, Work and Heat Transfer*, Longman, London.

Sonntag, R. E. and Van Wylen, G. J. (1991) *Introduction to Thermodynamics*, Wiley, New York.

Wark, K. (1989) *Thermodynamics*, 5th ed., McGraw-Hill, New York.

FLUID MECHANICS

General

Douglas, J. F., Gasiorek, J. M. and Swaffield, J. A. (1979) *Fluid Mechanics*, Pitman, Bath.

John, J. E. A. and Haberman, W. (1971) *Introduction to Fluid Mechanics*, Prentice–Hall, Englewood Cliffs, NJ.

Massey, B. S. (1989) *Mechanics of Fluids*, 6th ed., Chapman and Hall, London.

Streeter, V. L. (1971) *Fluid Mechanics*, 5th ed., McGraw-Hill, NY.

Gas Dynamics

Anderson, J. D. Jr. (1982), *Modern Compressible Flow*, McGraw-Hill, New York.

Daneshyar, H. (1976) *One-Dimensional Compressible Flow*, Pergamon, Oxford.

Emanuuel, G. (1986) *Gas Dynamics, Theory and Applications*, AIAA Education Series, USA.

John, J. E. A. (1969) *Gas Dynamics*, Allyn and Bacon, Boston.

Shapiro, A. H. (1953) *The Dynamics and Thermodynamcis of Compressible Fluid Flow, Vols 1 and 2*, Wiley, New York.

Zucrow, M. J. and Hoffman, J. D. (1977) *Gas Dynamics*, Wiley, New York.

Pressure waves and shock waves

Courant, R. and Friedrichs, K. O. (1967) *Supersonic Flow and Shock Waves*, Interscience, (John Wiley), New York.

Han, Z. and Yin, X. (1992) *Shock Dynamics*, Kluwer, Dordrecht, Netherlands.

Rudinger, G. (1969) *Nonsteady Duct Flow – Wave Diagram Analysis*, Dover, New York.

Turbulent flow

Bradshaw, P. (1971) *An Introduction to Turbulence and its Measurement*, Pergamon, Oxford.

Hinze, J. O. (1959) *Turbulence, An Introduction to its Mechanism and Theory*, McGraw-Hill, New York.

Tennekes, H. and Lumley, J. L. (1972) *A First Course in Turbulence*, MIT Press, Cambridge.

Boundary layers

Schlichting, H. (1979) *Boundary Layer Theory*, 7th ed., McGraw-Hill, New York.

Young, A. D. (1989) *Boundary Layers*, BSP Professional Books, Oxford.

Two phase flow

Collier, J. G. (1972) *Convective Boiling and Condensation*, McGraw-Hill, New York.

Wallis, G. B. (1969) *One-Dimensional Two-Phase Flow*, McGraw-Hill, New York.

Hetsroni, G. (ed) (1982) *Handbook of Multiphase Systems*, Hemisphere, Washington.

Computational fluid dynamics

Fletcher, C. A. J. (1988) *Computational Techniques for Fluid Dynamics, Vols 1 and 2*, Springer-Verlag, Berlin.

Peyret, R. and Taylor, T. D., *Computational Methods for Fluid Flow*, Springer-Verlag, Berlin.

TURBOMACHINERY

Dixon, S. L. (1966) *Fluidmechanics, Thermodynamics of Turbomachinery*, Pergamon, Oxford.

Ferguson, T. B. (1963) *The Centrifugal Compressor Stage*, Butterworths, London.

Gostelow, J. P. (1984) *Cascade Aerodynamics*, Pergamon, Oxford.

Horlock, J. H. (1958) *Axial Flow Compressors*, Butterworths, London.

Horlock, J. H. (1966) *Axial Flow Turbines*, Butterworths, London.

HEAT AND MASS TRANSFER

Holman, J. P. (1968) *Heat Transfer*, McGraw-Hill, New York.

Incropera, F. P. and De Witt, D. P. (1990) *Fundamentals of Heat and Mass Transfer*, Wiley, New York.

COMBUSTION

Kanury, A. M. (1975) *Introduction to Combustion Phenomena*, Gordon and Breach, New York.

Glassman, I. (1977) *Combustion*, Academic Press, New York.

Strehlow, R. A. (1976) *Fundamentals of Combustion in Aerospace Propulsion Systems*, AIAA Education Series, USA.

Strehlow, R. A. (1984) *Combustion Fundamentals*, McGraw-Hill, New York.

INTERNAL COMBUSTION ENGINE

Reciprocating engines

General

Arcoumanis, C. (ed) (1988), *Internal Combustion Engines*, Academic Press, London.

Benson, R. S., and Whitehouse, N. D. (1979) *Internal Combustion Engines, Vols 1 and 2*, Pergamon, Oxford.

Heywood, J. B. (1988) *Internal Combustion Engine Fundamentals*, McGraw-Hill, New York.

Lichty, L. C. (1967) *Internal Combustion Engines, 6th ed.*, McGraw-Hill, New York.

Markatos, N. C. (ed) (1989) *Computer Simulation for Fluid, Flow, Heat and Mass Transfer and Combustion in Reciprocating Engines*, Hemisphere, New York.

Obert, E. F. (1973) *Internal Combustion Engines and Air Pollution*, Intext, New York.

Schmidt, F. A. F. (1965) *The Internal Combustion Engine*, Chapman and Hall, London.

Stone, R. (1992) *Introduction to Internal Combustion Engines*, 2nd ed., Macmillan, London.

Taylor, C. F. (1977) *The Internal Combustion Engine in Theory and Practice*, Vols 1 and 2, MIT Press, Cambridge.

Specialist areas

Annand, W. J. D. and Roe, G. E. (1974) *Gas Flows in the Internal Combustion Engine*, Haessner, Newfoundland, NJ.

Watson, N. and Janota, M. S. (1982) *Turbocharging the Internal Combustion Engine*, Macmillan, London.

Springer, G. S. and Patterson, D. J. (1973) *Engine Emissions–Pollutant Formation and Control*, Plenum, New York.

Engine history

Cummins, C. L. (1976), *Internal Fire*, Carnot Press, Lake Oswago, Oreg.

Koln, I. (1972) *The Evolution of the Heat Engine*, Longman, London.

Historical interest

Chaloner, J., Dale, C., Lawrence, H., Paulin, C. and Walshaw, T. (1950) *Modern Oil Engine Practice*, 4th ed., Newnes, London.

Ricardo, H. (1993) *The High Speed Internal Combustion Engine*, Blackie, London.

Gas turbines (and aircraft engines in general)

General

Bathie, W. M. (1984) *Fundamentals of Gas Turbines*, Wiley, New York.

Cohen, H. H., Rogers, G. F. C. and Saravanamittoo, H. I. H. (1972) *Gas Turbine Theory*, 2nd ed., Longmans, Harlow.

Hill, P. G. and Peterson, C. R. (1992) *Mechanics and Thermodynamics of Propulsion*, 2nd ed., Addison-Wesley, Reading, MA.

Mattingly, J. D., Heiser, W. H. and Daley, D. H. (1987) *Aircraft Engine Design*, AIAA Education Series, USA.

Oates, G. C. (1984) *Aerothermodynamics of Gas Turbine and Rocket Propulsion*, AIAA Education Series, USA.

Oates, G. C. (1989) *Aircraft Propulsion Systems Technology and Design*, AIAA Education Series, USA.

Rolls Royce plc (1986) *The Jet Engine*, BPCC Ltd for Rolls Royce, Derby.

Whittle, F. (1981) *Gas Turbine Aero-Thermodynamics*, Pergamon, Oxford.

Wilson, D. (1984) *The Design of High-Efficiency Turbomachinery and Gas Turbines*, MIT Press, Cambridge.

Specialist areas

Archer, R. D. and Saarlas, M. (1995) *Elements of Aerospace Propulsion*, Prentice-Hall, Englewood Cliffs, NJ.

Foa, J. V. (1960) *Elements of Flight Propulsion*, John Wiley, New York.

Hesse, W. J. and Mumford, N. V. S. Jr (1964), *Jet Propulsion for Aerospace Applications*. 2nd ed., Pitman, London.

Lefebvre, A. H. (1981) *Gas Turbine Combustion*, Pergamon, Oxford.

Seddon, J. and Goldsmith, E. L. (1985) *Intake Aerodynamics*, AIAA Education Series, USA.

FUEL

Blackmore, D. R. and Thomas, A. (1979) *Fuel Economy and the Gasoline Engine: Fuel, Lubricant and Other Effects*, Wiley, New York.

Goodger, E. M. (1975) *Hydrocarbon Fuels*, Macmillan, London.

Goodger, E. M. (1982) *Alternative Fuels for Transport*, Cranfield Press, Cranfield.

TABLES AND CHARTS

Hottel, H., Williams, G. and Satterfield, C. (1936) *Thermodynamic Charts for Combustion Processes, Vols 1 and 2*, Wiley, New York.

JANAF Thermochemical Tables (1971) The Dow Chemical Company Thermal Research Laboratory, US National Board of Standards, NSRDS-NBS-37, Washington.

Keenan J. H. and Kaye J. (1983) *Gas Tables, International Version*, Wiley, New York.

Answers to problems

CHAPTER 1

1.1 7.54 hs for (an assumed) 50 l tank, fuel consumption 8.3 l/100 km (12 km l^{-1}).

1.2 (a) 38.4 MW; (b) 128 MW; (c) 2.98 kg s^{-1}, 3.92 ls^{-1}; approx 9 to 10 times petrol bowser flow rate; (d) 134 tonnes, 176 kl; (e) 19.3 tonnes.

1.3

Automobile

Fuel	Mass penalty (kg)	Volume penalty (l)
Methanol	56	65
Ethanol	20	27
CNG steel tank	233	212
CNG carb. fibre tank	48	212
LNG	15	30
Lead acid battery	85 (power only) 1373 (for range)	
New technology battery	6 (power only) 1373 (for range)	

Solar   Collection area 112 m^2, ($10.6 \text{ m} \times 10.6 \text{ m}$). Requires heavy storage systems for night, poor weather use

Aircraft

Fuel	Mass penalty (tonnes)	Volume penalty (kl)
Methanol	164	242
Ethanol	88	102
CNG steel tank	875	773
CNG carb. fibre tank	206	773
LNG	86	111

Aircraft (*Contd*)

Fuel	Mass penalty (tonnes)	Volume penalty (kl)
Lead acid battery	283 (power only) 15551 (for range)	
New technology battery	32 (power only) 7394 (for range)	
Solar	Collection area 261 224 m², (511 m × 511 m). Requires heavy storage systems for night, poor weather use	

1.7 Final temperature, 2220, 1708, 1401 K; Volume ratio, 2.7, 5.2, 8.5; Work from expansion 775, 1142, 1362.5 kJ kg^{-1}.

1.8 Estimated volumetric capacities, 5 l, 8 l, 3.5 l.

1.9 Core mass, volume flow rates = 31.2 kg s^{-1}, 40 m³ s^{-1}, $A = 0.2775$ m², $d = 0.594$ m; Overall mass, volume flow rate = 124.8 kg s^{-1}, 200 m³ s^{-1}, $A = 1.388$ m², $d = 1.33$ m.

1.10 Jet velocities, no bypass = 1762 m s^{-1}, with bypass = 560 m s^{-1}; Propulsive efficiency = 24%, 60%.

CHAPTER 2

2.1 Pressures, 100, 1000, 1000, 100 kPa; Temperatures, 300, 579.6, 1425.8, 738 K; Densities, 1.16, 6.01, 2.44, 0.472 kg m^{-3}; Compressor work input = 280.9, Turbine work output = 690.9, Net work = 410 kJ kg^{-1}; Thermal efficiency = 48.2%.

2.2 1146.2 K; 465 kPa; 4.66:1.

2.3 Pressure ratio 29.7; Pressures, 25, 742.5, 742.5, 25 kPa; Temperatures, 220, 579.6, 1425.8, 541 K; Net w = 527.6 kJ kg^{-1}; Thermal efficiency = 62%; Nozzle pressure ratio 10.7:1.

2.4 Pressure ratio 58.6; Pressures, 25, 1465, 1465, 25 kPa; Temperatures, 220, 703.8, 1550, 484.5 K; Net w = 584.3 kJ kg^{-1}; Thermal efficiency = 68.7%; Nozzle pressure ratio 15.8:1.

2.5 Pressure ratios, 15.3, 26.3, 37.7; Work output, 418.2, 527.2, 732.6 kJ kg^{-1}; Thermal efficiencies, 54.1, 60.7, 64.6%.

2.6 Cycle 1: 308.7 kJ kg^{-1}, 42.5%; Cycle 3: 586.5 kJ kg^{-1}, 54.4%.

2.7 $\eta_t = \eta_c = 63.8\%$.

2.8 Pressure ratio 127:1, thermal efficiency 59.4%, work output 460 kJ kg^{-1}. Compared with Problem 2.5, thermal efficiency, work output both reduced. Compared with Problem 2.6, thermal efficiency up, work output reduced.

2.9 11:1; 626 kJ kg^{-1}; 58%.

2.10 321.8 kJ kg^{-1}; 68.7%; 321.8 kJ kg^{-1}; 61.9%.

2.11 $w_c = 234.9$; $q_a = 1012.6$; $w_{net} = 456$ kJ kg^{-1}; Thermal efficiency $= 45\%$.

2.12 $w = 567.9$, kJ kg^{-1}; Thermal efficiency $= 70.7\%$; $w = 471.7$ kJ kg^{-1}; Thermal efficiency $= 58.8\%$; $w = 671.2$ kJ kg^{-1}; Thermal efficiency $= 76.\%$.

2.13 Heat input 2750 kJ kg^{-1}

For isentropic index 1.4:

Pressures $p2 = 1987, 2104, 1870$ kPa; Temperatures $T_2 = 763$ K;
Pressures $p3 = 11\,968, 12\,673, 11\,264$ kPa; Temperatures $T_3 = 4596$ K;
Pressures $p4 = 512, 542, 482$ kPa; Temperatures $T_4 = 1868$ K;
$w = 1632$ kJ kg^{-1}; Thermal efficiency $= 59.4\%$;
Work/cycle $= 3.12, 3.3, 2.94$ kJ;
Power $= 26, 82.6, 122.4$ kW;
MEP $= 1560, 1652, 1469$ kPa.

For isentropic index 1.3:

Pressures $p2 = 1587, 1680, 1493$ kPa; Temperatures $T_2 = 609$ K;
Pressures $p3 = 8397, 8889, 7900$ kPa; Temperatures $T_3 = 3222$ K;
Pressures $p4 = 450, 476, 423$, kPa; Temperatures $T_4 = 640$ K;
$w = 1228$ kJ kg^{-1}; Thermal efficiency $= 49.1\%$;
Work/cycle $= 2.34, 2.5, 2.21$ kJ;
Power $= 19.5, 62.1\,92$ kW;
MEP $= 1170, 1250, 1105$ kPa.

2.14 $w_p = 0.05, 0.04, 0.06$ kJ cycle^{-1}; $w = 2.29, 2.46, 2.15$, kJ cycle^{-1}; Power $= 19.1, 61.1, 89.5$ kW; Thermal efficiency $= 48.1, 48.3, 47.8\%$; Increased density of the charge.

2.15 Max. pressure and temperature, 10677 kPa, 3260 K; $w = 1211$ kJ kg^{-1}; 2.91 kJ/cycle; Power $= 121$ kW; Thermal efficiency $= 49.1\%$. Modified compression ratio, $r = 8.25:1$; $w = 1172$ kJ kg^{-1}; 2.82 kJ cycle^{-1}; Power $= 117.4$ kW; Thermal efficiency $= 46.9\%$. Intercooled: $w = 1172$ kJ kg^{-1}, 3.16 kJ cycle^{-1}; Power $= 132$ kW; Thermal efficiency $= 46.9\%$.

2.16 Pressures, 85, 3124.5, 3124.5, 394.6 kPa; Temperatures 310, 712.2, 2320, 1439 K; $w = 919.6$ kJ kg^{-1}, 10.54 kJ cycle^{-1}; Power $= 263.5$ kW; Thermal efficiency $= 46\%$, with reduced cut-off, 52.8%; Expansion ratios, 4.91, 9.82; End of expansion pressures, 394.6, 160.3 kPa; End of expansion temperatures, 1439, 585 K.

2.17 Max pressure and temperature, $5936.5\,\text{kPa}, 1598.5\,\text{K}$; $w = 543.6\,\text{kJ kg}^{-1}$; $11.6\,\text{kJ cycle}^{-1}$; Power $= 290\,\text{kW}$; Thermal efficiency $= 50.1\%$.

2.19 Original Otto cycle: Work output, 1228, 1218.4, 1183.4, 1090, $821.5\,\text{kJ kg}^{-1}$; Thermal efficiency, 49.1, 48.75, 47.31, 43.58, 32.85%. Original Diesel cycle: Work output, 920.7, 989.7, 1035, 1093.5, $1131.2\,\text{kJ kg}^{-1}$; Thermal efficiency, 46, 49.4, 51.7, 54.6, 56.5%.

CHAPTER 3

3.2 Best $w = 231\,667\text{kJk mol}^{-1}$, $\Delta g = -231\,667\,\text{kJ kmol}^{-1}$; $\Delta h = -288\,750\,\text{kJk mol}^{-1}$; Best efficiency $= 80.2\%$, current $= 45.1\%$; $q = -57,\ 083\ 45.1\,\text{kJk mol}^{-1}$; Δs for cell $= -163\,\text{kJk mol}^{-1}$, for atmos. $= 190.3\,\text{kJk mol}^{-1}$, overall $= +27.3\,\text{kJ kmol}^{-1}\,\text{K}^{-1}$; Irreversible.

3.3 $0.0015\,\text{kJ K}^{-1}$; $0.57\,\text{kJ}$; $1.55\,\text{kJ}$; $0.98\,\text{kJ}$; $0.94\,\text{kJ}$.

3.4 $-4.3\,\text{kJ kg}^{-1}$; $-84\,\text{kJ kg}^{-1}$, $-88.3\,\text{kJ kg}^{-1}$.

3.5 $9.515\,\text{kJ}$; $-3.111\,\text{kJ}$; $-6.804\,\text{kJ}$; 0 as availability is a property; 71.5%.

3.9 Actual, $945.8\,\text{kJ kg}^{-1}$; cold $880.9\,\text{kJ kg}^{-1}\,\text{K}^{-1}$; Required c_p 1.078; mean $1.083\,\text{kJ kg}^{-1}\,\text{K}^{-1}$; Actual $\Delta s = 1.5211\,\text{kJ kg}^{-1}\,\text{K}^{-1}$.

3.11 $u_2 - u_1 = -69.53\,\text{kJ kg}^{-1}$; $h_2 - h_1 = -93.65\,\text{kJ kg}^{-1}$; $w = -93.65\,\text{kJ kg}^{-1}$.

3.12 $T_1 = 298.2\,\text{K}$, $p_2, T_2 = 9.97\,\text{MPa}, 396.4\,\text{K}$; $w = 18.4$, $u_2 - u_1 = 194.9$; $q = 213.3\text{kJ kg}^{-1}$.

3.14 At 3000 K: oxygen $u/R = 9507$, $h/R = 12\,507$, $c_v/R = 3.4536$, $c_p/R = 4.4536$, $\gamma = 1.290$; nitrogen: $u/R = 9118$, $h/R = 12\,118$, $c_v/R = 3.4001$, $c_p/R = 4.4001$, $\gamma = 1.294$; hydrogen: $u/R = 8371$, $h/R = 11\,371$, $c_v/R = 3.1982$, $c_p/R = 4.1982$, $\gamma = 1.313$.

At 6000 K: oxygen $u/R = 19937$, $h/R = 25\,937$, $c_v/R = 3.4882$, $c_p/R = 4.4882$, $\gamma = 1.287$; nitrogen: $u/R = 19464$, $h/R = 25\,464$, $c_v/R = 3.4738$, $c_p/R = 4.4738$, $\gamma = 1.288$; hydrogen: $u/R = 18378$, $h/R = 24\,378$, $c_v/R = 3.4119$, $c_p/R = 4.4119$, $\gamma = 1.293$.

3.15 (a) 1.599 (b) 30%, 70%.

3.16 370 K; 356 K; 386 K.

3.17 10.70, 18.76, 70.54 kPa.

CHAPTER 4

4.1 $139.4\,\text{m s}^{-1}$; 74.4%.

4.2 $311\,\text{m s}^{-1}$; $0.0377\,\text{m}^2$; $0.0175\,\text{kJ kg}^{-1}\,\text{K}^{-1}$; 322 K; $329\,\text{m s}^{-1}$; $0.035\,\text{m}^2$. Non-isentropic flow transfers energy from flow (i.e. kinetic energy) to internal energy and increases temperature. The density and velocity differences mean that area is no longer compatible. For same area and isentropic flow, conditions are $0.366\,\text{MPa}$, 341 K, $265\,\text{m s}^{-1}$.

4.3 $316.9\,\mathrm{m\,s^{-1}}$; $2634\,\mathrm{m\,s^{-1}}$.

4.4 For air: $328.8\,\mathrm{m\,s^{-1}}$, $559.5\,\mathrm{m\,s^{-1}}$, 2.16, 4.7, $445\,\mathrm{mm^2}$, $2091\,\mathrm{mm^2}$; For helium: $915.8\,\mathrm{m\,s^{-1}}$, $1420.9\,\mathrm{m\,s^{-1}}$, 2.13, 1.7, 1, $127\,\mathrm{mm^2}$, $1880\,\mathrm{mm^2}$.

4.5 $839.6\,\mathrm{m\,s^{-1}}$; $11.31\,\mathrm{kN}$; $12.56\,\mathrm{kN}$.

4.6 3 min 28 s, to 8.5 MPa; 28 min 47 s to 184.5 kPa.

4.7 $M_1 = 1.42$; $M_2 = 0.7314$;
     Static conditions, 153 kPa, 329.6 K; Stagnation conditions, upstream 229 ka, 365 K; Stagnation conditions, downstream 218.4 kPa, 365 K.

4.8 $923.5\,\mathrm{kPa}$, $307\,\mathrm{m\,s^{-1}}$; $1.842\,\mathrm{kPa}$, $158\,\mathrm{m\,s^{-1}}$.

4.9 125.9 kPa; 127.9 kPa; Loss in stagnation pressure 128.4 kPa; $M = 0.15$; 42 kPa; 44.34 kPa; Loss in stagnation pressure 212 kPa; $M = 0.445$.

4.10 $M = 0.6544$; 1.871.

4.11 2.63 m; 2.72 m; 9.6%.

4.12 241 to 164 kPa.

4.13 $1178\,\mathrm{kJ\,kg^{-1}}$ (air)/fuel = 34 : 1; decrease = 3.7%.

4.14 Isentropic wave: Temperature ratios, 1.2996, 1.5845, 1.9320; Wave velocity 1.8391, 2.5510, 3.3370; Particle velocity, 0.6993, 1.2925, 1.9475. Shock wave: Temperature ratios, 1.3282, 1.7742, 2.6230; Wave velocity (Mach number), 1.5119, 2.1044, 2.9520; Particle velocity, 0.7087, 1.3577, 2.1778.

4.15 $252\,\mathrm{m\,s^{-1}}$; $676.5\,\mathrm{m\,s^{-1}}$; $0.0073\,\mathrm{s}$.

4.16 2.208:1; $662.7\,\mathrm{m\,s^{-1}}$; $442.4\,\mathrm{m\,s^{-1}}$.

4.17 $M = 1.86$; Stagnation conditions after shock, 473 K, 510.6 kPa; Stagnation conditions before shock, 260 K, 80 KPa.

4.18 $M = 1.656$; $V(\mathrm{abs}) = 316\,\mathrm{m\,s^{-1}}$; $p = 939\,\mathrm{kPa}$; $T = 585\,\mathrm{K}$.

4.19 $M = 2.37, 2.69, 3.6$; $p = 63.9, 82.8, 194.5\,\mathrm{kPa}$; $T = 604, 700, 1036\,\mathrm{K}$.

4.20 0.022s; Particle velocity 0, 34, 68, $102\,\mathrm{m\,s^{-1}}$; Acoustic velocity 340, 346.8, 353.6, $360.4\,\mathrm{m\,s^{-1}}$; Reflected particle velocity, zero for all cases; Wave 1 is acoustic: wave velocity $= 340\,\mathrm{m\,s^{-1}}$, pressure $= 100\,\mathrm{kPa}$; Wave 2: wave velocity $= 353.6\,\mathrm{m\,s^{-1}}$, pressure $= 131.6\,\mathrm{kPa}$.

CHAPTER 5

5.2 25.8°; 5251 m; 1.72 Nm; 7.2 kW.

5.2 2.2 Nm; 9.1 kW; 2.1 Nm; 8.9 kW.

5.3 28.3°; 5.8 kW; Blade angle as in Problem 5.2; Power as in 5.2(b) as extra power is required to generate initial whirl.

5.4 49°; 25°; 3.1°.

5.5 26.6°; 180 kW; $618.5\,\mathrm{m\,s^{-1}}$; 14°.

5.6 $143\,\mathrm{m\,s^{-1}}$; 32.6 and 34 kW; 51.6° towards rotation, and 65.4° with rotation; $259.4\,\mathrm{m\,s^{-1}}$.

5.7 Blade angles; inner, 36.9°, 20.6°; outer, 20.6°, 36.9°; Stator, 14° throughout; 10.7 MW.

5.8 86.4%; 87.4%.

5.10 (a) 87.3%, (b) 87.7%, (c) section 1, 89.1%, section 2, 91.1%, overall 87.8%.

5.11 Compressor, 85%; Turbine 92.5%.

CHAPTER 6

6.1 Air/fuel (volume) = 9.52, 23.8, 7.14, 59.5, 48.8; Stoichiometric air/fuel (mass) = 17.11, 15.6, 6.4, 15.0, 14.5. At 16:1: % air = -6.5 (insufficient), 2.56, 150, 6.7, 10.5 (excess); Equivalence ratio = 1.069, 0.975, 0.4, 0.9375, 0.905.

6.2 % theoretical air = 199.5, 64.5; Air/fuel (mass) = 34.1:1, 11.0:1; Equivalence ratio = 0.502, 1.555.

6.3 1:1.25:0.8125:9.165; C_4H_{10}.

6.4 Higher, 29670; lower 26804 kJ kg^{-1}; 1.2176 CO_2; 0.7824 CO; -24864 kJ kg^{-1} (from system).

6.5 1132 K; 2837 K; 1329 K; 2.66 atmospheres.

6.6 46.4, 49.9 MJ/kg fuel
water vapour, 1 kmole, condensed to liquid, 3 kmole
52 414 kJ/kg from the system.

6.7 3.3, 4.3, 2.725, 30.74, 23.24; 0.667; 1829 K.

6.8 $N_2, O_2, N, O, NO_2, N_2O, NO$; NO dominant; At 298 K: $2N_2, 1O_2$ zero NO; At 3200 K: 1.899 N_2, 0.899 O_2, 0.2017 NO; No change as pressure exponent is zero in this case.

6.9 0.6 by volume, 9.525 by mass; Equivalence ratio 0.833; $T = 3200$ K; $q = -64636$ kJ kmol^{-1} of hydrogen.

6.10 0.75 C_8H_{18}; 0.25 C_6H_6; 2245 K; $-195, 313$ kJ kmol^{-1} of fuel.

6.11 1.097 CO_2; 1.903 CO; 0.951 O_2; $4H_2O$; 18.8 N_2.

6.12 Equivalence ratio = 0.833; 700 K; 16 atmospheres; 2372 K; 0.9856 CO_2; 0.0146 CO; 0.4072 O_2; $2H_2O$; $9N_2$; CO = 0.11%

6.13 4.249 CO_2; 2.751 CO; 1.378 O_2; 5.515 H_2O; 0.493 H_2; 0.9798 OH.

CHAPTER 7

7.1 3019 K; 1.173 kJ kg^{-1}K^{-1}; 1421 K.

7.2 51.7 mm; 0.0178 kg; 0.0055 kg; 0.0123 kg.

7.3 52.8 cm s^{-1}, 11 cm s^{-1}; 63.8 cm s^{-1}.

7.4 2915 kPa; 765 K; 0.0056 kg; 0.00448 kg; 0.00112 kg; 8872 kPa; 2327.5 K; 5800 kPa; 236 m s^{-1}.

7.5 CO -65%; NO$_x$ 0%; UHC -12.1%; CO $+125\%$; NO$_x$ -36.8%; UHC $+32.8\%$; Transient CO $+15\%$; NO$_x$ -9.2%; UHC $+5.2\%$.

7.6 At 298 K, negligible; At 3500 K, 0.3510 = 7.4% of air.

CHAPTER 8

8.1 32.4 kW; 0.69 kW.

8.2 To head, walls, crown: Radiation, 1.876, 1.718, 1.374 kW; Convection 3.722, 3.409, 2.727 kW; Combined 5.598, 7.367, 4.101; Overall 17.01 kW, 28.3%

8.3 92 kW m^{-2}; 424 K; 414 K.

8.4 For 15°, 30° and 90° only: 174, 291, 393 W m^{-2} K^{-1}; 205 W, 1303 W, 6928 W.

8.5 For 15°, 30° and 90° only: 587, 832, 316 W m^{-2} K^{-1}; 726 W, 3919 W, 5581 W.

Index